常微分方程

Ordinary Differential Equations

主　编　孙书荣　韩振来　张　超
副主编　李　华　李西成　孙　莹

山东大学出版社

图书在版编目(CIP)数据

常微分方程 / 孙书荣，韩振来，张超主编. —济南：山东大学出版社，2018.11

ISBN 978-7-5607-6255-5

Ⅰ. ①常… Ⅱ. ①孙… ②韩… ③张… Ⅲ. ①常微分方程—教材 Ⅳ. ①O175.1

中国版本图书馆 CIP 数据核字(2018)第 284875 号

责任编辑：宋亚卿
封面设计：张　荔

出版发行：山东大学出版社
　　社　址　山东省济南市山大南路 20 号
　　邮　编　250100
　　电　话　市场部(0531)88363008
经　　销：新华书店
印　　刷：济南华林彩印有限公司
规　　格：787 毫米×1092 毫米　1/16
　　　　18.75 印张　342千字
版　　次：2018 年 11 月第 1 版
印　　次：2018 年 11 月第 1 次印刷
定　　价：42.00 元

Preface

"Ordinary Differential Equations" is a compulsory undergraduate course and also a required part of the graduate school entrance exam for the math majors. But the available English version textbooks are either engineering-oriented, or targeted at the math graduate students. Up till now, no such English textbook, which can satisfy the requirements of the bilingual "Ordinary Differential Equations" syllabus, has been published. However, "Ordinary Differential Equations" is currently the most common bilingual course for math majors in higher education institutions in China. Therefore, an ideal English version textbook is in urgent need.

This textbook is compiled on the basis of the syllabus, with references to the latest research progress on the academic study of ordinary differential equations and the bilingual teaching of "Ordinary Differential Equations" as a course. Reserving the advantages of the original English references, and meanwhile, corresponding to the outline and contents of the Chinese teaching materials, this is an ideal textbook for undergraduate math majors.

The book is divided into eight chapters.

Chapter 1 covers models, definitions, classifications, and simple illustrations of solutions of differential equations. The second chapter treats classical cases of first order differential equations that can be solved explicitly. Chapter 3 deals with existence and uniqueness of solutions, that is, we examine under what conditions a differential equation has solutions and how the solutions are determined, without solving the differential equation analytically. Picard's iteration method and its applications are introduced. Continuation of solutions, continuous dependence on initial conditions and parameters, differentiability of the solution on initial-value are also given. In Chapter 4, we study high-order equations. Some important methods to solve linear equations are developed. These methods include undetermined coefficients, variation of parameters, power series, and so on. Many methods of finding solutions of first-order ordinary differential system are summarized in Chapter 5. Matrix expression is an excellent way of finding solutions of first-order linear system with constant coefficients. In Chapter 6, we study the stability of critical points for general linear systems, and present results on how the stability of a critical point for an "almost linear system" can be determined by studying the corresponding linear system. Finally, Lyapunov's direct method, which is used to analyze the stability of a critical point is discussed. In Chapter 7, we introduce the concepts and basic results concerning fractional calculus, and study existence and uniqueness of solutions for fractional differential equations involving Riemann-Liouville and Caputo derivatives. In Chapter 8, we introduce the basic theory of time scale calculus and oscillation theory for dynamic equations on time scales.

We acknowledge with gratitude the support of the "Textbook Construction Project Funds of University of Jinan". We would like to express our appreciation to the support of School of Mathematical Sciences in University of Jinan.

Shurong, Sun

2017.11

Contents

Chapter 1

Introduction

In sciences, engineering and social sciences, mathematical models are developed to aid in the understanding of physical/natural/social phenomena. These models often yield an equation (or a system of equations) that involves an unknown function and its derivative(s). Such an equation is called a **differential equation.**

1.1 Differential Equations and Mathematical Models

The laws of the universe are written in the language of mathematics. Algebra is sufficient to solve many static problems, but the most interesting natural phenomena involve change and are described by equations that relate changing quantities.

Because the derivative $\mathrm{d}x/\mathrm{d}t=f'(t)$ of the function f is the rate at which the quantity $x=f(t)$ is changing with respect to the independent variable t, it is natural that equations involving derivatives are frequently used to describe the changing universe. An equation relating an unknown function and one or more of its derivatives is called a **differential equation.**

Example 1.1 Heat transfer

An object of uniform temperature T_0 (e.g., a potato) is placed in an oven of constant temperature T_e. It is observed that over time the potato heats up and eventually its temperature becomes that of the oven environment T_e. We want a model that governs the temperature $T(t)$ of the potato at any time t.

Newton's law of cooling may be stated in this way: The time rate of change (the rate of change with respect to time t) of the temperature $T(t)$ of a body is proportional to the difference between T and the temperature T_e of the surrounding medium. That is,

$$\frac{\mathrm{d}T}{\mathrm{d}t}=-k(T-T_e), \tag{1.1}$$

where k is a positive constant. Observe that if $T > T_e$, then $dT/dt < 0$, so the temperature is a decreasing function of t and the body is cooling. But if $T < T_e$, then $dT/dt > 0$, so that T is increasing.

Thus the physical law is translated into a differential equation. If we are given the values of k and T_e, we should be able to find an explicit formula for $T(t)$ satisfying equation (1. 1) using a simple change of variables method, and then—with the aid of this formula—we can predict the future temperature of the body. If we let $u = T - T_e$ be a new dependent variable, then $u' = T'$ and equation (1. 1) may be written as $u' = -ku$. The solution of this equation is $u = ce^{-kt}$. Therefore, $T - T_e = ce^{-kt}$, or $T = T_e + ce^{-kt}$, where c is an arbitrary constant. When we impose an initial condition $T(0) = T_0$, we find $c = T_0 - T_e$, giving the particular solution to the differential equation $T(t) = T_e + (T_0 - T_e)e^{-kt}$. We now see clearly that $T(t) \to T_e$ as $t \to \infty$. A plot of the solution showing how an object heats up is given in Figure 1. 1.

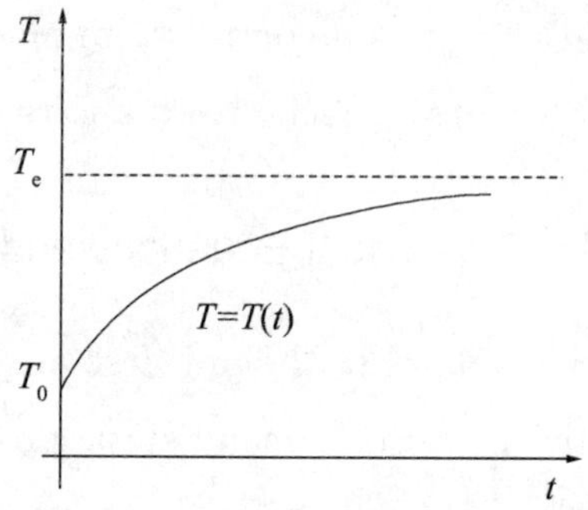

Figure 1. 1 The heating curve of an object

Temperature history in Newton's law of cooling shows how the temperature approaches the equilibrium temperature.

Example 1. 2 Dilution problems

Consider a tank which initially holds V_0 L of brine that contains a kg of salt. Another solution, containing b kg of salt per liter, is poured into the tank at the rate of e L/min while simultaneously, the well-stirred solution leaves the tank at the rate of f L/min. The problem is to find the amount of salt in the tank at any time t.

Let Q denote the mass of salt in the tank at any time. The time rate of change of Q, dQ/dt, equals the rate at which salt enters the tank minus the rate at which salt leaves the tank. Salt enters the tank at the rate of be L/min. To determine the rate at which salt leaves the tank, we first calculate the volume of brine in the tank at any time t, which is the initial volume plus the volume of brine added et minus the volume of brine removed ft. Thus, the volume of brine at any time is $V_0 + et - ft$. The concentration of salt in the tank at any time is $Q/(V_0 + et - ft)$, from which it follows that salt leaves the tank at the rate of $f[Q/(V_0 + et - f)]$ L/min. Thus,

$$dQ/dt = be - f[Q/(V_0 + et - f)]. \tag{1.2}$$

Example 1.3 Electrical circuits

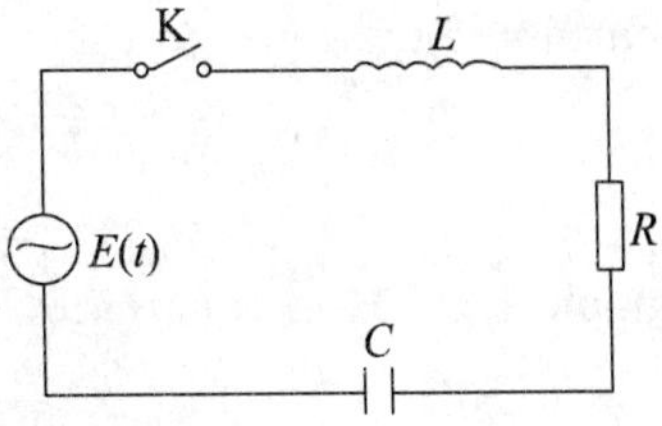

Figure 1.2 A simple electrical circuit

The RLC circuit shown on the right has a resistor, an inductor and a capacitor with a constant driving electro-motive force (emf) E connected in series (see Figure 1.2). A voltage V is applied when the switch K is closed. The voltage across the resistor is given by $V_R = Ri$. The voltage across the inductor is given by $V_L = L\frac{di}{dt}$. The voltage across the capacitor is given by $V_C = \frac{Q}{C}$. Kirchhoff's voltage law says that the directed sum of the voltages around a circuit must be zero. This result can be expressed in the following differential equation:

$$Ri + L\frac{di}{dt} + \frac{Q}{C} = E(t).$$

Differentiating, we have

$$L\frac{d^2 i}{dt^2} + R\frac{di}{dt} + \frac{1}{C}i = \frac{dE(t)}{dt}. \tag{1.3}$$

The charge $Q(t)$ satisfies the following differential equation:

$$L\frac{d^2 Q}{dt^2} + R\frac{dQ}{dt} + \frac{1}{C}Q = E(t). \tag{1.4}$$

Example 1.4 The simple pendulum

A simple pendulum consists of a mass m hanging from a string of length l and fixed at a pivot point P. When displaced to an initial angle and released, the pendulum will swing back and forth with periodic motion. By applying Newton's second law for rotational systems, the equation of motion for the pendulum may be obtained:

$$m\frac{dv}{dt} = -mg\sin\theta, \tag{1.5}$$

and rearranged as

$$\frac{d^2\theta}{dt^2} + \frac{g}{l}\sin\theta = 0, \tag{1.6}$$

where g is acceleration due to gravity, l is the length of the pendulum, and θ is the angular displacement. This is known as Mathieu's equation.

If the amplitude of angular displacement is small enough that the small angle approximation holds true, then the equation of motion reduces to the equation of simple harmonic motion:

$$\frac{d^2\theta}{dt^2}+\frac{g}{l}\theta=0. \tag{1.7}$$

Example 1.5 Heat equation

$$a^2\left(\frac{\partial^2 T}{\partial x^2}+\frac{\partial^2 T}{\partial y^2}+\frac{\partial^2 T}{\partial z^2}\right)=\frac{\partial T}{\partial t}, \tag{1.8}$$

where a is a constant and $T=T(x, y, z, t)$ is the temperature of a point (x, y, z) at time t.

Example 1.6 Volterra's prey-predator equation

$$\begin{cases}\dfrac{dx}{dt}=x(a-by),\\ \dfrac{dy}{dt}=-y(c-dx),\end{cases} \tag{1.9}$$

where a, b, c, d are positive constants and $x=x(t)$, $y=y(t)$ are the prey population and predator population respectively at time t.

1.2 Basic Concept

To begin our study of differential equations, we need some common terminology. If an equation involves the derivative of one variable with respect to another, then the former is called a **dependent variable** and the latter an **independent variable.**

A differential equation involving only ordinary derivatives with respect to a single independent variable is called an **ordinary differential equation.** A differential equation involving partial derivatives with respect to more than one independent variable is a **partial differential equation.**

Equations (1.1)-(1.7) are ordinary differential equations, and equation (1.8) is a partial differential equation.

In what follows, we will consider ordinary differential equations (simply called ODE) only.

The **order** of a differential equation is the order of the highest-order derivatives present in the equation.

Equations (1.1), (1.2) and (1.5) are first-order ordinary differential equations.

Equations (1.3), (1.4), (1.6) and (1.7) are second-order ordinary differential equations. Equation (1.8) is a second-order partial differential equation.

It will be useful to classify ordinary differential equations as being either linear or nonlinear. Remember that lines (in two dimensions) and planes (in three dimensions) are especially easy to visualize, when compared to nonlinear objects such as cubic curves or quadric surfaces. For example, all the points on a line can be found if we know just two of them. Correspondingly, linear differential equations are more amenable to solution than nonlinear ones. Now the equations for lines $ax+by=c$ and planes $ax+by+cz=d$ have the feature that the variables appear in additive combinations of their first powers only. By analogy a linear differential equation is one in which the dependent variable y and its derivatives appear in additive combinations of their first powers.

More precisely, a differential equation is **linear** if it has the form

$$a_n(x)y^{(n)}+a_{n-1}(x)y^{(n-1)}+\ldots+a_1(x)y'+a_0(x)y=f(x),$$

where $a_n(x)$, $a_{n-1}(x)$, ..., $a_1(x)$, $a_0(x)$ and $f(x)$ depend only on the independent variable x. If an ordinary differential equation is not linear, then we call it **nonlinear.**

In particular, if $f(x)=0$, the equation is called a **homogeneous** linear one; otherwise it is a **non-homogeneous** one.

An nth-order ordinary differential equation is an equality relating the independent variable to the nth derivative (and usually lower-order derivatives as well) of the dependent variable.

Examples are as follows:

$x^2\frac{d^2y}{dx^2}+x\frac{dy}{dx}+y=x^3$ (second-order, x is independent variable, y is dependent variable);

$\sqrt{1-\frac{d^2y}{dt^2}}-y=0$ (second-order, t is independent variable, y is dependent variable);

$\frac{d^4x}{dt^4}=xt$ (fourth-order, t is independent variable, x is dependent variable).

Thus, a general form for an nth-order equation with x independent, y dependent, can be expressed as

$$F\left(x, y, \frac{dy}{dx}, \ldots, \frac{d^ny}{dx^n}\right)=0, \tag{1.10}$$

where F is a function that depends on x, y, and the derivatives of y up to order n, that is, $x, y, \ldots, \mathrm{d}^n y/\mathrm{d}x^n$. We assume that the equation holds for all x in an open interval I ($a<x<b$, where a or b could be infinite). In many cases, we can isolate the highest-order term $\mathrm{d}^n y/\mathrm{d}x^n$ and write equation (1.10) as

$$\frac{\mathrm{d}^n y}{\mathrm{d}x^n}=f\left(x, y, \frac{\mathrm{d}y}{\mathrm{d}x}, \ldots, \frac{\mathrm{d}^{n-1}y}{\mathrm{d}x^{n-1}}\right), \tag{1.11}$$

which is often preferable to equation (1.10) for theoretical and computational purposes.

Explicit Solution

Definition 1.1 A function $\varphi(x)$ that when substituted for y in equation (1.10) [or (1.11)] satisfies the equation for all x in the interval I is called an **explicit solution** to the equation on I.

Example 1.7 Show that $\varphi(x)=x^2-x^{-1}$ is an explicit solution to the linear equation

$$\frac{\mathrm{d}^2 y}{\mathrm{d}x^2}-\frac{2}{x^2}y=0, \tag{1.12}$$

but $\psi(x)=x^3$ is not.

Solution The functions $\varphi(x)=x^2-x^{-1}$, $\varphi'(x)=2x+x^{-2}$, and $\varphi''(x)=2-2x^{-3}$ are defined for all $x\neq 0$. Substitution of $\varphi(x)$ for y in equation (1.12) gives

$$(2-2x^{-3})-\frac{2}{x^2}(x^2-x^{-1})=(2-2x^{-3})-(2-2x^{-3})=0.$$

Since this is valid for any $x\neq 0$, the function $\varphi(x)=x^2-x^{-1}$ is an explicit solution to equation (1.12) on $(-\infty, 0)$ and also on $(0, +\infty)$.

For $\psi(x)=x^3$, we have $\psi'(x)=3x^2$, $\psi''(x)=6x$, and substitution into equation (1.12) gives

$$6x-\frac{2}{x^2}x^3=4x=0,$$

which is valid only at the point $x=0$ and not on an interval. Hence, $\psi(x)$ is not a solution. ◆

Example 1.8 Show that for any choice of the constants c_1 and c_2, the function

$$\varphi(x)=c_1 \mathrm{e}^{-x}+c_2 \mathrm{e}^{2x}$$

is an explicit solution to the linear equation

$$y''-y'-2y=0. \tag{1.13}$$

Solution We compute $\varphi'(x)=-c_1\mathrm{e}^{-x}+2c_2\mathrm{e}^{2x}$ and $\varphi''(x)=c_1\mathrm{e}^{-x}+4c_2\mathrm{e}^{2x}$. Substitution of φ, φ' and φ'' for y, y' and y'' in equation (1.13) yields

$$(c_1 e^{-x}+4c_2 e^{2x})-(-c_1 e^{-x}+2c_2 e^{2x})-2(c_1 e^{-x}+c_2 e^{2x})$$
$$=(c_1+c_1-2c_1)e^{-x}+(4c_2-2c_2-2c_2)e^{2x}=0.$$

Since equality holds for all x in $(-\infty,+\infty)$, then $\varphi(x)=c_1 e^{-x}+c_2 e^{2x}$ is an explicit solution to equation (1.13) on the interval $(-\infty,+\infty)$ for any choice of the constants c_1 and c_2. ◆

As we will see in Chapter 2, the methods for solving differential equations do not always yield an explicit solution for the equation. We may have to settle for a solution that is defined implicitly. Consider the following example.

Example 1.9 Show that the relation

$$y^2-x^3+8=0 \tag{1.14}$$

implicitly defines a solution to the nonlinear equation

$$\frac{dy}{dx}=\frac{3x^2}{2y} \tag{1.15}$$

on the interval $(2,+\infty)$.

Solution When we solve equation(1.14) for y, we obtain $y=\pm\sqrt{x^3-8}$. Let's try $\varphi(x)=\sqrt{x^3-8}$ to see if it is an explicit solution. Since $d\varphi/dx=3x^2/(2\sqrt{x^3-8})$, both φ and $d\varphi/dx$ are defined on $(2,+\infty)$. Substituting them into equation(1.15) yields

$$\frac{3x^2}{2\sqrt{x^3-8}}=\frac{3x^2}{2(\sqrt{x^3-8})},$$

which is indeed valid for all x in $(2,+\infty)$. [You can check that $\psi(x)=-\sqrt{x^3-8}$ is also an explicit solution to equation(1.15).] ◆

Implicit Solution

Definition 1.2 A relation $G(x,y)=0$ is said to be an **implicit solution** to equation (1.10) on the interval I if it defines one or more explicit solutions on I.

Example 1.10 Show that

$$x+y+e^{xy}=0 \tag{1.16}$$

is an implicit solution to the nonlinear equation

$$(1+xe^{xy})\frac{dy}{dx}+1+ye^{xy}=0. \tag{1.17}$$

Solution First, we observe that we are unable to solve equation (1.16) directly for y in terms of x alone. However, for (1.16) to hold, we realize that any change in x requires a change in y, so we expect the relation (1.16) to define implicitly at least one function $y(x)$. This is difficult to show directly but can be rigorously verified using the implicit function theorem of advanced calculus, which guarantees that such a function $y(x)$ exists that is also differentiable.

Once we know that y is a differentiable function of x, we can use the technique of implicit differentiation. Indeed, from equation (1.16) we obtain on differentiating with respect to x and applying the product and chain rules,

$$\frac{\mathrm{d}}{\mathrm{d}x}(x+y+\mathrm{e}^{xy})=1+\frac{\mathrm{d}y}{\mathrm{d}x}+\mathrm{e}^{xy}\left(y+x\frac{\mathrm{d}y}{\mathrm{d}x}\right)=0$$

or

$$(1+x\mathrm{e}^{xy})\frac{\mathrm{d}y}{\mathrm{d}x}+1+y\mathrm{e}^{xy}=0,$$

which is identical to the differential equation (1.17). Thus, relation (1.16) is an implicit solution on some interval guaranteed by the implicit function theorem. ◆

Example 1.11 Verify that for every constant c the relation $4x^2-y^2=c$ is an implicit solution to

$$y\frac{\mathrm{d}y}{\mathrm{d}x}-4x=0. \tag{1.18}$$

Graph the solution curves for $c=0, \pm 1, \pm 4$. (We call the collection of all such solutions a **one-parameter family of solutions.**)

Solution When we implicitly differentiate the equation $4x^2-y^2=c$ with respect to x, we find

$$8x-2y\frac{\mathrm{d}y}{\mathrm{d}x}=0,$$

which is equivalent to equation (1.18). In Figure 1.3, we have sketched the implicit solutions for $c=0, \pm 1, \pm 4$. The curves are hyperbolas with common asymptotes $y=\pm 2x$. Notice that the implicit solution curves (with c arbitrary) fill the entire plane and are nonintersecting for $c\neq 0$. For $c=0$, the implicit solution gives rise to the two explicit solutions $y=2x$ and $y=-2x$, both of which pass through the origin. ◆

For brevity, we hereafter use the term solution to mean either an explicit or an implicit solution.

In the beginning of Section 1.1, we saw that the solution of the equation of simple harmonic motion (1.7) invoked two arbitrary constants of integration c_1, c_2:

$$h(t)=\frac{-gt^2}{2}+c_1t+c_2,$$

whereas the solution of the first-order radioactive decay equation contained a single constant c:

$$A(t)=ce^{-kt}.$$

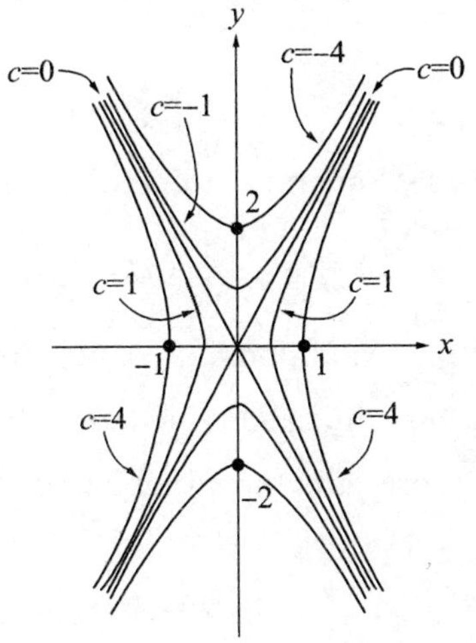

Figure 1.3 Implicit solutions curves of $4x^2-y^2=c$

It is clear that integration of the simple fourth-order equation

$$\frac{d^4 y}{dx^4}=0$$

brings in four undetermined constants:

$$y(x)=c_1x^3+c_2x^2+c_3x+c_4.$$

It will be shown later in the text that in general the methods for solving nth-order differential equations evoke n arbitrary constants. In most cases, we will be able to evaluate these constants if we know n initial values $y(x_0)$, $y'(x_0)$, ..., $y^{(n-1)}(x_0)$.

Initial Value Problem

Definition 1.3 By an **initial value problem** for an nth-order differential equation

$$F\left(x, y, \frac{dy}{dx}, \ldots, \frac{d^n y}{dx^n}\right)=0,$$

we mean: Find a solution to the differential equation on an interval I that satisfies at x_0 the following n initial conditions:

$y(x_0)=y_0$,

$\frac{dy}{dx}(x_0)=y_1$,

...

$$\frac{d^{n-1}}{dx^{n-1}}(x_0)=y_{n-1},$$

where $x_0 \in I$ and $y_0, y_1, \dots, y_{n-1}$ are given constants.

In the case of a first-order equation, the initial conditions reduce to the single requirement

$$y(x_0)=y_0,$$

and in the case of a second-order equation, the initial conditions have the form

$$y(x_0)=y_0, \quad \frac{dy}{dx}(x_0)=y_1.$$

The terminology "initial conditions" comes from mechanics, where the independent variable x represents time and is customarily symbolized as t. Then if t_0 is the starting time, $y(t_0)=y_0$ represents the initial location of an object and $y'(t_0)$ gives its initial velocity.

Example 1.12 Show that $\varphi(x)=\sin x-\cos x$ is a solution to the initial value problem

$$\frac{d^2y}{dx^2}+y=0, \quad y(0)=-1, \quad \frac{dy}{dx}(0)=1. \tag{1.19}$$

Solution Observe that $\varphi(x)=\sin x-\cos x$, $d\varphi/dx=\cos x+\sin x$, and $d^2\varphi/dx^2=-\sin x+\cos x$ are all defined on $(-\infty,+\infty)$. Substituting them into the differential equation gives

$$(-\sin x+\cos x)+(\sin x-\cos x)=0,$$

which holds for all $x\in(-\infty,+\infty)$. Hence, $\varphi(x)$ is a solution to the differential equation (1.19) on $(-\infty,+\infty)$. When we check the initial conditions, we find

$$\begin{cases} \varphi(0)=\sin 0-\cos 0=-1, \\ \dfrac{d\varphi}{dx}(0)=\cos 0+\sin 0=1, \end{cases}$$

which meets the requirements of equation (1.19). Therefore, $\varphi(x)$ is a solution to the given initial value problem. ◆

Example 1.13 As shown in Example 1.8, the function $\varphi(x)=c_1e^{-x}+c_2e^{2x}$ is a solution to

$$\frac{d^2x}{dx^2}-\frac{dy}{dx}-2y=0$$

for any choice of the constants c_1 and c_2. Determine c_1 and c_2 so that the initial conditions

$$y(0)=2 \quad \text{and} \quad \frac{dy}{dx}(0)=-3$$

are satisfied.

Solution To determine the constants c_1 and c_2, we first compute $d\varphi/dx$ to get $d\varphi/dx=-c_1e^{-x}+2c_2e^{2x}$. Substituting in our initial conditions gives the following system of equations:

$$\begin{cases}\varphi(0)=c_1e^0+c_2e^0=2,\\ \dfrac{d\varphi}{dx}(0)=-c_1e^0+2c_2e^0=-3,\end{cases} \quad \text{or} \quad \begin{cases}c_1+c_2=2,\\ -c_1+2c_2=-3.\end{cases}$$

Adding the last two equations yields $3c_2=-1$, so $c_2=-1/3$. Since $c_1+c_2=2$, we find $c_1=7/3$. Hence, the solution to the initial value problem is $\varphi(x)=(7/3)e^{-x}-(1/3)e^{2x}$. ◆

Exercises 1.1

In Problems 1-12, a differential equation is given along with the field or problem area in which it arises. Classify each as an ordinary differential equation (ODE) or a partial differential equation (PDE), give the order, and indicate the independent and dependent variables. If the equation is an ordinary differential equation, indicate whether the equation is linear or nonlinear.

1. $\dfrac{d^2y}{dx^2}-2x\dfrac{dy}{dx}+2y=0.$

(Hermite's equation, quantum-mechanical harmonic oscillator)

2. $5\dfrac{d^2x}{dt^2}+4\dfrac{dx}{dt}+9x=2\cos 3t.$

(mechanical vibrations, electrical circuits, seismology)

3. $\dfrac{\partial^2u}{\partial x^2}+\dfrac{\partial^2u}{\partial y^2}=0.$

(Laplace's equation, potential theory, electricity, heat, aerodynamics)

4. $\dfrac{dy}{dx}=\dfrac{y(2-3x)}{x(1-3y)}.$

(competition between two species, ecology)

5. $\dfrac{dx}{dt}=k(4-x)(1-x)$, where k is a constant.

(chemical reaction rates)

6. $y\left[1+\left(\dfrac{dy}{dx}\right)^2\right]=C$, where C is a constant.

(brachistochrone problem, calculus of variations)

7. $\sqrt{1-y}\dfrac{d^2y}{dx^2}+2x\dfrac{dy}{dx}=0$.

(Kidder's equation, flow of gases through a porous medium)

8. $\dfrac{dp}{dt}=kp(P-p)$, where k and P are constants.

(logistic curve, epidemiology, economics)

9. $8\dfrac{d^4y}{dx^4}=x(1-x)$.

(deflection of beams)

10. $x\dfrac{d^2y}{dx^2}+\dfrac{dy}{dx}+xy=0$.

(aerodynamics, stress analysis)

11. $\dfrac{\partial N}{\partial t}=\dfrac{\partial^2 N}{\partial r^2}+\dfrac{1}{r}\dfrac{\partial N}{\partial r}+kN$, where k is a constant.

(nuclear fission)

12. $\dfrac{d^2y}{dx^2}-0.1(1-y^2)\dfrac{dy}{dx}+9y=0$.

(van der Pol's equation, triode vacuum tube)

In Problems 13-16, write a differential equation that fits the physical description.

13. The rate of change of the population p of bacteria at time t is proportional to the population at time t.

14. The velocity at time t of a particle moving along a straight line is proportional to the fourth power of its position x.

15. The rate of change in the temperature T of coffee at time t is proportional to the difference between the temperature M of the air at time t and the temperature of the coffee at time t.

16. The rate of change of the mass A of salt at time t is proportional to the square of the mass of salt present at time t.

17. **Drag Race.** Two drivers, Alison and Kevin, are participating in a drag race. Beginning from a standing start, they each proceed with a constant acceleration. Alison covers the last of the distance in 3 seconds, whereas Kevin covers the last of the distance in 4 seconds. Who wins and by how much time?

18. (1) Show that $y^2+x-3=0$ is an implicit solution to $dy/dx=-1/(2y)$ on the interval $(-\infty, 3)$.

(2) Show that $xy^3-xy^3\sin x=1$ is an implicit solution to

$$\frac{dy}{dx}=\frac{(x\cos x+\sin x-1)y}{3(x-x\sin x)}$$

on the interval $(0, \pi/2)$.

19. (1) Show that $\varphi(x)=x^2$ is an explicit solution to

$$x\frac{dy}{dx}=2y$$

on the interval $(-\infty, +\infty)$.

(2) Show that $\varphi(x)=e^x-x$ is an explicit solution to

$$\frac{dy}{dx}+y^2=e^{2x}+(1-2x)e^x+x^2-1$$

on the interval $(-\infty, +\infty)$.

(3) Show that $\varphi(x)=x^2-x^{-1}$ is an explicit solution to $x^2 d^2y/dx^2=2y$ on the interval $(0, +\infty)$.

In Problems 20-25, determine whether the given function is a solution to the given differential equation.

20. $x=2\cos t-3\sin t$, $\quad x''+x=0$.

21. $y=\sin x+x^2$, $\quad \dfrac{d^2y}{dx^2}+y=x^2+2$.

22. $x=\cos 2t$, $\quad \dfrac{dx}{dt}+tx=\sin 2t$.

23. $\theta=2e^{3t}-e^{2t}$, $\quad \dfrac{d^2\theta}{dt^2}-\theta\dfrac{d\theta}{dt}+3\theta=-2e^{2t}$.

24. $y=3\sin 2x+e^{-x}$, $\quad y''+4y=5e^{-x}$.

25. $y=e^{2x}-3e^{-x}$, $\quad \dfrac{d^2y}{dx^2}-\dfrac{dy}{dx}-2y=0$.

In Problems 26-30, determine whether the given relation is an implicit solution to the given differential equation. Assume that the relationship does define y implicitly as a function of x and use implicit differentiation.

26. $y-\ln y=x^2+1$, $\quad \dfrac{dy}{dx}=\dfrac{2xy}{y-1}$.

27. $x^2+y^2=4$, $\quad \dfrac{dy}{dx}=\dfrac{x}{y}$.

28. $e^{xy}+y=x-1$, $\quad \dfrac{dy}{dx}=\dfrac{e^{-xy}-y}{e^{-xy}+x}$.

29. $x^2-\sin(x+y)=1$, $\quad \dfrac{dy}{dx}=2x\sec(x+y)-1$.

30. $\sin y+xy-x^3=2$, $\quad y''=\dfrac{6xy'+(y')^3\sin y-2(y')^2}{3x^2-y}$.

31. Show that $\varphi(x)=c_1\sin x+c_2\cos x$ is a solution to $d^2y/dx^2+y=0$ for any choice of the constants c_1 and c_2. Thus, $c_1\sin x+c_2\cos x$ is a two-parameter family of solutions to the differential equation.

32. Verify that $\varphi(x)=2/(1-ce^x)$, where c is an arbitrary constant, is a one-parameter family of solutions to

$$\frac{dy}{dx}=\frac{y(y-2)}{2},$$

and graph the solution curves corresponding to $c=0, \pm 1, \pm 2$ using the same coordinate axes.

33. Verify that $x^2+cy^2=1$, where c is an arbitrary nonzero constant, is a one-parameter family of implicit solutions to

$$\frac{dy}{dx}=\frac{xy}{x^2-1},$$

and graph several of the solution curves using the same coordinate axes.

34. Show that $\varphi(x)=ce^{3x}+1$ is a solution to $dy/dx-3y=-3$ for any choice of the constant c. Thus, $ce^{3x}+1$ is a one-parameter family of solutions to the differential equation. Graph several of the solution curves using the same coordinate axes.

35. Determine for which values of m the function $\varphi(x)=e^{mx}$ is a solution to the given equation.

(1) $\dfrac{d^2y}{dx^2}+6\dfrac{dy}{dx}+5y=0$.

(2) $\dfrac{d^3y}{dx^3}+3\dfrac{d^2y}{dx^2}+2\dfrac{dy}{dx}=0$.

36. Determine for which values of m the function $\varphi(x)=x^m$ is a solution to the given equation.

(1) $3x^2\dfrac{d^2y}{dx^2}+11x\dfrac{dy}{dx}-3y=0$.

(2) $x^2\dfrac{d^2y}{dx^2}-x\dfrac{dy}{dx}-5y=0$.

37. Verify that the function $\varphi(x)=c_1 e^x + c_2 e^{-2x}$ is a solution to the linear equation

$$\frac{d^2 y}{dx^2}+\frac{dy}{dx}-2y=0$$

for any choice of the constants c_1 and c_2. Determine c_1 and c_2 so that each of the following initial conditions is satisfied.

(1) $y(0)=2$, $y'(0)=1$.

(2) $y(1)=1$, $y'(1)=0$.

1.3 Direction Fields

The existence and uniqueness theorem certainly has great value, but it stops short of telling us anything about the nature of the solution to a differential equation. For practical reasons, we may need to know the value of the solution at a certain point, or the intervals where the solution is increasing, or the points where the solution attains a maximum value. Certainly, knowing an explicit representation (a formula) for the solution would be a considerable help in answering these questions. However, for many of the differential equations that we are likely to encounter in real-world applications, it will be impossible to find such a formula. Moreover, even if we are lucky enough to obtain an implicit solution, using this relationship to determine an explicit form may be difficult. Thus, we must rely on other methods to analyze or approximate the solution.

One technique that is useful in visualizing (graphing) the solutions to a first-order differential equation is to sketch the direction field for the equation. To describe this method, we need to make a general observation. Namely, a first-order equation

$$\frac{dy}{dx}=f(x,y)$$

specifies a slope at each point in the xy-plane where f is defined. In other words, it gives the direction that a graph of a solution to the equation must have at each point. Consider, for example, the equation

$$\frac{dy}{dx}=x^2-y. \tag{1.20}$$

The graph of a solution to equation (1.20) that passes through the point $(-2,1)$ must have slope $(-2)^2-1=3$ at that point, and a solution through $(-1,1)$ has zero slope at that point.

A plot of short line segments drawn at various points in the xy-plane showing the slope of the solution curve there is called a **direction field** for the differential equation. Because the direction field gives the "flow of solutions", it facilitates the drawing of any particular solution (such as the solution to an initial value problem). In Figure 1.4(a) we have sketched the direction field for equation (1.20) and in Figure 1.4(b) have drawn several solution curves.

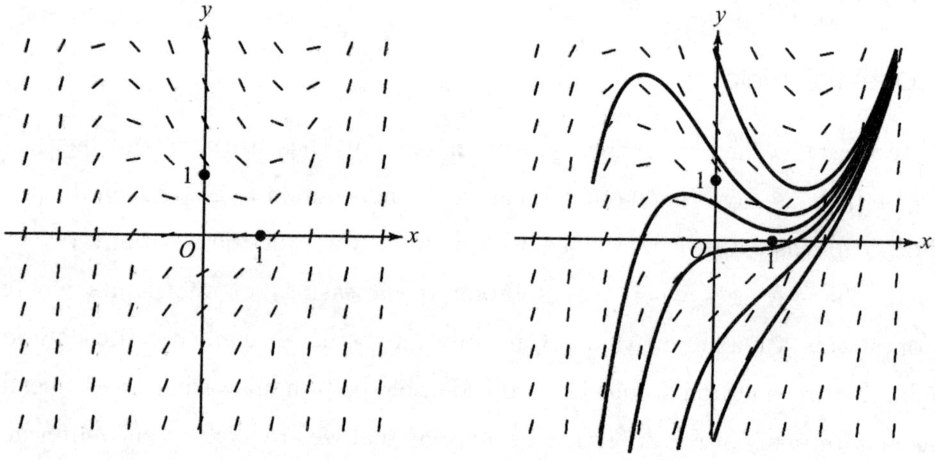

(a) Direction field for $dy/dx=x^2-y$ (b) Solutions to $dy/dx=x^2-y$

Figure 1.4

The logistic equation for the population p (in thousands) at time t of a certain species is given by

$$\frac{dp}{dt}=p(2-p). \tag{1.21}$$

Of course, p is nonnegative.

From the direction field sketched in Figure 1.5, answer the following questions:

(i) If the initial population is 3000 [that is, $p(0)=3$], what can you say about the limiting population $\lim\limits_{t\to+\infty}p(t)$?

(ii) Can a population of 1000 ever decline to 500?

(iii) Can a population of 1000 ever increase to 3000?

Solution (i) The direction field indicates that all solution curves [other than $p(t)\equiv 0$] will approach the horizontal line $p=2$ as $t\to+\infty$, that is, this line is an asymptote for all positive solutions. Thus, $\lim\limits_{t\to+\infty}p(t)=2$.

(ii) The direction field further shows that populations greater than 2000 will steadily decrease, whereas those less than 2000 will increase. In particular, a population of 1000 can never decline to 500.

(iii) As mentioned in part (ii), a population of 1000 will increase with time. But the direction field indicates it can never reach 2000 or any larger value; i. e., the solution curve cannot cross the line $p=2$. Indeed, the constant function $p(t)\equiv 2$ is a solution to equation (1. 21). ◆

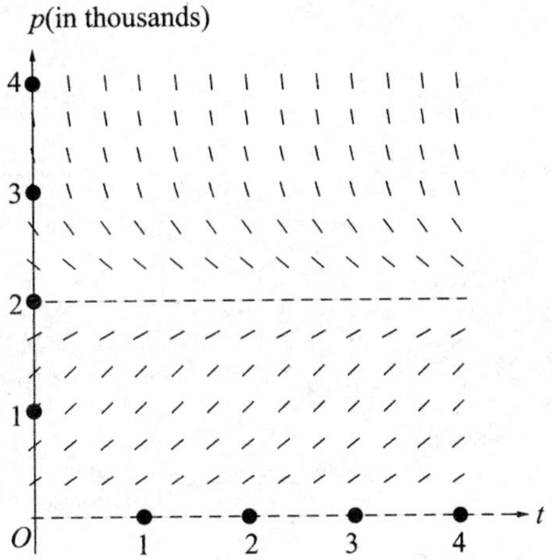

Figure 1. 5 Direction field for logistic equation

Notice that the direction field in Figure 1. 5 has the nice feature that the slopes do not depend on t; that is, the slopes are the same along each horizontal line. This is the key property of so-called **autonomous equations** $y'=f(y)$, where the right-hand side is a function of the dependent variable only.

Hand sketching the direction field for a differential equation is often tedious. Fortunately, several software programs have been developed to obviate this task. When hand sketching is necessary, however, the method of isoclines can be helpful in reducing the work.

The Method of Isoclines

An **isocline** for the differential equation

$$y'=f(x,y)$$

is a set of points in the xy-plane where all the solutions have the same slope $\mathrm{d}y/\mathrm{d}x$; thus, it is a level curve for the function $f(x,y)$. For example, if

$$y'=f(x,y)=x+y, \tag{1. 22}$$

the isoclines are simply the curves (straight lines) $x+y=c$ or $y=-x+c$. Here c is an arbitrary constant. But c can be interpreted as the numerical value of the slope

dy/dx of every solution curve as it crosses the isocline. (Note that c is not the slope of the isocline itself; the latter is, obviously, -1.) Figure 1.6(a) depicts the isoclines for equation (1.22).

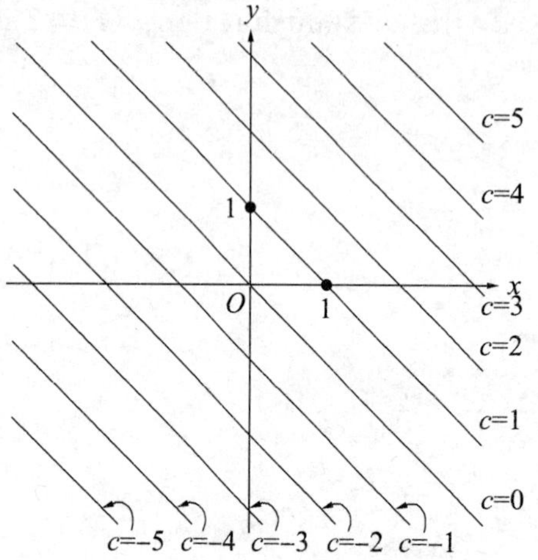

(a) Isoclines for $y'=x+y$

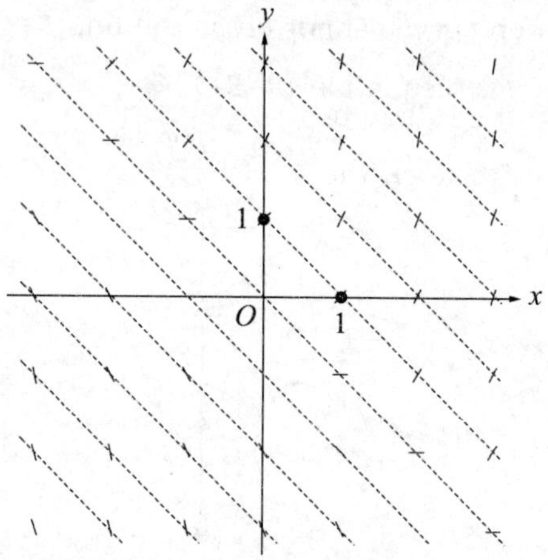

(b) Direction field for $y'=x+y$

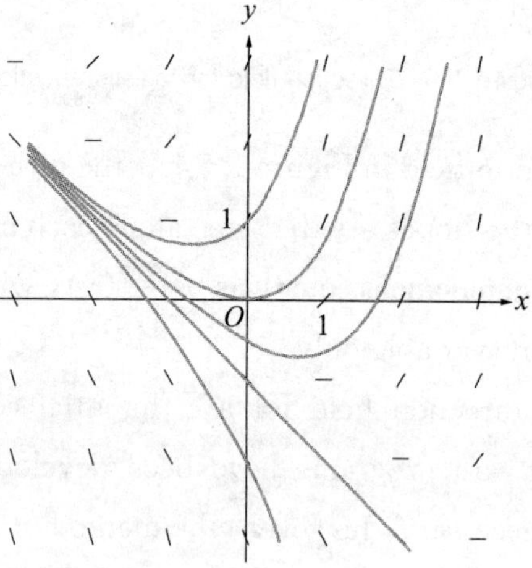

(c) Solutions to $y'=x+y$

Figure 1.6

To implement the method of isoclines for sketching direction fields, we draw hash marks with slope c along the isocline $f(x,y)=c$ for a few selected values of c. If we then erase the underlying isocline curves, the hash marks constitute a part of the direction field for the differential equation. Figure 1.6(b) depicts this process for the isoclines shown in Figure 1.6(a), and Figure 1.6(c) displays some solution curves.

Remark The isoclines themselves are not always straight lines. For equation (1.20) at the beginning of this section, they are parabolas $x^2-y=c$. When the isocline curves are complicated, this method is not practical.

Exercises 1. 2

1. The direction field for $dy/dx=2x+y$ is shown in Figure 1. 7.

(1) Sketch the solution curve that passes through $(0, -2)$. From this sketch, write the equation for the solution.

(2) Sketch the solution curve that passes through $(-1, 3)$.

(3) What can you say about the solution in part (2) as $x \to +\infty$? How about $x \to -\infty$?

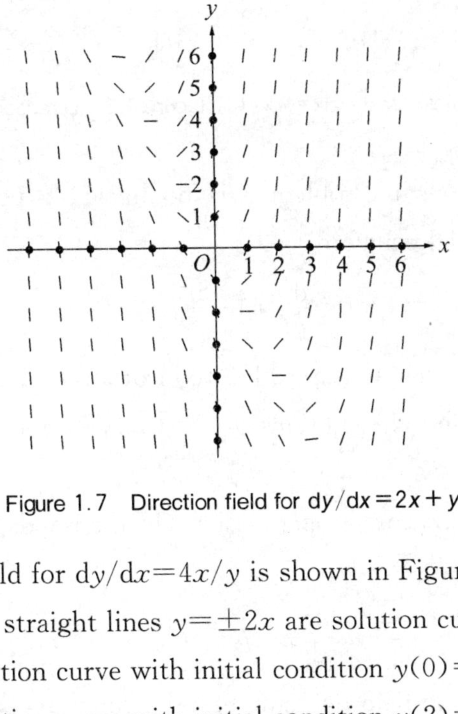

Figure 1.7 Direction field for $dy/dx = 2x + y$

2. The direction field for $dy/dx=4x/y$ is shown in Figure 1. 8.

(1) Verify that the straight lines $y=\pm 2x$ are solution curves, provided $x \neq 0$.

(2) Sketch the solution curve with initial condition $y(0)=2$.

(3) Sketch the solution curve with initial condition $y(2)=1$.

(4) What can you say about the behavior of the above solutions as $x \to +\infty$? How about $x \to -\infty$?

3. A model for the velocity v at time t of a certain object falling under the influence of gravity in a viscous medium is given by the equation

$$\frac{dv}{dt}=1-\frac{v}{8}.$$

From the direction field shown in Figure 1. 9, sketch the solution curves with the initial conditions $v(0)=5$, 8, and 15. Why is the value $v=8$ called the **terminal velocity**?

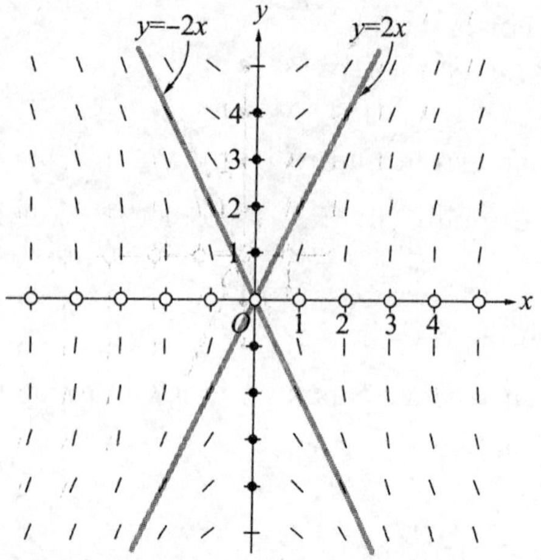

Figure 1.8 Direction field for $dy/dx=4x/y$

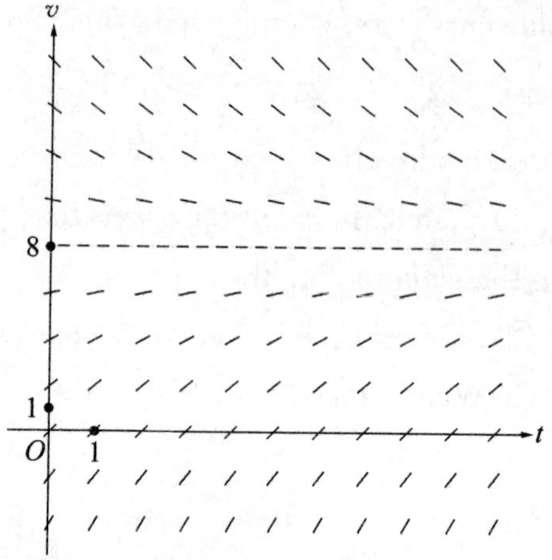

Figure 1.9 Direction field for $\frac{dv}{dt}=1-\frac{v}{8}$

4. If the viscous force in Problem 3 is nonlinear, a possible model would be provided by the differential equation

$$\frac{dv}{dt}=1-\frac{v^3}{8}.$$

Redraw the direction field in Figure 1.9 to incorporate this v^3 dependence. Sketch the solution curves with initial conditions $v(0)=0$, 1, 2, and 3. What is the terminal velocity in this case?

5. The logistic equation for the population (in thousands) of a certain species is given by

$$\frac{dp}{dt}=3p-2p^2.$$

(1) Sketch the direction field by using either a computer software package or the method of isoclines.

(2) If the initial population is 3000 [that is, $p(0)=3$], what can you say about the limiting population $\lim\limits_{t\to+\infty} p(t)$?

(3) If $p(0)=0.8$, what is $\lim\limits_{t\to+\infty} p(t)$?

(4) Can a population of 2000 ever decline to 800?

6. Consider the differential equation

$$\frac{dy}{dx}=x+\sin y.$$

(1) A solution curve passes through the point $(1,\pi/2)$. What is its slope at this

point?

(2) Argue that every solution curve is increasing for $x>1$.

(3) Show that the second derivative of every solution satisfies

$$\frac{d^2 y}{dx^2}=1+x\cos y+\frac{1}{2}\sin 2y.$$

(4) A solution curve passes through (0,0). Prove that this curve has a relative minimum at (0,0).

7. Consider the differential equation

$$\frac{dp}{dt}=p(p-1)(2-p)$$

for the population p (in thousands) of a certain species at time t.

(1) Sketch the direction field by using either a computer software package or the method of isoclines.

(2) If the initial population is 4000 [that is, $p(0)=4$], what can you say about the limiting population $\lim\limits_{t\to+\infty} p(t)$?

(3) If $p(0)=1.7$, what is $\lim\limits_{t\to+\infty} p(t)$?

(4) If $p(0)=0.8$, what is $\lim\limits_{t\to+\infty} p(t)$?

(5) Can a population of 900 ever increase to 1100?

8. The motion of a set of particles moving along the x-axis is governed by the differential equation

$$\frac{dx}{dt}=t^3-x^3,$$

where $x(t)$ denotes the position at time t of the particle.

(1) If a particle is located at $x=1$ when $t=2$, what is its velocity at this time?

(2) Show that the acceleration of a particle is given by

$$\frac{d^2 x}{dt^2}=3t^2-3t^3x^2+3x^5.$$

(3) If a particle is located at $x=2$ when $t=2.5$, can it reach the location $x=1$ at any later time?

[Hint: $t^3-x^3=(t-x)(t^2+xt+x^2)$]

9. Let $\varphi(x)$ denote the solution to the initial value problem

$$\frac{dy}{dx}=x-y,\quad y(0)=1.$$

(1) Show that $\varphi''(x)=1-\varphi'(x)=1-x+\varphi(x)$.

(2) Argue that the graph of φ is decreasing for x near zero and that as x increases from zero, $\varphi(x)$ decreases until it crosses the line $y=x$, where its derivative is zero.

(3) Let x^* be the abscissa of the point where the solution curve $y=\varphi(x)$ crosses the line $y=x$. Consider the sign of $\varphi''(x^*)$ and argue that φ has a relative minimum at x^*.

(4) What can you say about the graph of $y=\varphi(x)$ for $x>x^*$?

(5) Verify that $y=x-1$ is a solution to $dy/dx=x-y$ and explain why the graph of $\varphi(x)$ always stays above the line $y=x-1$.

(6) Sketch the direction field for $dy/dx=x-y$ by using the method of isoclines or a computer software package.

(7) Sketch the solution curve $y=\varphi(x)$ using the direction field in part (6).

10. Use a computer software package to sketch the direction field for the following differential equations. Sketch some of the solution curves.

(1) $dy/dx=\sin x$.

(2) $dy/dx=\sin y$.

(3) $dy/dx=\sin x \sin y$.

(4) $dy/dx=x^2+2y^2$.

(5) $dy/dx=x^2-2y^2$.

In Problems 11-16, draw the isoclines with their direction markers and sketch several solution curves, including the curve satisfying the given initial conditions.

11. $dy/dx=-x/y$, $y(0)=4$.

12. $dy/dx=y$, $y(0)=1$.

13. $dy/dx=2x$, $y(0)=-1$.

14. $dy/dx=x/y$, $y(0)=-1$.

15. $dy/dx=2x^2-y$, $y(0)=0$.

16. $dy/dx=x+2y$, $y(0)=1$.

17. From a sketch of the direction field, what can one say about the behavior as $x\to+\infty$ of a solution to the following?

$$\frac{dy}{dx}=3-y+\frac{1}{x}.$$

18. From a sketch of the direction field, what can one say about the behavior as $x\to+\infty$ of a solution to the following?

$$\frac{dy}{dx}=-y.$$

19. By rewriting the differential equation $dy/dx = -y/x$ in the form

$$\frac{1}{y}dy = \frac{-1}{x}dx,$$

integrate both sides to obtain the solution $y = c/x$ for an arbitrary constant c.

20. A bar magnet is often modeled as a magnetic dipole with one end labeled the north pole N and the opposite end labeled the south pole S. The magnetic field for the magnetic dipole is symmetric with respect to rotation about the axis passing lengthwise through the center of the bar. Hence we can study the magnetic field by restricting ourselves to a plane with the bar magnet centered on the x-axis.

For a point P that is located a distance r from the origin, where r is much greater than the length of the magnet, the **magnetic field lines** satisfy the differential equation

$$\frac{dy}{dx} = \frac{3xy}{2x^2 - y^2} \tag{1.23}$$

and the **equipotential lines** satisfy the equation

$$\frac{dy}{dx} = \frac{y^2 - 2x^2}{3xy}. \tag{1.24}$$

(1) Show that the two families of curves are perpendicular where they intersect. [Hint: Consider the slopes of the tangent lines of the two curves at a point of intersection.]

(2) Sketch the direction field for equation (1.23) for $-5 \leqslant x \leqslant 5, -5 \leqslant y \leqslant 5$. You can use a software package to generate the direction field or use the method of isoclines. The direction field should remind you of the experiment where iron filings are sprinkled on a sheet of paper that is held above a bar magnet. The iron filings correspond to the hash marks.

(3) Use the direction field found in part (2) to help sketch the magnetic field lines that are solutions to equation (1.23).

(4) Apply the statement of part (1) to the curves in part (3) to sketch the equipotential lines that are solutions to equation (1.23). The magnetic field lines and the equipotential lines are examples of orthogonal trajectories (see Problem 32 in Exercises 2.4).

Chapter Summary

In this chapter, we introduced some basic terminology for differential equations. The **order** of a differential equation is the order of the highest derivative present. The subject of this text is **ordinary differential equations**, which involve derivatives with respect to a single independent variable. Such equations are classified as **linear** or **nonlinear.**

An **explicit solution** of a differential equation is a function of the independent variable that satisfies the equation on some interval. An **implicit solution** is a relation between the dependent and independent variables that implicitly defines a function that is an explicit solution. A differential equation typically has infinitely many solutions. In contrast, some theorems ensure that a unique solution exists for certain **initial value problems** in which one must find a solution to the differential equation that also satisfies given initial conditions. For an nth-order equation, these conditions refer to the values of the solution and its first $n-1$ derivatives at some point.

Even if one is not successful in finding explicit solutions to a differential equation, several techniques can be used to help analyze the solutions. One such method for first-order equations views the differential equation $\mathrm{d}y/\mathrm{d}x = f(x, y)$ as specifying directions (slopes) at points on the plane. The conglomerate of such slopes is the **direction field** for the equation. Knowing the "flow of solutions" is helpful in sketching the solution curve to an initial value problem. Furthermore, carrying out this method algebraically leads to numerical approximations to the desired solution. This numerical process is called **Euler's method.**

Chapter 2

First-Order Differential Equations

2.1 Introduction: Motion of a Falling Body

An object falls through the air toward Earth. Assuming that the only forces acting on the object are gravity and air resistance, determine the velocity of the object as a function of time.

Newton's second law states that force is equal to mass times acceleration. We can express this by the equation

$$m\frac{dv}{dt}=F,$$

where F represents the total force on the object, m is the mass of the object, and dv/dt is the acceleration, expressed as the derivative of velocity with respect to time. It will be convenient in the future to define v as positive when it is directed downward.

Near Earth's surface, the force due to gravity is just the weight of the objects and is also directed downward. This force can be expressed by mg, where g is the acceleration due to gravity. No general law precisely models the air resistance acting on the object, since this force seems to depend on the velocity of the object, the density of the air, and the shape of the object, among other things. However, in some instances, air resistance can be reasonably represented by bv, where b is a positive constant depending on the density of the air and the shape of the object. We use the negative sign because air resistance is a force that opposes the motion. The forces acting on the object are depicted in Figure 2.1.

Applying Newton's law, we obtain the first-order differential equation

$$m\frac{dv}{dt}=mg-bv. \tag{2.1}$$

To solve this equation, we exploit a technique called **separation of variables**, which was used to analyze the heat transfer model in Section 1.1 and will be developed in full detail in Section 2. 2. Treating dv and dt as differentials, we rewrite equation (2. 1) so as to isolate the variables v and t on opposite sides of the equation:

$$\frac{\mathrm{d}v}{mg-bv}=\frac{\mathrm{d}t}{m}.$$

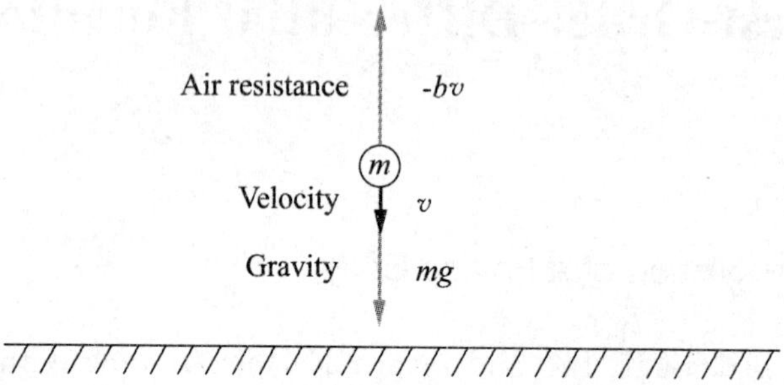

Figure 2. 1 Forces on falling object

Next we integrate the separated equation

$$\int\frac{\mathrm{d}v}{mg-bv}=\int\frac{\mathrm{d}t}{m}\tag{2. 2}$$

and derive

$$-\frac{1}{b}\ln|mg-bv|=\frac{t}{m}+c.\tag{2. 3}$$

Therefore,

$$|mg-bv|=\mathrm{e}^{-bc}\,\mathrm{e}^{-bt/m}$$

or

$$mg-by=A\mathrm{e}^{-bt/m},$$

where the new constant A has magnitude e^{-bc} and the same sign ($\pm$) as ($mg-bv$). Solving for v, we obtain

$$v=\frac{mg}{b}-\frac{A}{b}\mathrm{e}^{-bt/m},\tag{2. 4}$$

which is called a **general solution** to the differential equation because, as we will see in Section 2. 3, every solution to equation (2. 1) can be expressed in the form given in equation (2. 4).

In a specific case, we would be given the values of m, g, and b. To determine the constant A in the general solution, we can use the initial velocity of the object v_0. That is, we solve the initial value problem

$$m\frac{\mathrm{d}v}{\mathrm{d}t}=mg-bv,\quad v(0)=v_0.$$

Substituting $v = v_0$ and $t = 0$ into the general solution to the differential equation, we can solve for A. With this value for A, the solution to the initial value problem is

$$v = \frac{mg}{b} + \left(v_0 - \frac{mg}{b}\right)e^{-bt/m}. \tag{2.5}$$

The preceding formula gives the velocity of the object falling through the air as a function of time if the initial velocity of the object is v_0. In Figure 2.2, we have sketched the graph of $v(t)$ for various values of v_0. It appears from Figure 2.2 that the velocity $v(t)$ approaches mg/b regardless of the initial velocity v_0. [This is easy to see from formula (2.5) by letting $t \to +\infty$.] The constant mg/b is referred to as the **limiting** or **terminal velocity** of the object.

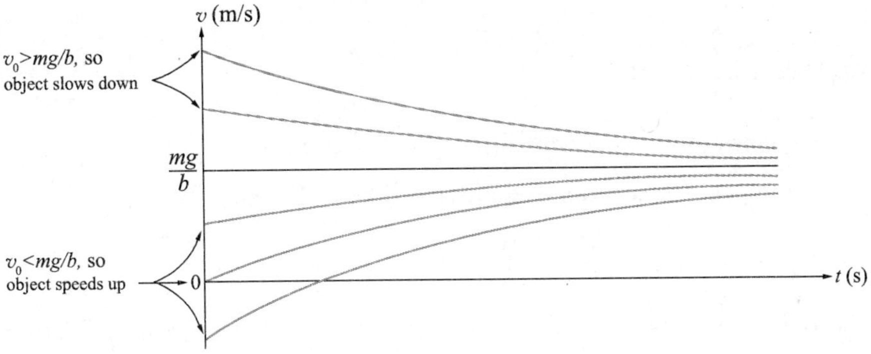

Figure 2.2 Graph of $v(t)$ for six different initial velocities v_0 ($g = 9.8\ \text{m/s}^2$, $m/b = 5$ s)

From this model for a falling body, we can make certain observations. Because $e^{-bt/m}$ rapidly tends to zero, the velocity is approximately the weight, mg, divided by the coefficient of air resistance, b. Thus, in the presence of air resistance, the heavier the object, the faster it will fall, assuming shapes and sizes are the same. Also, when air resistance is lessened (b is made smaller), the object will fall faster. These observations certainly agree with our experience.

Many other physical problems,① when formulated mathematically, lead to first-order differential equations or initial value problems. Several of these will be discussed in Chapter 3. In this chapter, we learn how to recognize and obtain solutions for some special types of first-order equations. We begin by studying separable equations, then

① The physical problem just discussed has other mathematical models. For example, one could take into account the variations in the gravitational field of Earth and the more general equations for air resistance.

linear equations, and then exact equations. The methods for solving these are the most basic. In the last two sections, we illustrate how devices such as integrating factors, substitutions, and transformations can be used to transform certain equations into either separable, exact, or linear equations that we can solve. Through our discussion of these special types of equations, you will gain insight into the behavior of solutions to more general equations and the possible difficulties in finding these solutions.

A word of warning is in order: In solving differential equations, integration plays an essential role. In particular, the separable equations in Section 2.2 always entail integration, as demonstrated in equations (2.2) and (2.3) above.

2.2 Separable Equations

A simple class of first-order differential equations that can be solved using integration is the class of **separable equations.** These are equations

$$\frac{\mathrm{d}y}{\mathrm{d}x}=f(x,y), \tag{2.6}$$

that can be rewritten to isolate the variables x and y (together with their differentials $\mathrm{d}x$ and $\mathrm{d}y$) on opposite sides of the equation, as in

$$h(y)\mathrm{d}y=g(x)\mathrm{d}x.$$

So the original right-hand side $f(x,y)$ must have the factored form

$$f(x,y)=g(x)\cdot\frac{1}{h(y)}.$$

More formally, we write $p(y)=1/h(y)$ and present the following definition.

Separable Equation

Definition 2.1 If the right-hand side of the equation

$$\frac{\mathrm{d}y}{\mathrm{d}x}=f(x,y)$$

can be expressed as a function $g(x)$ that depends only on x times a function $p(y)$ that depends only on y, then the differential equation is called **separable.** ①

① Historical footnote: A procedure for solving separable equations was discovered implicitly by Gottfried Leibniz in 1691. The explicit technique called **separation of variables** was formalized by John Bernoulli in 1694.

In other words, a first-order equation is separable if it can be written in the form

$$\frac{dy}{dx}=g(x)p(y).$$

For example, the equation

$$\frac{dy}{dx}=\frac{2x+xy}{y^2+1}$$

is separable, since (if one is sufficiently alert to detect the factorization)

$$\frac{2x+xy}{y^2+1}=x\frac{2+y}{y^2+1}=g(x)p(y).$$

However, the equation

$$\frac{dy}{dx}=1+xy$$

admits no such factorization of the right-hand side and so is not separable.

Informally speaking, one solves separable equations by performing the separation and then integrating each side.

Method for Solving Separable Equations

To solve the equation

$$\frac{dy}{dx}=g(x)p(y) \tag{2.7}$$

multiply by dx and by $h(y):=1/p(y)$ ① to obtain

$$h(y)dy=g(x)dx.$$

Then integrate both sides:

$$\int h(y)dy=\int g(x)dx,$$

$$H(y)=G(x)+C, \tag{2.8}$$

where we have merged the two constants of integration into a single symbol C. The last equation gives an implicit solution to the differential equation.

Caution: Constant functions $y\equiv C$ such that $p(C)=0$ are also solutions to equation (2.7), which may or may not be included in equation (2.8) (as we shall see in Example 2.3).

We will look at the mathematical justification of this "streamlined" procedure shortly, but first we study some examples.

Example 2.1 Solve the nonlinear equation

$$\frac{dy}{dx}=\frac{x-5}{y^2}.$$

---

① The symbol ":=" means "is defined to be".

Solution Following the streamlined approach, we separate the variables and rewrite the equation in the form

$$y^2 \mathrm{d}y = (x-5)\mathrm{d}x.$$

Integrating, we have

$$\int y^2 \mathrm{d}y = \int (x-5)\mathrm{d}x,$$

$$\frac{y^3}{3} = \frac{x^2}{2} - 5x + C,$$

and solving this last equation for y gives

$$y = \left(\frac{3x^2}{2} - 15x + 3C\right)^{1/3}.$$

Since C is a constant of integration that can be any real number, $3C$ can also be any real number. Replacing $3C$ by the single symbol K, we then have

$$y = \left(\frac{3x^2}{2} - 15x + K\right)^{1/3}.$$

If we wish to abide by the custom of letting C represent an arbitrary constant, we can go one step further and use C instead of K in the final answer. This solution family is graphed in Figure 2. 3. ◆

As Example 2. 1 attests, separable equations are among the easiest to solve. However, the procedure does require a facility for computing integrals. Many of the procedures to be discussed in the text also require a familiarity with the techniques of integration.

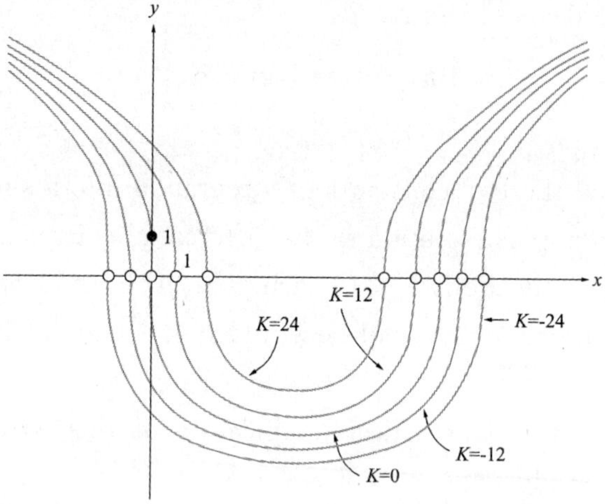

Figure 2. 3 Family of solutions for Example 2. 1①

① The gaps in the curves reflect the fact that in the original differential equation, y appears in the denominator, so that $y=0$ must be excluded.

Example 2.2 Solve the initial value problem

$$\frac{dy}{dx}=\frac{y-1}{x+3}, \quad y(-1)=0. \tag{2.9}$$

Solution Separating the variables and integrating gives

$$\frac{dy}{y-1}=\frac{dx}{x+3},$$

$$\int\frac{dy}{y-1}=\int\frac{dx}{x+3},$$

$$\ln|y-1|=\ln|x+3|+C. \tag{2.10}$$

At this point, we can either solve for y explicitly (retaining the constant C) or use the initial condition to determine C and then solve explicitly for y. Let's try the first approach.

Exponentiating equation (2.10), we have

$$e^{\ln|y-1|}=e^{\ln|x+3|+C}=e^{C}e^{\ln|x+3|},$$

$$|y-1|=e^{C}|x+3|=C_1|x+3|, \tag{2.11}$$

where $C_1:=e^{C}$. Now, depending on the values of y, we have $|y-1|=\pm(y-1)$; and similarly, $|x+3|=\pm(x+3)$. Thus, equation (2.11) can be written as

$$y-1=\pm C_1(x+3) \text{ or } y=1\pm C_1(x+3),$$

where the choice of sign depends (as we said) on the values of x and y. Because C_1 is a positive constant (recall that $C_1=e^{C}>0$), we can replace $\pm C_1$ by K, where K now represents an arbitrary nonzero constant. We then obtain

$$y=1+K(x+3). \tag{2.12}$$

Finally, we determine K such that the initial condition $y(-1)=0$ is satisfied. Putting $x=-1$ and $y=0$ in equation (2.12) gives

$$0=1+K(-1+3)=1+2K,$$

and so $K=-1/2$. Thus, the solution to the initial value problem is

$$y=1-\frac{1}{2}(x+3)=-\frac{1}{2}(x+1). \tag{2.13}$$

Alternative Approach. The second approach is to first set $x=-1$ and $y=0$ in equation (2.10) and solve for C. In this case, we obtain

$$\ln|0-1|=\ln|-1+3|+C,$$

$$0=\ln 1=\ln 2+C,$$

and so $C=-\ln 2$. Thus, from equation (2.10), the solution y is given implicitly by

$$\ln(1-y)=\ln(x+3)-\ln 2.$$

Here we have replaced $|y-1|$ by $(1-y)$ and $|x+3|$ by $(x+3)$, since we are interested in x and y near the initial values $x=-1$, $y=0$ (for such values, $y-1<0$ and $x+3>0$). Solving for y, we find

$$\ln(1-y)=\ln(x+3)-\ln 2=\ln\left(\frac{x+3}{2}\right),$$

$$1-y=\frac{x+3}{2},$$

$$y=1-\frac{1}{2}(x+3)=-\frac{1}{2}(x+1),$$

which agrees with the solution (2.13) found by the first method. ◆

Example 2.3 Solve the nonlinear equation

$$\frac{\mathrm{d}y}{\mathrm{d}x}=\frac{6x^5-2x+1}{\cos y+\mathrm{e}^y}. \tag{2.14}$$

Solution Separating variables and integrating, we find

$$(\cos y+\mathrm{e}^y)\mathrm{d}y=(6x^5-2x+1)\mathrm{d}x,$$

$$\int(\cos y+\mathrm{e}^y)\mathrm{d}y=\int(6x^5-2x+1)\mathrm{d}x,$$

$$\sin y+\mathrm{e}^y=x^6-x^2+x+C.$$

At this point, we reach an impasse. We would like to solve for y explicitly, but we cannot. This is often the case in solving nonlinear first-order equations. Consequently, when we say "solve the equation", we must on occasion be content if only an implicit form of the solution has been found. ◆

The separation of variables technique, as well as several other techniques discussed in this book, entails rewriting a differential equation by performing certain algebraic operations on it.

"Rewriting $\mathrm{d}y/\mathrm{d}x=g(x)p(y)$ as $h(y)\mathrm{d}y=g(x)\mathrm{d}x$" amounts to dividing both sides by $p(y)$. You may recall from your algebra days that doing this can be treacherous. For example, the equation $x(x-2)=4(x-2)$ has two solutions: $x=2$ and $x=4$. But if we "rewrite" the equation as $x=4$ by dividing both sides by $(x-2)$, we lose track of the root $x=2$. Thus, we should record the zeros of $(x-2)$ itself before dividing by this factor.

By the same token, we must take note of the zeros of $p(y)$ in the separable equation $\mathrm{d}y/\mathrm{d}x=g(x)p(y)$ prior to dividing. After all, if (say) $g(x)p(y)=(x-2)^2(y-13)$, then observe that the constant function $y(x)\equiv 13$ solves the differential equation $dy/dx=g(x)p(y)$:

$$\frac{dy}{dx}=\frac{d(13)}{dx}=0,$$

$$g(x)p(y)=(x-2)^2(13-13)=0.$$

Indeed, in solving the equation of Example 2. 2,

$$\frac{dy}{dx}=\frac{y-1}{x+3},$$

we obtained $y=1+K(x+3)$ as the set of solutions, where K was a nonzero constant (since K replaced $\pm e^C$). But notice that the constant function $y\equiv 1$ (which in this case corresponds to $K=0$) is also a solution to the differential equation. The reason we lost this solution can be traced back to a division by $(y-1)$ in the separation process. (See Problem 30 for an example of where a solution is lost and cannot be retrieved by setting the constant $K=0$.)

Formal Justification of Method. We close this section by reviewing the separation of variables procedure in a more rigorous framework. The original differential equation (2. 7) is rewritten in the form

$$h(y)\frac{dy}{dx}=g(x), \tag{2. 15}$$

where $h(y):=1/p(y)$. Letting $H(y)$ and $G(x)$ denote antiderivatives (indefinite integrals) of $h(y)$ and $g(x)$, respectively—that is,

$$H'(y)=h(y), \quad G'(x)=g(x),$$

we recast equation (2. 15) as

$$H'(y)\frac{dy}{dx}=G'(x).$$

By the chain rule for differentiation, the left-hand side is the derivative of the composite function $H(y(x))$:

$$\frac{d}{dx}H(y(x))=H'(y(x))\frac{dy}{dx}.$$

Thus, if $y(x)$ is a solution to equation (2. 7), then $H(y(x))$ and $G(x)$ are two functions of x that have the same derivative. Therefore, they differ by a constant:

$$H(y(x))=G(x)+C. \tag{2. 16}$$

Equation (2. 16) agrees with equation (2. 8), which was derived informally, and we have thus verified that the latter can be used to construct implicit solutions.

Exercises 2.1

In Problems 1-6, determine whether the given differential equation is separable.

1. $\frac{dy}{dx}-\sin(x+y)=0.$

2. $\frac{dy}{dx}=4y^2-3y+1.$

3. $\frac{ds}{dt}=t\ln(s^{2t})+8t^2.$

4. $\frac{dy}{dx}=\frac{ye^{x+y}}{x^2+2}.$

5. $(xy^2+3y^2)dy-2xdx=0.$

6. $s^2+\frac{ds}{dt}=\frac{s+1}{st}$

In Problems 7-16, solve the equation.

7. $\frac{dx}{dt}=3xt^2.$

8. $x\frac{dy}{dx}=\frac{1}{y^3}.$

9. $\frac{dy}{dx}=\frac{x}{y^2\sqrt{1+x}}.$

10. $\frac{dx}{dt}=\frac{t}{xe^{t+2x}}.$

11. $\frac{dy}{dx}=\frac{\sec^2 y}{1+x^2}.$

12. $x\frac{dv}{dx}=\frac{1-4v^2}{3v}.$

13. $\frac{dx}{dt}-x^3=x.$

14. $\frac{dy}{dx}=3x^2(1+y^2)^{3/2}.$

15. $y^{-1}dy+ye^{\cos x}\sin x\,dx=0.$

16. $(x+xy^2)dx+e^{x^2}y\,dy=0.$

In Problems 17-26, solve the initial value problem.

17. $y'=x^3(1-y),\quad y(0)=3.$

18. $\frac{dy}{dx}=(1+y^2)\tan x,\quad y(0)=\sqrt{3}.$

19. $\frac{1}{2}\frac{dy}{dx}=\sqrt{y+1}\cos x$, $y(\pi)=0$.

20. $x^2\frac{dy}{dx}=\frac{4x^2-x-2}{(x+1)(y+1)}$, $y(1)=1$.

21. $\frac{1}{\theta}\frac{dy}{d\theta}=\frac{y\sin\theta}{y^2+1}$, $y(\pi)=1$.

22. $x^2dx+2ydy=0$, $y(0)=2$.

23. $t^{-1}\frac{dy}{dt}=2\cos^2 y$, $y(0)=\pi/4$.

24. $\frac{dy}{dx}=8x^3e^{-2y}$, $y(1)=0$.

25. $\frac{dy}{dx}=x^2(1+y)$, $y(0)=3$.

26. $\sqrt{y}dx+(1+x)dy=0$, $y(0)=1$.

27. Sketch the solution to the initial value problem

$$\frac{dy}{dt}=2y-2yt,\quad y(0)=3$$

and determine its maximum value.

28. As stated in this section, the separation of equation (2.7) requires division by $p(y)$, and this may disguise the fact that the roots of the equation $p(y)=0$ are actually constant solutions to the differential equation.

(1) To explore this further, separate the equation

$$\frac{dy}{dx}=(x-3)(y+1)^{2/3}$$

to derive the solution,

$$y=-1+(x^2/6-x+C)^3.$$

(2) Show that $y\equiv 1$ satisfies the original equation

$$dy/dx=(x-3)(y+1)^{2/3}.$$

(3) Show that there is no choice of the constant C that will make the solution in part (1) yield the solution $y\equiv -1$. Thus, we lost the solution $y\equiv -1$ when we divided by $(y+1)^{2/3}$.

29. **Interval of Definition.** By looking at an initial value problem $dy/dx=f(x, y)$ with $y(x_0)=y_0$, it is not always possible to determine the domain of the solution $y(x)$ or the interval over which the function $y(x)$ satisfies the differential equation.

(1) Solve the equation $dy/dx=xy^3$.

(2) Give explicitly the solutions to the initial value problem with $y(0)=1$, $y(0)=1/2$, $y(0)=2$.

(3) Determine the domains of the solutions in part (2).

(4) As found in part (3), the domains of the solutions depend on the initial conditions. For the initial value problem $dy/dx=xy^3$ with $y(0)=a(a>0)$, show that as a approaches zero from the right the domain approaches the whole real line $(-\infty, +\infty)$ and as a approaches $+\infty$ the domain shrinks to a single point.

(5) Sketch the solutions to the initial value problem $dy/dx=xy^3$ with $y(0)=a$ for $a=\pm 1/2$, ± 1, and ± 2.

30. Analyze the solution $y=\varphi(x)$ to the initial value problem

$$\frac{dy}{dx}=y^2-3y+2, \quad y(0)=1.5$$

using approximation methods and then compare with its exact form as follows.

(1) Sketch the direction field of the differential equation and use it to guess the value of $\lim_{x\to\infty}\varphi(x)$.

(2) Use Euler's method with a step size of 0.1 to find an approximation of $\varphi(1)$.

(3) Find a formula for $\varphi(x)$ and graph $\varphi(x)$ on the direction field from part (1).

(4) What is the exact value of $\varphi(1)$? Compare with your approximation in part (2).

(5) Using the exact solution obtained in part (3), determine $\lim_{x\to\infty}\varphi(x)$ and compare with your guess in part (1).

31. **Mixing.** Suppose a brine containing 0.3 kilogram (kg) of salt per liter (L) runs into a tank initially filled with 400 L of water containing 2 kg of salt. If the brine enters at 10 L/min, the mixture is kept uniform by stirring, and the mixture flows out at the same rate. Find the mass of salt in the tank after 10 min (see Figure 2.4). [Hint: Let A denote the number of kilograms of salt in the tank at t min after the process begins and use the fact that rate of increase in A=rate of input−rate of exit.

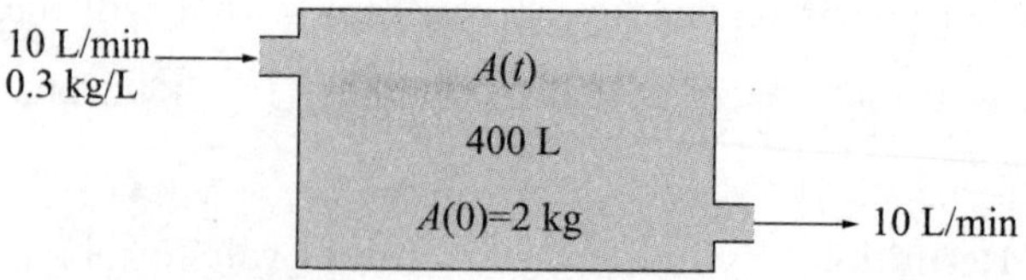

Figure 2.4 Schematic representation of a mixing problem

32. **Newton's Law of Cooling.** According to Newton's law of cooling, if an object

at temperature T is immersed in a medium having the constant temperature M, then the rate of change of T is proportional to the difference of temperature $M-T$. This gives the differential equation

$$dT/dt=k(M-T).$$

(1) Solve the differential equation for T.

(2) A thermometer reading 100 ℃ is placed in a medium having a constant temperature of 70 ℃. After 6 min, the thermometer reads 80 ℃. What is the reading after 20 min?

33. Blood plasma is stored at 4 ℃. Before the plasma can be used, it must be at 32 ℃. When the plasma is placed in an oven at 48 ℃, it takes 45 min for the plasma to warm to 32 ℃. Assume Newton's law of cooling (see Problem 32) applies. How long will it take for the plasma to warm to 32 ℃ if the oven temperature is set at (1) 37 ℃, (2) 60 ℃, and (3) 26 ℃?

34. A pot of boiling water at 100 ℃ is removed from a stove at time $t=0$ and left to cool in the kitchen. After 5 min, the water temperature has decreased to 80 ℃, and another 5 min later it has dropped to 65 ℃. Assuming Newton's law of cooling (Problem 32) applies, determine the (constant) temperature of the kitchen.

35. **Compound Interest.** If $P(t)$ is the amount of dollars in a savings bank account that pays a yearly interest rate of $r\%$ compounded continuously, then

$$\frac{dP}{dt}=\frac{r}{100}P, \quad t \text{ in years.}$$

Assume the interest is 5% annually, $P(0)=\$1000$, and no monies are withdrawn.

(1) How much will be in the account after 2 years?

(2) When will the account reach \$4000?

(3) If \$1000 is added to the account every 12 months, how much will be in the account after $3\frac{1}{2}$ years?

36. **Free Fall.** In Section 2.1, we discussed a model for an object falling toward Earth. Assuming that only air resistance and gravity are acting on the object, we found that the velocity v must satisfy the equation

$$m\frac{dv}{dt}=mg-bv,$$

where m is the mass, g is the acceleration due to gravity, and $b>0$ is a constant (see

Figure 2.1). If $m=100$ kg, $g=9.8\ \mathrm{m/s^2}$, $b=5$ kg/s, and $v(0)=10$ m/s, solve for $v(t)$. What is the limiting (i.e., terminal) velocity of the object?

37. **Grand Prix Race.** Driver A had been leading archrival B for a while by a steady 4.8 kilometres. Only 3.2 kilomiles from the finish, driver A ran out of gas and decelerated thereafter at a rate proportional to the square of his remaining speed. 1.6 kilometres later, driver A's speed was exactly halved. If driver B's speed remained constant, who won the race?

38. The atmospheric pressure (force per unit area) on a surface at an altitude z is due to the weight of the column of air situated above the surface. Therefore, the difference in air pressure p between the top and bottom of a cylindrical volume element of height Δz and cross-section area A equals the weight of the air enclosed (density ρ times volume $V=A\Delta z$ times gravity g), per unit area:

$$p(z+\Delta z)-p(z)=-\frac{p(z)(A\Delta z)g}{A}=-p(z)g\Delta z.$$

Let $\Delta z\to 0$ to derive the differential equation $\mathrm{d}p/\mathrm{d}z=-\rho g$. To analyze this further, we must postulate a formula that relates pressure and density. The perfect gas law relates pressure, volume, mass m, and absolute temperature T according to $pV=mRT/M$, where R is the universal gas constant and M is the molar mass of the air. Therefore, density and pressure are related by $\rho:=m/V=Mp/RT$.

(1) Derive the equation $\dfrac{\mathrm{d}p}{\mathrm{d}z}=-\dfrac{Mg}{RT}p$ and solve it for the "isothermal" case where T is constant to obtain the barometric pressure equation $p(z)=p(z_0)\exp[-Mg(z-z_0)/RT]$.

(2) If the temperature also varies with altitude $T=T(z)$, derive the solution

$$p(z)=p(z_0)\exp\left[-\frac{Mg}{R}\int_{z_0}^{z}\frac{\mathrm{d}\zeta}{T(\zeta)}\right].$$

(3) Suppose an engineer measures the barometric pressure at the top of a building to be 99,000 Pa (pascals), and 101,000 Pa at the base ($z=z_0$). If the absolute temperature varies as $T(z)=288-0.0065(z-z_0)$, determine the height of the building. Take $R=8.31$ J/(mol · K), $M=0.029$ kg/mol, and $g=9.8\ \mathrm{m/s^2}$. (An amusing story concerning this problem can be found at http://www.snopes.com/college/exam/barometer.asp.)

2.3 Linear Equations

A type of first-order differential equation that occurs frequently in applications is the linear equation. A linear first-order equation is an equation that can be expressed in the form

$$a_1(x)\frac{dy}{dx}+a_0(x)y=b(x), \tag{2.17}$$

where $a_1(x)$, $a_0(x)$, and $b(x)$ depend only on the independent variable x, not on y.

For example, the equation

$$x^2\sin x-(\cos x)y=(\sin x)\frac{dy}{dx}$$

is linear, because it can be rewritten in the form

$$(\sin x)\frac{dy}{dx}+(\cos x)y=x^2\sin x.$$

However, the equation

$$y\frac{dy}{dx}+(\sin x)y^3=e^x+1$$

is not linear; it cannot be put in the form of equation (2.17) due to the presence of the y^3 and $y\,dy/dx$ terms.

There are two situations for which the solution of a linear differential equation is quite immediate. The first arises if the coefficient $a_0(x)$ is identically zero, for then equation (2.17) reduces to

$$a_1(x)\frac{dy}{dx}=b(x), \tag{2.18}$$

which is equivalent to

$$y(x)=\int\frac{b(x)}{a_1(x)}dx+C$$

[as long as $a_1(x)$ is not zero].

The second is less trivial. Note that if $a_0(x)$ happens to equal the derivative of $a_1(x)$ —that is, $a_0(x)=a_1'(x)$ —then the two terms on the left-hand side of equation (2.17) simply comprise the derivative of the product $a_1(x)y$:

$$a_1(x)y'+a_0(x)y=a_1(x)y'+a_1'(x)y=\frac{d}{dx}[a_1(x)y].$$

Therefore, equation (2.17) becomes

$$\frac{d}{dx}[a_1(x)y]=b(x) \tag{2.19}$$

and the solution is again elementary:

$$a_1(x)y=\int b(x)dx+C,$$

$$y(x)=\frac{1}{a_1(x)}\left[\int b(x)\mathrm{d}x+C\right].$$

One can seldom rewrite a linear differential equation so that it reduces to a form as simple as equation (2.18). However, the form (2.19) can be achieved through multiplication of the original equation (2.17) by a well-chosen function $\mu(x)$. Such a function $\mu(x)$ is then called an **integrating factor** for equation (2.17). The easiest way to see this is first to divide the original equation (2.17) by $a_1(x)$ and put it into standard form

$$\frac{\mathrm{d}y}{\mathrm{d}x}+P(x)y=Q(x), \tag{2.20}$$

where $P(x)=a_0(x)/a_1(x)$ and $Q(x)=b(x)/a_1(x)$.

Next we wish to determine $\mu(x)$ so that the left-hand side of the multiplied equation

$$\mu(x)\frac{\mathrm{d}y}{\mathrm{d}x}+\mu(x)P(x)y=\mu(x)Q(x) \tag{2.21}$$

is just the derivative of the product $\mu(x)y$:

$$\mu(x)\frac{\mathrm{d}y}{\mathrm{d}x}+\mu(x)P(x)y=\frac{\mathrm{d}}{\mathrm{d}x}[\mu(x)y]=\mu(x)\frac{\mathrm{d}y}{\mathrm{d}x}+\mu'(x)y.$$

Clearly, this requires that μ satisfy

$$\mu'=\mu P. \tag{2.22}$$

To find such a function, we recognize that equation (2.22) is a separable differential equation, which we can write as $(1/\mu)\,\mathrm{d}\mu=P(x)\,\mathrm{d}x$. Integrating both sides gives

$$\mu(x)=\mathrm{e}^{\int P(x)\mathrm{d}x}. \tag{2.23}$$

With this choice[①] for $\mu(x)$, equation (2.21) becomes

$$\frac{\mathrm{d}}{\mathrm{d}x}[\mu(x)y]=\mu(x)Q(x),$$

which has the solution

$$y(x)=\frac{1}{\mu(x)}\left[\int\mu(x)Q(x)\mathrm{d}x+C\right]. \tag{2.24}$$

Here C is an arbitrary constant, so equation (2.24) gives a one-parameter family of solutions to equation (2.20). This form is known as the **general solution** to equation (2.20).

We can summarize the method for solving linear equations as follows.

① Any choice of the integration constant in $\int P(x)\mathrm{d}x$ will produce a suitable $\mu(x)$.

Method for Solving Linear Equations

(i) Write the equation in the standard form

$$\frac{dy}{dx}+P(x)y=Q(x).$$

(ii) Calculate the integrating factor $\mu(x)$ by the formula

$$\mu(x)=e^{\int P(x)dx}.$$

(iii) Multiply the equation in standard form by $\mu(x)$ and, recalling that the left-hand side is just $\frac{d}{dx}[\mu(x)y]$, we can obtain

$$\underbrace{\mu(x)\frac{dy}{dx}+P(x)\mu(x)y}_{\frac{d}{dx}[\mu(x)y]}=\mu(x)Q(x),$$

$$\frac{d}{dx}[\mu(x)y] \qquad =\mu(x)Q(x).$$

(iv) Integrate the last equation and solve for y by dividing by $\mu(x)$ to obtain equation (2.24).

Example 2.4 Find the general solution to

$$\frac{1}{x}\frac{dy}{dx}-\frac{2y}{x^2}=x\cos x, \quad x>0. \tag{2.25}$$

Solution To put this linear equation in standard form, we multiply by x to obtain

$$\frac{dy}{dx}-\frac{2}{x}y=x^2\cos x. \tag{2.26}$$

Here $P(x)=-2/x$, so

$$\int P(x)dx=\int\frac{-2}{x}dx=-2\ln|x|.$$

Thus, an integrating factor is

$$\mu(x)=e^{-2\ln|x|}=e^{\ln(x^{-2})}=x^{-2}.$$

Multiplying equation (2.26) by $\mu(x)$ yields

$$\underbrace{x^{-2}\frac{dy}{dx}-2x^{-3}y}_{\frac{d}{dx}(x^{-2}y)}=\cos x,$$

$$\frac{d}{dx}(x^{-2}y) \qquad =\cos x.$$

We now integrate both sides and solve for y to find

$$x^{-2}y=\int\cos x\,dx=\sin x+C,$$

$$y=x^2\sin x+Cx^2. \tag{2.27}$$

It is easily checked that this solution is valid for all $x>0$. In Figure 2. 5, we have sketched solutions for various values of the constant C in equation (2. 27). ◆

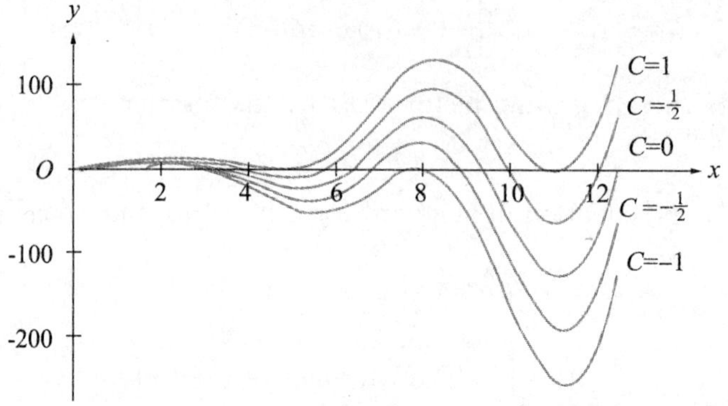

Figure 2.5 Graph of $y=x^2 \sin x+Cx^2$ for five values of the constant C

In the next example, we encounter a linear equation that arises in the study of the radioactive decay of an isotope.

Example 2. 5 A rock contains two radioactive isotopes, RA_1 and RA_2, that belong to the same radioactive series; that is, RA_1 decays into RA_2, which then decays into stable atoms. Assume that the rate at which RA_1 decays into RA_2 is $50e^{-10t}$ kg/s. Because the rate of decay of RA_2 is proportional to the mass $y(t)$ of RA_2 present, the rate of change in RA_2 is

$$\frac{dy}{dt}=\text{rate of creation}-\text{rate of decay},$$

$$\frac{dy}{dt}=50e^{-10t}-ky, \tag{2. 28}$$

where $k>0$ is the decay constant. If $k=2/\text{s}$ and initially $y(0)=40$ kg, find the mass $y(t)$ of RA_2 for $t\geqslant 0$.

Solution Equation (2. 28) is linear, so we begin by writing it in standard form

$$\frac{dy}{dt}+2y=50e^{-10t}, \quad y(0)=40, \tag{2. 29}$$

where we have substituted $k=2$ and displayed the initial condition. We now see that $P(t)=2$, so $\int P(t)\,dt=\int 2\,dt=2t$. Thus, an integrating factor is $\mu(t)=e^{2t}$. Multiplying equation (2. 29) by $\mu(t)$ yields

$$\underbrace{e^{2t}\frac{dy}{dt}+2e^{2t}y}=50e^{-10t+2t}=50e^{-8t},$$

$$\frac{d}{dt}(e^{2t}y)=50e^{-8t}.$$

Integrating both sides and solving for y, we find

$$e^{2t}y=-\frac{25}{4}e^{-8t}+C,$$

$$y=-\frac{25}{4}e^{-10t}+Ce^{-2t}.$$

Substituting $t=0$ and $y(0)=40$ gives

$$40=-\frac{25}{4}e^{0}+Ce^{0}=-\frac{25}{4}+C,$$

so $C=40+25/4=185/4$. Thus, the mass $y(t)$ of RA_2 at time t is given by

$$y(t)=\left(\frac{185}{4}\right)e^{-2t}-\left(\frac{25}{4}\right)e^{-10t}\quad t\geqslant 0. \blacklozenge \tag{2.30}$$

Example 2.6 For the initial value problem

$$y'+y=\sqrt{1+\cos^2 x},\quad y(1)=4,$$

find the value of $y(2)$.

Solution The integrating factor for the differential equation is, from equation (2.23),

$$\mu(x)=e^{\int 1dx}=e^{x}.$$

The general solution form (2.24) thus reads

$$y(x)=e^{-x}\left(\int e^{x}\sqrt{1+\cos^2 x}\,dx+C\right).$$

However, this indefinite integral cannot be expressed in finite terms with elementary functions. We revert to the form (2.21), which in this case reads

$$\frac{d}{dx}(e^{x}y)=e^{x}\sqrt{1+\cos^2 x},$$

and take the definite integral from the initial value $x=1$ to the desired value $x=2$:

$$e^{x}y\Big|_{x=1}^{x=2}=e^{2}y(2)-e^{1}y(1)=\int_{x=1}^{x=2}e^{x}\sqrt{1+\cos^2 x}\,dx.$$

Inserting the given value of $y(1)$ and solving, we express

$$y(2)=e^{-2+1}(4)+e^{-2}\int_{1}^{2}e^{x}\sqrt{1+\cos^2 x}\,dx.$$

Using Simpson's rule, we find that the definite integral is approximately 4.841, so

$$y(2)\approx 4e^{-1}+4.841e^{-2}\approx 2.127. \blacklozenge$$

Because we have established explicit formulas for the solutions to linear first-order differential equations, we get as a dividend a direct proof of the following theorem.

Existence and Uniqueness of Solution

Theorem 2.1 Suppose $P(x)$ and $Q(x)$ are continuous on an interval (a,b) that contains the point x_0. Then for any choice of initial value y_0, there exists a unique solution $y(x)$ on (a,b) to the initial value problem

$$\frac{dy}{dx}+P(x)y=Q(x), \quad y(x_0)=y_0. \tag{2.31}$$

In fact, the solution is given by equation (2.24) for a suitable value of C.

Exercises 2.2

In Problems 1-6, determine whether the given equation is separable, linear, neither, or both.

1. $\frac{dx}{dt}+xt=e^x$.

2. $x^2\frac{dy}{dx}+\sin x-y=0$.

3. $3t=e^t\frac{dy}{dt}+y\ln t$.

4. $(t^2+1)\frac{dy}{dt}=yt-y$.

5. $3r=\frac{dr}{d\theta}-\theta^3$.

6. $x\frac{dx}{dt}+t^2x=\sin t$.

In Problems 7-16, obtain the general solution to the equation.

7. $\frac{dy}{dx}=\frac{y}{x}+2x+1$.

8. $\frac{dy}{dx}-y-e^{3x}=0$.

9. $x\frac{dy}{dx}+2y=x^{-3}$.

10. $\frac{dr}{d\theta}+r\tan\theta=\sec\theta$.

11. $(t+y+1)dt-dy=0$.

12. $\frac{dy}{dx}=x^2e^{-4x}-4y$.

13. $y\frac{dx}{dy}+2x=5y^3$.

14. $x\frac{dy}{dx}+3(y+x^2)=\frac{\sin x}{x}$.

15. $(x^2+1)\frac{dy}{dx}+xy-x=0$.

16. $(1-x^2)\frac{dy}{dx}-x^2y=(1+x)\sqrt{1-x^2}$.

In Problems 17-22, solve the initial value problem.

17. $\frac{dy}{dx}-\frac{y}{x}=xe^x$, $y(1)=e-1$.

18. $\frac{dy}{dx}+4y-e^{-x}=0$, $y(0)=\frac{4}{3}$.

19. $t^2\frac{dx}{dt}+3tx=t^4\ln t+1$, $x(1)=0$.

20. $\frac{dy}{dx}+\frac{3y}{x}+2=3x$, $y(1)=1$.

21. $\cos x\frac{dy}{dx}+y\sin x=2x\cos^2 x$, $y\left(\frac{\pi}{4}\right)=\frac{-15\sqrt{2}\pi^2}{32}$.

22. $\sin x\frac{dy}{dx}+y\cos x=x\sin x$, $y\left(\frac{\pi}{2}\right)=2$.

23. **Radioactive Decay.** In Example 2.5, assume that the rate at which RA_1 decays into RA_2 is $40e^{-20t}$ kg/s and the decay constant for RA_2 is $k=5/s$. Find the mass $y(t)$ of RA_2 for $t\geqslant 0$ if initially $y(0)=10$ kg.

24. In Example 2.5, the decay constant for isotope RA_1 is 10/s, which expresses itself in the exponent of the rate term $50e^{-10t}$ kg/s. When the decay constant for RA_2 is $k=2/s$, we see that in formula (2.30) for y the term $(185/4)e^{-2t}$ eventually dominates (has greater magnitude for t large).

(1) Redo Example 2.5 taking $k=20/s$. Now which term in the solution eventually dominates?

(2) Redo Example 2.5 taking $k=10/s$.

25. Using definite integration, show that the solution to the initial value problem

$$\frac{dy}{dx}+2xy=1,\quad y(2)=1,$$

can be expressed as

$$y(x)=e^{-x^2}\left(e^4+\int_2^x e^{t^2}\,dt\right).$$

26. **Constant Multiples of Solutions.**

(1) Show that $y=e^{-x}$ is a solution of the linear equation

$$\frac{dy}{dx}+y=0, \tag{2.32}$$

and $y=x^{-1}$ is a solution of the nonlinear equation

$$\frac{dy}{dx}+y^2=0. \tag{2.33}$$

(2) Show that for any constant C, the function Ce^{-x} is a solution of equation (2.32), while Cx^{-1} is a solution of equation (2.33) only when $C=0$ or 1.

(3) Show that for any linear equation of the form

$$\frac{dy}{dx}+P(x)y=0,$$

if $\hat{y}(x)$ is a solution, then for any constant C the function $C\hat{y}(x)$ is also a solution.

27. Use your ingenuity to solve the equation

$$\frac{dy}{dx}=\frac{1}{e^{4y}+2x}.$$

[Hint: The roles of the independent and dependent variables may be reversed.]

28. **Bernoulli Equations.** The equation

$$\frac{dy}{dx}+2y=xy^{-2} \tag{2.34}$$

is an example of a Bernoulli equation. (Further discussion of Bernoulli equations is in Section 2.6.)

(1) Show that the substitution $v=y^3$ reduces equation (2.34) to the equation

$$\frac{dv}{dx}+6v=3x. \tag{2.35}$$

(2) Solve equation (2.35) for v. Then make the substitution $v=y^3$ to obtain the solution to equation (2.34).

29. **Discontinuous Coefficients.** As we will see in Chapter 3, occasions arise when the coefficient $P(x)$ in a linear equation fails to be continuous because of jump discontinuities. Fortunately, we may still obtain a "reasonable" solution. For example, consider the initial value problem

$$\frac{dy}{dx}+P(x)y=x, \quad y(0)=1,$$

where

$$P(x):=\begin{cases}1, & 0\leqslant x\leqslant 2,\\ 3, & x>2.\end{cases}$$

(1) Find the general solution for $0\leqslant x\leqslant 2$.

(2) Choose the constant in the solution of part (1) so that the initial condition is satisfied.

(3) Find the general solution for $x>2$.

(4) Now choose the constant in the general solution from part (3) so that the solution from part (2) and the solution from part (3) agree at $x=2$. By patching the two solutions together, we can obtain a continuous function that satisfies the differential equation except at $x=2$, where its derivative is undefined.

(5) Sketch the graph of the solution from $x=0$ to $x=5$.

30. **Discontinuous Forcing Terms.** There are occasions when the forcing term $Q(x)$ in a linear equation fails to be continuous because of jump discontinuities. Fortunately, we may still obtain a reasonable solution imitating the procedure discussed in Problem 29. Use this procedure to find the continuous solution to the initial value problem

$$\frac{dy}{dx}+2y=Q(x), \quad y(0)=0,$$

where

$$Q(x):=\begin{cases} 2, & 0\leqslant x\leqslant 3, \\ -2, & x>3. \end{cases}$$

Sketch the graph of the solution from $x=0$ to $x=7$.

31. **Singular Points.** Those values of x for which $P(x)$ in equation (2.20) is not defined are called **singular points** of the equation. For example, $x=0$ is a singular point of the equation $xy'+2y=3x$, since when the equation is written in the standard form, $y'+(2/x)y=3$, we see that $P(x)=2/x$ is not defined at $x=0$. On an interval containing a singular point, the questions of the existence and uniqueness of a solution are left unanswered, since Theorem 2.1 does not apply. To show the possible behavior of solutions near a singular point, consider the following equations.

(1) Show that $xy'+2y=3x$ has only one solution defined at $x=0$. Then show that the initial value problem for this equation with initial condition $y(0)=y_0$ has a unique solution when $y_0=0$ and no solution when $y_0\neq 0$.

(2) Show that $xy'-2y=3x$ has an infinite number of solutions defined at $x=0$. Then show that the initial value problem for this equation with initial condition $y(0)=0$ has an infinite number of solutions.

32. **Existence and Uniqueness.** Under the assumptions of Theorem 2.1, we will

prove that equation (2. 24) gives a solution to equation (2. 20) on (a,b). We can then choose the constant C in equation (2. 24) so that the initial value problem (2. 31) is solved.

(1) Show that since $P(x)$ is continuous on (a,b), then $\mu(x)$ defined in equation (2. 23) is a positive, continuous function satisfying $\mathrm{d}\mu/\mathrm{d}x=P(x)\mu(x)$ on (a,b).

(2) Since

$$\frac{\mathrm{d}}{\mathrm{d}x}\int \mu(x)Q(x)\,\mathrm{d}x=\mu(x)Q(x),$$

verify that y given in equation (2. 24) satisfies equation (2. 20) by differentiating both sides of equation (2. 24).

(3) Show that when we let $\int \mu(x)Q(x)\,\mathrm{d}x$ be the antiderivative whose value at x_0 is 0 (i. e., $\int_{x_0}^{x} \mu(t)Q(t)\,\mathrm{d}t$) and choose C to be $y_0\mu(x_0)$, the initial condition $y(x_0)=y_0$ is satisfied.

(4) Start with the assumption that $y(x)$ is a solution to the initial value problem (2. 31) and argue that the discussion leading to equation (2. 24) implies that $y(x)$ must obey equation (2. 24). Then argue that the initial condition in (2. 31) determines the constant C uniquely.

33. **Mixing.** Suppose a brine containing 0. 2 kg of salt per liter runs into a tank initially filled with 500 L of water containing 5 kg of salt. The brine enters the tank at a rate of 5 L/min. The mixture, kept uniform by stirring, is flowing out at the rate of 5 L/min (see Figure 2. 6).

(1) Find the concentration, in kilograms per liter, of salt in the tank after 10 min. [Hint: Let A denote the number of kilograms of salt in the tank at t min after the process begins and use the fact that rate of increase in A=rate of input−rate of exit.]

(2) After 10 min, a leak develops in the tank and an additional liter per minute of mixture flows out of the tank (see Figure 2. 7). What will be the concentration, in kilograms per liter, of salt in the tank 20 min after the leak develops?

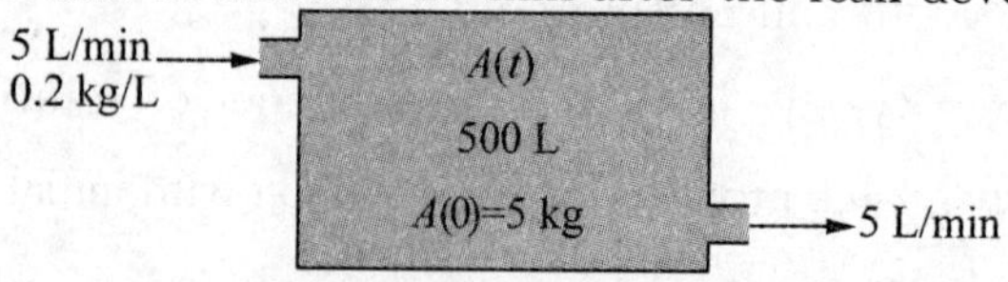

Figure 2.6 Mixing problem with equal flow rates

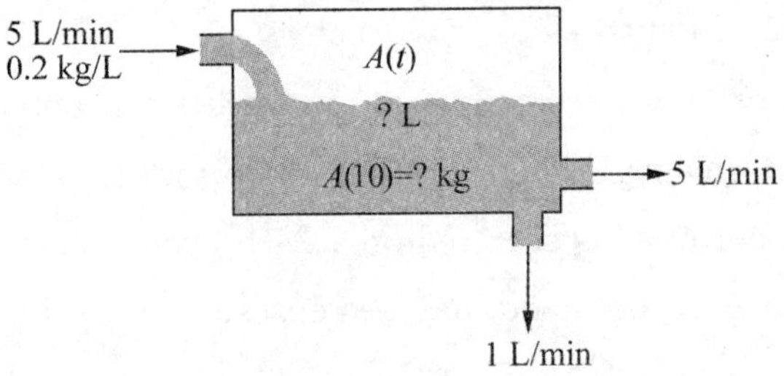

Figure 2.7 Mixing problem with unequal flow rates

34. **Variation of Parameters.** Here is another procedure for solving linear equations that is particularly useful for higher-order linear equations. This method is called **variation of parameters.** It is based on the idea that just by knowing the form of the solution, we can substitute into the given equation and solve for any unknowns. Here we illustrate the method for first-order equations (see Sections 4. 6 and 6. 4 for the generalization to higher-order equations).

(1) Show that the general solution to

$$\frac{dy}{dx}+P(x)y=Q(x) \tag{2.36}$$

has the form

$$y(x)=Cy_h(x)+y_p(x),$$

where $y_h(\not\equiv 0)$ is a solution to equation (2. 36) when $Q(x)\equiv 0$, C is a constant, and $y_p(x)=v(x)y_h(x)$ for a suitable function $v(x)$. [Hint: Show that we can take $y_h=\mu^{-1}(x)$ and then use equation (2. 24).]

We can in fact determine the unknown function y_h by solving a separable equation. Then direct substitution of vy_h in the original equation will give a simple equation that can be solved for v.

Use this procedure to find the general solution to

$$\frac{dy}{dx}+\frac{3}{x}y=x^2, \quad x>0, \tag{2.37}$$

by completing the following steps.

(2) Find a nontrivial solution y_h to the separable equation

$$\frac{dy}{dx}+\frac{3}{x}y=0, \quad x>0. \tag{2.38}$$

(3) Assuming equation (2. 37) has a solution of the form $y_p(x)=v(x)y_h(x)$, substitute this into equation (2. 37), and simplify to obtain $v'(x)=x^2/y_h(x)$.

(4) Now integrate to get $v(x)$.

(5) Verify that $y(x)=Cy_h(x)+v(x)y_h(x)$ is a general solution to equation (2.37).

35. **Secretion of Hormones.** The secretion of hormones into the blood is often a periodic activity. If a hormone is secreted on a 24 h cycle, then the rate of change of the level of the hormone in the blood may be represented by the initial value problem

$$\frac{\mathrm{d}x}{\mathrm{d}t}=\alpha-\beta\cos\frac{\pi t}{12}-kx,\quad x(0)=x_0,$$

where $x(t)$ is the amount of the hormone in the blood at time t, α is the average secretion rate, β is the amount of daily variation in the secretion, and k is a positive constant reflecting the rate at which the body removes the hormone from the blood. If $\alpha=\beta=1$, $k=2$, and $x_0=10$, solve for $x(t)$.

36. Use the separation of variables technique to derive the solution (2.23) to the differential equation (2.22).

37. The temperature T (in units of 100 ℃) of a university classroom on a cold winter day varies with time t (in hours) as

$$\frac{\mathrm{d}T}{\mathrm{d}t}=\begin{cases}1-T, & \text{if heating unit is ON;}\\ -T, & \text{if heating unit is OFF.}\end{cases}$$

Suppose $T=0$ at 9:00 a.m., the heating unit is ON from 9 to 10 a.m., OFF from 10 to 11 a.m., ON again from 11 a.m. to noon, and so on for the rest of the day. How warm will the classroom be at noon? At 5:00 p.m.?

2.4 Exact Equations

Suppose the mathematical function $F(x,y)$ represents some physical quantity, such as temperature, in a region of the xy-plane. Then the level curves of F, where $F(x,y)=C$, could be interpreted as isotherms on a weather map, as depicted in Figure 2.8.

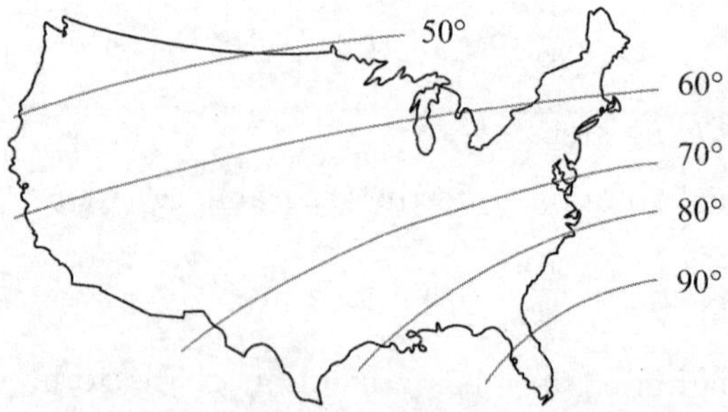

Figure 2.8 Level curves of $F(x,y)$

How does one calculate the slope of the tangent to a level curve? It is accomplished by implicit differentiation. One takes the derivative, with respect to x, of both sides of the equation $F(x,y)=C$, taking into account that y depends on x along the curve:

$$\frac{\mathrm{d}}{\mathrm{d}x}F(x,y)=\frac{\mathrm{d}}{\mathrm{d}x}(C) \quad \text{or}$$

$$\frac{\partial F}{\partial x}+\frac{\partial F}{\partial y}\frac{\mathrm{d}y}{\mathrm{d}x}=0, \tag{2.39}$$

and solves for the slope:

$$\frac{\mathrm{d}y}{\mathrm{d}x}=f(x,y)=-\frac{\partial F/\partial x}{\partial F/\partial y}. \tag{2.40}$$

The expression obtained by formally multiplying the left-hand member of equation (2.39) by $\mathrm{d}x$ is known as the total differential of F, written as $\mathrm{d}F$:

$$\mathrm{d}F:=\frac{\partial F}{\partial x}\mathrm{d}x+\frac{\partial F}{\partial y}\mathrm{d}y,$$

and our procedure for obtaining the equation for the slope $f(x,y)$ of the level curve $F(x,y)=C$ can be expressed as setting the total differential $\mathrm{d}F=0$ and solving.

Because equation (2.40) has the form of a differential equation, we should be able to reverse this logic and come up with a very easy technique for solving some differential equations. After all, any first-order differential equation $\mathrm{d}y/\mathrm{d}x=f(x,y)$ can be rewritten in the (differential) form

$$M(x,y)\mathrm{d}x+N(x,y)\mathrm{d}y=0 \tag{2.41}$$

(in a variety of ways). Now, if the left-hand side of equation (2.41) can be identified as a total differential,

$$M(x,y)\mathrm{d}x+N(x,y)\mathrm{d}y=\frac{\partial F}{\partial x}\mathrm{d}x+\frac{\partial F}{\partial y}\mathrm{d}y=\mathrm{d}F(x,y),$$

then its solutions are given (implicitly) by the level curves

$$F(x,y)=C$$

for an arbitrary constant C.

Example 2.7 Solve the differential equation

$$\frac{\mathrm{d}y}{\mathrm{d}x}=-\frac{2xy^2+1}{2x^2y}.$$

Solution Some of the choices of differential forms corresponding to this equation are

$$(2xy^2+1)\mathrm{d}x+2x^2y\mathrm{d}y=0,$$

$$\frac{2xy^2+1}{2x^2y}\mathrm{d}x+\mathrm{d}y=0,$$

$$\mathrm{d}x+\frac{2x^2y}{2xy^2+1}\mathrm{d}y=0, \text{ etc.}$$

However, the first form is best for our purposes because it is a total differential of the function $F(x,y)=x^2y^2+x$:

$$\begin{aligned}(2xy^2+1)\mathrm{d}x+2x^2y\mathrm{d}y &= \mathrm{d}(x^2y^2+x)\\ &= \frac{\partial}{\partial x}(x^2y^2+x)\mathrm{d}x+\frac{\partial}{\partial y}(x^2y^2+x)\mathrm{d}y.\end{aligned}$$

Thus, the solutions are given implicitly by the formula $x^2y^2+x=C$ (see Figure 2.9). ◆

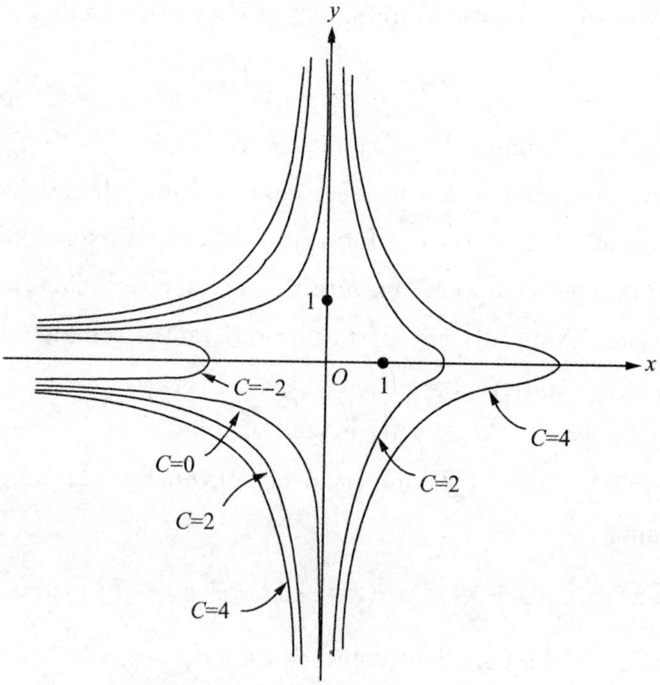

Figure 2.9 Solutions of Example 2.7

Next we introduce some terminology.

Exact Differential Form

Definition 2.2 The differential form $M(x,y)\,\mathrm{d}x+N(x,y)\,\mathrm{d}y$ is said to be **exact** in a rectangle R if there is a function $F(x,y)$ such that

$$\frac{\partial F}{\partial x}(x,y)=M(x,y) \quad \text{and} \quad \frac{\partial F}{\partial y}(x,y)=N(x,y) \tag{2.42}$$

for all (x,y) in R. That is, the total differential of $F(x,y)$ satisfies

$$dF(x,y)=M(x,y)dx+N(x,y)dy.$$

If $M(x,y)dx+N(x,y)dy$ is an exact differential form, then the equation

$$M(x,y)dx+N(x,y)dy=0$$

is called an **exact equation.**

As you might suspect, in applications a differential equation is rarely given to us in exact differential form. However, the solution procedure is so quick and simple for such equations that we devote this section to it. From Example 2.7, we see that what is needed is (i) a test to determine if a differential form $M(x,y)dx+N(x,y)dy$ is exact and, if so, (ii) a procedure for finding the function $F(x,y)$ itself.

The test for exactness arises from the following observation. If

$$M(x,y)dx+N(x,y)dy=\frac{\partial F}{\partial x}dx+\frac{\partial F}{\partial y}dy,$$

then the calculus theorem concerning the equality of continuous mixed partial derivatives

$$\frac{\partial}{\partial y}\frac{\partial F}{\partial x}=\frac{\partial}{\partial x}\frac{\partial F}{\partial y}$$

would dictate a "compatibility condition" on the functions M and N:

$$\frac{\partial}{\partial y}M(x,y)=\frac{\partial}{\partial x}N(x,y).$$

In fact, Theorem 2.2 states that the compatibility condition is also sufficient for the differential form to be exact.

Test for Exactness

Theorem 2.2 Suppose the first partial derivatives of $M(x,y)$ and $N(x,y)$ are continuous in a rectangle R. Then

$$M(x,y)dx+N(x,y)dy=0$$

is an exact equation in R if and only if the compatibility condition

$$\frac{\partial M}{\partial y}(x,y)=\frac{\partial N}{\partial x}(x,y) \tag{2.43}$$

holds for all (x,y) in R. ①

Before we address the proof of Theorem 2.2, note that in Example 2.7 the differential form that led to the total differential was

① Historical footnote: This theorem was proven by Leonhard Euler in 1734.

$$(2xy^2+1)\mathrm{d}x+(2x^2y)\mathrm{d}y=0.$$

The compatibility conditions are easily confirmed:

$$\frac{\partial M}{\partial y}=\frac{\partial}{\partial y}(2xy^2+1)=4xy,$$

$$\frac{\partial N}{\partial x}=\frac{\partial}{\partial x}(2x^2y)=4xy.$$

Also clear is the fact that the other differential forms considered,

$$\frac{2xy^2+1}{2x^2y}\mathrm{d}x+\mathrm{d}y=0,\quad \mathrm{d}x+\frac{2x^2y}{2xy^2+1}\mathrm{d}y=0,$$

do not meet the compatibility conditions.

Proof There are two parts to the theorem: Exactness implies compatibility, and compatibility implies exactness. First, we have seen that if the differential equation is exact, then the two members of equation (2.43) are simply the mixed second partials of a function $F(x,y)$. As such, their equality is ensured by the theorem of calculus that states that mixed second partials are equal if they are continuous. Because the hypothesis of Theorem 2.2 guarantees the latter condition, equation (2.43) is validated.

Rather than proceed directly with the proof of the second part of the theorem, let's derive a formula for a function $F(x,y)$ that satisfies $\partial F/\partial x=M$ and $\partial F/\partial y=N$. Integrating the first equation with respect to x yields

$$F(x,y)=\int M(x,y)\mathrm{d}x+g(y). \tag{2.44}$$

Notice that instead of using C to represent the constant of integration, we have written $g(y)$. This is because y is held fixed while integrating with respect to x, and so our "constant" may well depend on y. To determine $g(y)$, we differentiate both sides of equation (2.44) with respect to y to obtain

$$\frac{\partial F}{\partial y}(x,y)=\frac{\partial}{\partial y}\int M(x,y)\mathrm{d}x+\frac{\partial}{\partial y}g(y). \tag{2.45}$$

As g is a function of y alone, we can write $\partial g/\partial y=g'(y)$, and solving equation (2.45) for $g'(y)$ gives

$$g'(y)=\frac{\partial F}{\partial y}(x,y)-\frac{\partial}{\partial y}\int M(x,y)\mathrm{d}x.$$

Since $\partial F/\partial y=N$, this last equation becomes

$$g'(y)=N(x,y)-\frac{\partial}{\partial y}\int M(x,y)\mathrm{d}x. \tag{2.46}$$

Notice that although the right-hand side of equation (2.46) indicates a possible dependence on x, the appearances of this variable must cancel because the left-hand side, $g'(y)$, depends only on y. By integrating equation (2.46), we can determine $g(y)$ up to a numerical constant, and therefore we can determine the function $F(x,y)$ up to a numerical constant from the functions $M(x,y)$ and $N(x,y)$.

To finish the proof of Theorem 2.2, we need to show that the condition (2.43) implies that $M\mathrm{d}x+N\mathrm{d}y=0$ is an exact equation. This we do by actually exhibiting a function $F(x,y)$ that satisfies $\partial F/\partial x=M$ and $\partial F/\partial y=N$. Fortunately, we needn't look too far for such a function. The discussion in the first part of the proof suggests equation (2.44) as a candidate, where $g'(y)$ is given by equation (2.46). Namely, we define $F(x,y)$ by

$$F(x,y):=\int_{x_0}^{x} M(t,y)\mathrm{d}t+g(y), \tag{2.47}$$

where (x_0, y_0) is a fixed point in the rectangle R and $g(y)$ is determined, up to a numerical constant, by the equation

$$g'(y):=N(x,y)-\frac{\partial}{\partial y}\int_{x_0}^{x} M(t,y)\mathrm{d}t. \tag{2.48}$$

Before proceeding we must address an extremely important question concerning the definition of $F(x,y)$. That is, how can we be sure (in this portion of the proof) that $g'(y)$, as given in equation (2.48), is really a function of just y alone? To show that the right-hand side of equation (2.48) is independent of x (that is, the appearances of the variable x cancel), all we need to do is to show that its partial derivative with respect to x is zero. This is where condition (2.43) is utilized. We leave to the reader this computation and the verification that $F(x, y)$ satisfies conditions (2.42) (see Problems 35 and 36 in Exercises 2.3). ◆

The construction in the proof of Theorem 2.2 actually provides an explicit procedure for solving exact equations. Let's recap and look at some examples.

Method for solving exact equations is as follows:

(i) If $M\,\mathrm{d}x+N\,\mathrm{d}y=0$ is exact, then $\partial F/\partial x=M$. Integrate this last equation with respect to x to get

$$F(x,y)=\int M(x,y)\mathrm{d}x+g(y). \tag{2.49}$$

(ii) To determine $g(y)$, take the partial derivative with respect to y of both sides of equation (2.49) and substitute N for $\partial F/\partial y$. We can now solve for $g'(y)$.

(iii) Integrate $g'(y)$ to obtain $g(y)$ up to a numerical constant. Substituting $g(y)$ into equation (2.49) gives $F(x,y)$.

(iv) The solution to $M\mathrm{d}x+N\mathrm{d}y=0$ is given implicitly by

$$F(x,y)=C.$$

(Alternatively, starting with $\partial F/\partial y=N$, the implicit solution can be found by first integrating with respect to y; see Example 2.9.)

Example 2.8 Solve

$$(2xy-\sec^2 x)\mathrm{d}x+(x^2+2y)\mathrm{d}y=0. \tag{2.50}$$

Solution Here $M(x,y)=2xy-\sec^2 x$ and $N(x,y)=x^2+2y$. Because

$$\frac{\partial M}{\partial y}=2x=\frac{\partial N}{\partial x},$$

equation (2.50) is exact. To find $F(x,y)$, we begin by integrating M with respect to x:

$$\begin{aligned}F(x,y)&=\int(2xy-\sec^2 x)\mathrm{d}x+g(y)\\&=x^2y-\tan x+g(y).\end{aligned} \tag{2.51}$$

Next we take the partial derivative of equation (2.52) with respect to y and substitute x^2+2y for N:

$$\frac{\partial F}{\partial y}(x,y)=N(x,y),$$

$$x^2+g'(y)=x^2+2y.$$

Thus, $g'(y)=2y$, and since the choice of the constant of integration is not important, we can take $g(y)=y^2$. Hence, from equation (2.51), we have $F(x,y)=x^2y-\tan x+y^2$, and the solution to equation (2.50) is given implicitly by $x^2y-\tan x+y^2=C$. ◆

Remark The procedure for solving exact equations requires several steps. As a check on our work, we observe that when we solve for $g'(y)$, we must obtain a function that is independent of x. If this is not the case, then we have erred either in our computation of $F(x,y)$ or in computing $\partial M/\partial y$ or $\partial N/\partial x$.

In the construction of $F(x,y)$, we can first integrate $N(x,y)$ with respect to y to get

$$F(x,y)=\int N(x,y)\mathrm{d}y+h(x) \tag{2.52}$$

and then proceed to find $h(x)$. We illustrate this alternative method in the next example.

Example 2.9 Solve

$$(1+e^x y+xe^x y)dx+(xe^x+2)dy=0. \tag{2.53}$$

Solution Here $M=1+e^x y+xe^x y$ and $N=xe^x+2$. Because

$$\frac{\partial M}{\partial y}=e^x+xe^x=\frac{\partial N}{\partial x},$$

equation (2.53) is exact. If we now integrate $N(x,y)$ with respect to y, we obtain

$$F(x,y)=\int(xe^x+2)dy+h(x)=xe^x y+2y+h(x).$$

When we take the partial derivative with respect to x and substitute for M, we get

$$\frac{\partial F}{\partial x}(x,y)=M(x,y),$$

$$e^x y+xe^x y+h'(x)=1+e^x y+xe^x y.$$

Thus, $h'(x)=1$, so we take $h(x)=x$. Hence, $F(x,y)=xe^x y+2y+x$, and the solution to equation (2.53) is given implicitly by $xe^x y+2y+x=C$. In this case, we can solve explicitly for y to obtain $y=(C-x)/(2+xe^x)$. ◆

Remark Since we can use either procedure for finding $F(x,y)$, it may be worthwhile to consider each of the integrals $\int M(x,y)\,dx$ and $\int N(x,y)\,dy$. If one is easier to evaluate than the other, this would be sufficient reason for us to use one method over the other. [The skeptical reader should try solving equation (2.53) by first integrating $M(x,y)$.]

Example 2.10 Show that

$$(x+3x^3\sin y)dx+(x^4\cos y)dy=0 \tag{2.54}$$

is not exact but that multiplying this equation by the factor x^{-1} yields an exact equation. Use this fact to solve equation (2.54).

Solution In equation (2.54), $M=x+3x^3\sin y$ and $N=x^4\cos y$. Because

$$\frac{\partial M}{\partial y}=3x^3\cos y\neq 4x^3\cos y=\frac{\partial N}{\partial x},$$

equation (2.54) is not exact. When we multiply equation (2.54) by the factor x^{-1}, we obtain

$$(1+3x^2\sin y)dx+(x^3\cos y)dy=0. \tag{2.55}$$

For this new equation, $M=1+3x^2\sin y$ and $N=x^3\cos y$. If we test for exactness, we now find that

$$\frac{\partial M}{\partial y}=3x^2\cos y=\frac{\partial N}{\partial x},$$

and hence equation (2.55) is exact. Upon solving equation (2.55), we find that the solution is given implicitly by $x+x^3\sin y=C$. Since equations (2.54) and (2.55) differ only by a factor of x, then any solution to one will be a solution for the other whenever $x\neq 0$. Hence the solution to equation (2.54) is given implicitly by $x+x^3\sin y=C$. ◆

In Section 2.5, we will discuss methods for finding factors that, like x^{-1} in Example 2.10, change inexact equations into exact equations.

Exercises 2.3

In Problems 1-8, classify the equation as separable, linear, exact, or none of these. Notice that some equations may have more than one classification.

1. $(x^{10/3}-2y)\mathrm{d}x+x\mathrm{d}y=0$.
2. $(x^2y+x^4\cos x)\mathrm{d}x-x^3\mathrm{d}y=0$.
3. $\sqrt{-2y-y^2}\,\mathrm{d}x+(3+2x-x^2)\mathrm{d}y=0$.
4. $(y\mathrm{e}^{xy}+2x)\mathrm{d}x+(x\mathrm{e}^{xy}-2y)\mathrm{d}y=0$.
5. $xy\mathrm{d}x+\mathrm{d}y=0$.
6. $y^2\,\mathrm{d}x+(2xy+\cos y)\mathrm{d}y=0$.
7. $[2x+y\cos(xy)]\mathrm{d}x+[x\cos(xy)-2y]\mathrm{d}y=0$.
8. $\theta\mathrm{d}r+(3r-\theta-1)\mathrm{d}\theta=0$.

In Problems 9-20, determine whether the equation is exact. If it is, then solve it.

9. $(2x+y)\mathrm{d}x+(x-2y)\mathrm{d}y=0$.
10. $(2xy+3)\mathrm{d}x+(x^2-1)\mathrm{d}y=0$.
11. $(\cos x\cos y+2x)\mathrm{d}x-(\sin x\sin y+2y)\mathrm{d}y=0$.
12. $(\mathrm{e}^x\sin y-3x^2)\mathrm{d}x+(\mathrm{e}^x\cos y+y^{-2/3}/3)\mathrm{d}y=0$.
13. $(t/y)\mathrm{d}y+(1+\ln y)\mathrm{d}y=0$.
14. $\mathrm{e}^t(y-t)\mathrm{d}t+(1+\mathrm{e}^t)\mathrm{d}y=0$.
15. $\cos\theta\,\mathrm{d}r-(r\sin\theta-\mathrm{e}^\theta)\mathrm{d}\theta=0$.
16. $(y\mathrm{e}^{xy}-1/y)\mathrm{d}x+(x\mathrm{e}^{xy}+x/y^2)\mathrm{d}y=0$.
17. $(1/y)\mathrm{d}x-(3y-x/y^2)\mathrm{d}y=0$.
18. $[2x+y^2-\cos(x+y)]\mathrm{d}x+[2xy-\cos(x+y)-\mathrm{e}^y]\mathrm{d}y=0$.

19. $\left(2x+\frac{y}{1+x^2y^2}\right)dx+\left(\frac{x}{1+x^2y^2}-2y\right)dy=0$.

20. $\left[\frac{2}{\sqrt{1-x^2}}+y\cos(xy)\right]dx+[x\cos(xy)-y^{-1/3}]dy=0$.

In Problems 21-26, solve the initial value problem.

21. $(1/x+2y^2x)dx+(2yx^2-\cos y)dy=0$, $y(1)=\pi$.

22. $(ye^{xy}-1/y)dx+(xe^{xy}+x/y^2)dy=0$, $y(1)=1$.

23. $(e^ty+te^ty)dt+(te^t+2)dt=0$, $y(0)=-1$.

24. $(e^tx+1)dt+(e^t-1)dx=0$, $x(1)=1$.

25. $(y^2\sin x)dx+(1/x-y/x)dy=0$, $y(\pi)=1$.

26. $(\tan y-2)dx+(x\sec^2 y+1/y)dy=0$, $y(0)=1$.

27. For each of the following equations, find the most general function $M(x,y)$ so that the equation is exact.

(1) $M(x,y)dx+(\sec^2 y-x/y)dy=0$.

(2) $M(x,y)dx+(\sin x\cos y-xy-e^{-y})dy=0$.

28. For each of the following equations, find the most general function $N(x,y)$ so that the equation is exact.

(1) $[y\cos(xy)+e^x]dx+N(x,y)dy=0$.

(2) $(ye^{xy}-4x^3y+2)dx+N(x,y)dy=0$.

29. Consider the equation

$$(y^2+2xy)dx-x^2\,dy=0.$$

(1) Show that this equation is not exact.

(2) Show that multiplying both sides of the equation by y^{-2} yields a new equation that is exact.

(3) Use the solution of the resulting exact equation to solve the original equation.

(4) Were any solutions lost in the process?

30. Consider the equation

$$(5x^2y+6x^3y^2+4xy^2)dx+(2x^3+3x^4y+3x^2y)dy=0.$$

(1) Show that the equation is not exact.

(2) Multiply the equation by x^ny^m and determine values for n and m that make the resulting equation exact.

(3) Use the solution of the resulting exact equation to solve the original equation.

31. Argue that in the proof of Theorem 2. 2 the function g can be taken as

$$g(y)=\int_{y_0}^{y} N(x,t)\mathrm{d}t-\int_{y_0}^{y}\left[\frac{\partial}{\partial t}\int_{x_0}^{x} M(s,t)\mathrm{d}s\right]\mathrm{d}t,$$

which can be expressed as

$$g(y)=\int_{y_0}^{y} N(x,t)\mathrm{d}t-\int_{x_0}^{x} M(s,y)\mathrm{d}s+\int_{x_0}^{x} M(s,y_0)\mathrm{d}s.$$

This leads ultimately to the representation

$$F(x,y)=\int_{y_0}^{y} N(x,t)\mathrm{d}t+\int_{x_0}^{x} M(s,y_0)\mathrm{d}s. \tag{2.56}$$

Evaluate this formula directly with $x_0=0$, $y_0=0$ to rework

(1) Example 2. 7.

(2) Example 2. 8.

(3) Example 2. 9.

32. **Orthogonal Trajectories.** A geometric problem occurring often in engineering is that of finding a family of curves (orthogonal trajectories) that intersects a given family of curves orthogonally at each point. For example, we may be given the lines of force of an electric field and want to find the equation for the equipotential curves. Consider the family of curves described by $F(x,y)=k$, where k is a parameter. Recall from the discussion of equation (2. 40) that for each curve in the family, the slope is given by

$$\frac{\mathrm{d}y}{\mathrm{d}x}=-\frac{\partial F}{\partial x}\Big/\frac{\partial F}{\partial y}.$$

(1) Recall that the slope of a curve that is orthogonal (perpendicular) to a given curve is just the negative reciprocal of the slope of the given curve. Using this fact, show that the curves orthogonal to the family $F(x,y)=k$ satisfy the differential equation

$$\frac{\partial F}{\partial y}(x,y)\mathrm{d}x-\frac{\partial F}{\partial x}(x,y)\mathrm{d}y=0.$$

(2) Using the preceding differential equation, show that the orthogonal trajectories to the family of circles $x^2+y^2=k$ are just straight lines through the origin (see Figure 2. 10).

(3) Show that the orthogonal trajectories to the family of hyperbolas $xy=k$ are the hyperbolas $x^2-y^2=k$ (see Figure 2. 11).

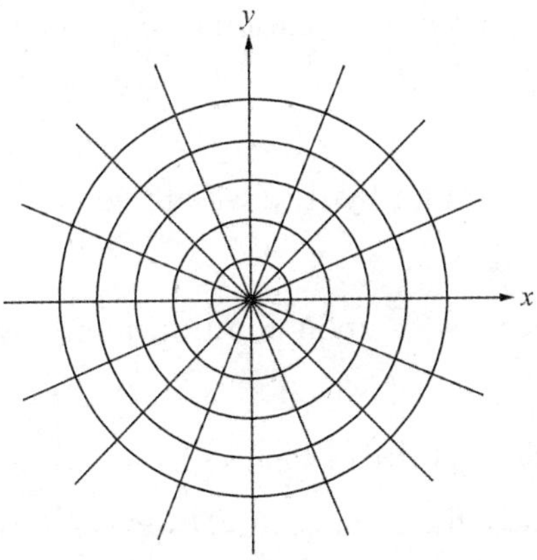

Figure 2.10 Orthogonal trajectories for concentric circles are lines through the center

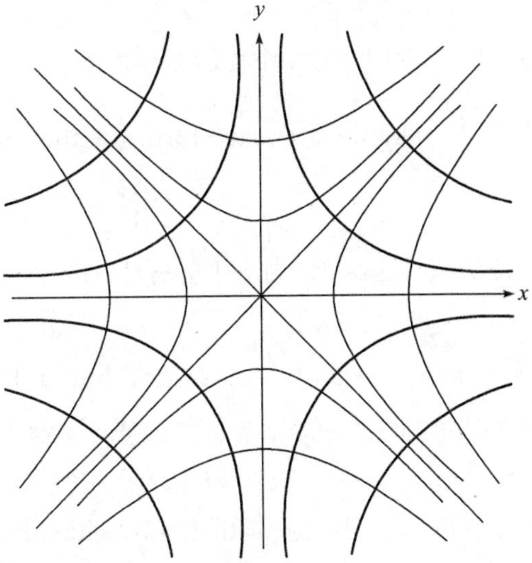

Figure 2.11 Families of orthogonal hyperbolas

33. Use the method in Problem 32 to find the orthogonal trajectories for each of the given families of curves, where k is a parameter.

(1) $2x^2+y^2=k$.

(2) $y=kx^4$.

(3) $y=e^{kx}$.

(4) $y^2=kx$.

[Hint: First express the family in the form $F(x,y)=k$.]

34. Use the method described in Problem 32 to show that the orthogonal trajectories to the family of curves $x^2+y^2=kx$, k is a parameter, satisfy

$$(2yx^{-1})\mathrm{d}x+(y^2x^{-2}-1)\mathrm{d}y=0.$$

Find the orthogonal trajectories by solving the above equation. Sketch the family of curves, along with their orthogonal trajectories. [Hint: Try multiplying the equation by $x^m y^n$ as in Problem 30.]

35. Using condition (2.43), show that the right-hand side of equation (2.48) is independent of x by showing that its partial derivative with respect to x is zero. [Hint: Since the partial derivatives of M are continuous, Leibniz's theorem allows you to interchange the operations of integration and differentiation.]

36. Verify that $F(x, y)$ as defined by equations (2.47) and (2.48) satisfies condition (2.42).

2.5 Special Integrating Factors

If we take the standard form for the linear differential equation of Section 2.3,

$$\frac{\mathrm{d}y}{\mathrm{d}x}+P(x)y=Q(x),$$

and rewrite it in differential form by multiplying through by $\mathrm{d}x$, we obtain

$$[P(x)y-Q(x)]\mathrm{d}x+\mathrm{d}y=0.$$

This form is certainly not exact, but it becomes exact upon multiplication by the integrating factor $\mu(x)=\mathrm{e}^{\int P(x)\mathrm{d}x}$. We have

$$[\mu(x)P(x)y-\mu(x)Q(x)]\mathrm{d}x+\mu(x)\mathrm{d}y=0$$

as the form, and the compatibility condition is precisely the identity $\mu(x)P(x)=\mu'(x)$.

This leads us to generalize the notion of an integrating factor.

Integrating Factor

Definition 2.3 If the equation

$$M(x,y)\mathrm{d}x+N(x,y)\mathrm{d}y=0 \tag{2.57}$$

is not exact, but the equation

$$\mu(x,y)M(x,y)\mathrm{d}x+\mu(x,y)N(x,y)\mathrm{d}y=0, \tag{2.58}$$

which results from multiplying equation (2.57) by the function $\mu(x,y)$, is exact, then $\mu(x,y)$ is called an **integrating factor**① of the equation (2.57).

Example 2.11 Show that $\mu(x,y)=xy^2$ is an integrating factor for

$$(2y-6x)\mathrm{d}x+(3x-4x^2y^{-1})\mathrm{d}y=0. \tag{2.59}$$

Use this integrating factor to solve the equation.

Solution We leave it to you to show that equation (2.59) is not exact. Multiplying equation (2.59) by $\mu(x,y)=xy^2$, we obtain

$$(2xy^3-6x^2y^2)\mathrm{d}x+(3x^2y^2-4x^3y)\mathrm{d}y=0. \tag{2.60}$$

For this equation, we have $M=2xy^3-6x^2y^2$ and $N=3x^2y^2-4x^3y$. Because

$$\frac{\partial M}{\partial y}(x,y)=6xy^2-12x^2y=\frac{\partial N}{\partial x}(x,y),$$

equation (2.60) is exact. Hence, $\mu(x,y)=xy^2$ is indeed an integrating factor of equation (2.59).

---

① Historical footnote: A general theory of integrating factors was developed by Alexis Clairaut in 1739. Leonhard Euler also studied classes of equations that could be solved using a specific integrating factor.

Let's now solve equation (2. 60) using the procedure of Section 2. 4. To find $F(x,y)$, we begin by integrating M with respect to x:

$$F(x,y)=\int (2xy^3-6x^2y^2)\,dx+g(y)=x^2y^3-2x^3y^2+g(y).$$

When we take the partial derivative with respect to y and substitute for N, we find

$$\frac{\partial F}{\partial y}(x,y)=N(x,y),$$

$$3x^2y^2-4x^3y+g'(y)=3x^2y^2-4x^3y.$$

Thus, $g'(y)=0$, so we can take $g(y)\equiv 0$. Hence, $F(x,y)=x^2y^3-2x^3y^2$, and the solution to equation (2. 60) is given implicitly by

$$x^2y^3-2x^3y^2=C.$$

Although equations (2. 59) and (2. 60) have essentially the same solutions, it is possible to lose or gain solutions when multiplying by $\mu(x,y)$. In this case, $y\equiv 0$ is a solution of equation (2. 60) but not of equation (2. 59). The extraneous solution arises because, when we multiply equation (2. 59) by $\mu(x,y)=xy^2$ to obtain equation (2. 60), we are actually multiplying both sides of equation (2. 59) by zero if $y\equiv 0$. This gives us $y\equiv 0$ as a solution to equation (2. 60), but it is not a solution to equation (2. 59).

Generally speaking, when using integrating factors, you should check whether any solutions to $\mu(x,y)=0$ are in fact solutions to the original differential equation.

How do we find an integrating factor? If $\mu(x,y)$ is an integrating factor of equation (2. 57) with continuous first partial derivatives, then testing equation (2. 58) for exactness, we must have

$$\frac{\partial}{\partial y}[\mu(x,y)M(x,y)]=\frac{\partial}{\partial x}[\mu(x,y)N(x,y)].$$

By use of the product rule, this reduces to the equation

$$M\frac{\partial \mu}{\partial y}-N\frac{\partial \mu}{\partial x}=\left(\frac{\partial N}{\partial x}-\frac{\partial M}{\partial y}\right)\mu. \tag{2.61}$$

But solving the partial differential equation (2. 61) for μ is usually more difficult than solving the original equation (2. 57). There are, however, two important exceptions.

Let's assume that equation (2. 57) has an integrating factor that depends only on x; that is, $\mu=\mu(x)$. In this case equation (2. 61) reduces to the separable equation

$$\frac{d\mu}{dx}=\left(\frac{\partial M/\partial y-\partial N/\partial x}{N}\right)\mu, \tag{2.62}$$

where $(\partial M/\partial y-\partial N/\partial x)/N$ is (presumably) just a function of x. In a similar fashion, if equation (2. 57) has an integrating factor that depends only on y, then equation (2. 61) reduces to the separable equation

$$\frac{d\mu}{dy}=\left(\frac{\partial N/\partial x-\partial M/\partial y}{M}\right)\mu, \tag{2.63}$$

where $(\partial N/\partial x-\partial M/\partial y)/M$ is just a function of y.

We can reverse the above argument. In particular, if $(\partial M/\partial y-\partial N/\partial x)/N$ is a function that depends only on x, then we can solve the separable equation (2. 62) to obtain the integrating factor

$$\mu(x)=\exp\left[\int\left(\frac{\partial M/\partial y-\partial N/\partial x}{N}\right)dx\right]$$

for equation (2. 57).

We summarize these observations in the following theorem.

Special Integrating Factors

Theorem 2. 3 If $(\partial M/\partial y-\partial N/\partial x)/N$ is continuous and depends only on x, then

$$\mu(x)=\exp\left[\int\left(\frac{\partial M/\partial y-\partial N/\partial x}{N}\right)dx\right] \tag{2.64}$$

is an integrating factor for equation (2. 57).

If $(\partial N/\partial x-\partial M/\partial y)/M$ is continuous and depends only on y, then

$$\mu(y)=\exp\left[\int\left(\frac{\partial N/\partial x-\partial M/\partial y}{M}\right)dy\right] \tag{2.65}$$

is an integrating factor for equation (2. 57).

Theorem 2. 3 suggests the following procedure.

Method for finding special integrating factors is as follows:

If $M\,dx+N\,dy=0$ is neither separable nor linear, compute $\partial M/\partial y$ and $\partial N/\partial x$. If $\partial M/\partial y=\partial N/\partial x$, then the equation is exact. If it is not exact, consider

$$\frac{\partial M/\partial y-\partial N/\partial x}{N}. \tag{2.66}$$

If (2. 66) is a function of just x, then an integrating factor is given by formula (2. 64). If not, consider

$$\frac{\partial N/\partial x-\partial M/\partial y}{M}. \tag{2.67}$$

If (2. 67) is a function of just y, then an integrating factor is given by formula (2. 65).

Example 2.12 Solve

$$(2x^2+y)dx+(x^2y-x)dy=0. \tag{2.68}$$

Solution A quick inspection shows that equation (2.68) is neither separable nor linear. We also note that

$$\frac{\partial M}{\partial y}=1\neq(2xy-1)=\frac{\partial N}{\partial x}.$$

Because equation (2.68) is not exact, we compute

$$\frac{\partial M/\partial y-\partial N/\partial x}{N}=\frac{1-(2xy-1)}{x^2y-x}=\frac{2(1-xy)}{-x(1-xy)}=\frac{-2}{x}.$$

We obtain a function of only x, so an integrating factor for equation (2.68) is given by formula (2.64). That is,

$$\mu(x)=e^{\int\frac{-2}{x}dx}=x^{-2}.$$

When we multiply equation (2.68) by $\mu(x)=x^{-2}$, we get the exact equation

$$(2+yx^{-2})dx+(y-x^{-1})dy=0.$$

Solving this equation, we ultimately derive the implicit solution

$$2x-yx^{-1}+\frac{y^2}{2}=C. \tag{2.69}$$

Notice that the solution $x\equiv0$ was lost in multiplying by $\mu(x)=x^{-2}$. Hence, equation (2.69) and $x\equiv0$ are solutions to equation (2.68). ◆

There are many differential equations that are not covered by Theorem 2.3 but for which an integrating factor nevertheless exists. The major difficulty, however, is in finding an explicit formula for these integrating factors, which in general will depend on both x and y.

Exercises 2.4

In Problems 1-6, identify the equation as separable, linear, exact, or having an integrating factor that is a function of either x alone or y alone.

1. $(2x+yx^{-1})dx+(xy-1)dy=0$.
2. $(2y^3+2y^2)dx+(3y^2x+2xy)dy=0$.
3. $(2x+y)dx+(x-2y)dy=0$.
4. $(y^2+2xy)dx-x^2\,dy=0$.
5. $(x^2\sin x+4y)dx+x\,dy=0$.
6. $(2y^2x-y)dx+x\,dy=0$.

In Problems 7-12, solve the equation.

7. $(2xy)\mathrm{d}x+(y^2-3x^2)\mathrm{d}y=0$.

8. $(3x^2+y)\mathrm{d}x+(x^2y-x)\mathrm{d}y=0$.

9. $(x^4-x+y)\mathrm{d}x-x\,\mathrm{d}y=0$.

10. $(2y^2+2y+4x^2)\mathrm{d}x+(2xy+x)\mathrm{d}y=0$.

11. $(y^2+2xy)\mathrm{d}x-x^2\,\mathrm{d}y=0$.

12. $(2xy^3+1)\mathrm{d}x+(3x^2y^2-y^{-1})\mathrm{d}y=0$.

In Problems 13 and 14, find an integrating factor of the form $x^n y^m$ and solve the equation.

13. $(2y^2-6xy)\mathrm{d}x+(3xy-4x^2)\mathrm{d}y=0$.

14. $(12+5xy)\mathrm{d}x+(6xy^{-1}+3x^2)\mathrm{d}y=0$.

15. (1) Show that if $(\partial N/\partial x-\partial M/\partial y)/(xM-yN)$ depends only on the product xy, that is,

$$\frac{\partial N/\partial x-\partial M/\partial y}{xM-yN}=H(xy),$$

then the equation $M(x,y)\mathrm{d}x+N(x,y)\mathrm{d}y=0$ has an integrating factor of the form $\mu(xy)$. Give the general formula for $\mu(xy)$.

(2) Use your answer to part (1) to find an implicit solution to

$$(3y+2xy^2)\mathrm{d}x+(x+2x^2y)\mathrm{d}y=0,$$

satisfying the initial condition $y(1)=1$.

16. (1) Prove that $M\mathrm{d}x+N\mathrm{d}y=0$ has an integrating factor that depends only on the sum $x+y$ if and only if the expression

$$\frac{\partial N/\partial x-\partial M/\partial y}{M-N}$$

depends only on $x+y$.

(2) Use part (1) to solve the equation $(3+y+xy)\mathrm{d}x+(3+x+xy)\mathrm{d}y=0$.

17. (1) Find a condition on M and N that is necessary and sufficient for $M\mathrm{d}x+N\mathrm{d}y=0$ to have an integrating factor that depends only on the product x^2y.

(2) Use part (1) to solve the equation

$$(2x+2y+2x^3y+4x^2y^2)\mathrm{d}x+(2x+x^4+2x^3y)\mathrm{d}y=0.$$

18. If $xM(x,y)+yN(x,y)\equiv 0$, find the solution to the equation $M(x,y)\mathrm{d}x+N(x,y)\mathrm{d}y=0$.

19. **Fluid Flow.** The streamlines associated with a certain fluid flow are represented by the family of curves $y=x-1+k\mathrm{e}^{-x}$. The velocity potentials of the

flow are just the orthogonal trajectories of this family.

(1) Use the method described in Problem 32 of Exercises 2. 3 to show that the velocity potentials satisfy $dx+(x-y)dy=0$.

[Hint: First express the family $y=x-1+ke^{-x}$ in the form $F(x,y)=k$.]

(2) Find the velocity potentials by solving the equation obtained in part (1).

20. Verify that when the linear differential equation $[P(x)y-Q(x)]dx+dy=0$ is multiplied by $\mu(x)=e^{\int P(x)dx}$, the result is exact.

2.6 Substitutions and Transformations

When the equation

$$M(x,y)dx+N(x,y)dy=0$$

is not a separable, exact, or linear equation, it may still be possible to transform it into one that we know how to solve. This was in fact our approach in Section 2. 5, where we used an integrating factor to transform our original equation into an exact equation.

In this section, we study four types of equations that can be transformed into either a separable or linear equation by means of a suitable substitution or transformation.

Substitution procedure is as follows:

(i) Identify the type of equation and determine the appropriate substitution or transformation.

(ii) Rewrite the original equation in terms of new variables.

(iii) Solve the transformed equation.

(iv) Express the solution in terms of the original variables.

Homogeneous Equations

Definition 2. 4 If the right-hand side of the equation

$$\frac{dy}{dx}=f(x,y) \tag{2.70}$$

can be expressed as a function of the ratio y/x alone, then we say the equation is **homogeneous.**

For example, the equation

$$(x-y)dx+x\,dy=0 \tag{2.71}$$

can be written in the form

$$\frac{\mathrm{d}y}{\mathrm{d}x}=\frac{y-x}{x}=\frac{y}{x}-1.$$

Since we have expressed $(y-x)/x$ as a function of the ratio y/x [that is, $(y-x)/x=G(y/x)$, where $G(v):=v-1$], then equation (2.71) is homogeneous.

The equation

$$(x-2y+1)\mathrm{d}x+(x-y)\mathrm{d}y=0 \tag{2.72}$$

can be written in the form

$$\frac{\mathrm{d}y}{\mathrm{d}x}=\frac{x-2y+1}{y-x}=\frac{1-2(y/x)+(1/x)}{(y/x)-1}.$$

Here the right-hand side cannot be expressed as a function of y/x alone because of the term $1/x$ in the numerator. Hence, equation (2.72) is not homogeneous.

One test for the homogeneity of equation (2.70) is to replace x by tx and y by ty. Then equation (2.70) is homogeneous if and only if

$$f(tx,ty)=f(x,y)$$

for all $t\neq 0$ [see Problem 43(1) in Exercises 2.5].

To solve a homogeneous equation, we make a rather obvious substitution. Let

$$v=\frac{y}{x}.$$

Our homogeneous equation now has the form

$$\frac{\mathrm{d}y}{\mathrm{d}x}=G(v), \tag{2.73}$$

and all we need is to express $\mathrm{d}y/\mathrm{d}x$ in terms of x and v. Since $v=y/x$, then $y=vx$. Keeping in mind that both v and y are functions of x, we use the product rule for differentiation to deduce from $y=vx$ that

$$\frac{\mathrm{d}y}{\mathrm{d}x}=v+x\frac{\mathrm{d}v}{\mathrm{d}x}.$$

We then substitute the above expression for $\mathrm{d}y/\mathrm{d}x$ into equation (2.73) to obtain

$$v+x\frac{\mathrm{d}v}{\mathrm{d}x}=G(v). \tag{2.74}$$

The new equation (2.74) is separable, and we can obtain its implicit solution from

$$\int\frac{1}{G(v)-v}\mathrm{d}v=\int\frac{1}{x}\mathrm{d}x.$$

All that remains to do is to express the solution in terms of the original variables x and y.

Example 2.13 Solve

$$(xy+y^2+x^2)\mathrm{d}x-x^2\,\mathrm{d}y=0. \tag{2.75}$$

Solution A check will show that equation (2.75) is not separable, exact, or linear. If we express equation (2.75) in the derivative form

$$\frac{dy}{dx}=\frac{xy+y^2+x^2}{x^2}=\frac{y}{x}+\left(\frac{y}{x}\right)^2+1, \tag{2.76}$$

then we see that the right-hand side of equation (2.76) is a function of just y/x. Thus, equation (2.75) is homogeneous.

Now let $v=y/x$ and recall that $dy/dx=v+x(dv/dx)$. With these substitutions, equation (2.76) becomes

$$v+x\frac{dv}{dx}=v+v^2+1.$$

The above equation is separable, and separating the variables and integrating, we obtain

$$\int\frac{1}{v^2+1}dv=\int\frac{1}{x}dx,$$

$$\arctan v=\ln|x|+C.$$

Hence,

$$v=\tan(\ln|x|+C).$$

Finally, we substitute y/x for v and solve for y to get

$$y=x\tan(\ln|x|+C)$$

as an explicit solution to equation (2.75). Also note that $x\equiv 0$ is a solution. ◆

Equations of the Form $dy/dx=G(ax+by)$

When the right-hand side of the equation $dy/dx=f(x,y)$ can be expressed as a function of the combination $ax+by$, where a and b are constants, that is,

$$\frac{dy}{dx}=G(ax+by),$$

then the substitution

$$z=ax+by$$

transforms the equation into a separable one. The method is illustrated in the next example.

Example 2.14 Solve

$$\frac{dy}{dx}=y-x-1+(x-y+2)^{-1}. \tag{2.77}$$

Solution The right-hand side can be expressed as a function of $x-y$, that is,

$$y-x-1+(x-y+2)^{-1}=-(x-y)-1+[(x-y)+2]^{-1},$$

so let $z=x-y$. To solve for dy/dx, we differentiate $z=x-y$ with respect to x to obtain $dz/dx=1-dy/dx$, and so $dy/dx=1-dz/dx$. Substituting into equation (2.77) yields

$$1-\frac{dz}{dx}=-z-1+(z+2)^{-1},$$

or

$$\frac{dz}{dx}=(z+2)-(z+2)^{-1}.$$

Solving this separable equation, we obtain

$$\int\frac{z+2}{(z+2)^2-1}dz=\int dx,$$

$$\frac{1}{2}\ln|(z+2)^2-1|=x+C_1,$$

from which it follows that

$$(z+2)^2=Ce^{2x}+1.$$

Finally, replacing z by $x-y$ yields

$$(x-y+2)^2=Ce^{2x}+1$$

as an implicit solution to equation (2.77). ◆

Bernoulli Equations

Definition 2.5 A first-order equation that can be written in the form

$$\frac{dy}{dx}+P(x)y=Q(x)y^n, \tag{2.78}$$

where $P(x)$ and $Q(x)$ are continuous on an interval (a, b) and n is a real number, is called a **Bernoulli equation.** ①

Notice that when $n=0$ or 1, equation (2.78) is also a linear equation and can be solved by the method discussed in Section 2.3. For other values of n, the substitution

$$v=y^{1-n}$$

transforms the Bernoulli equation into a linear equation, as we now show.

Dividing equation (2.78) by y^n yields

---

① Historical footnote: This equation was proposed for solution by James Bernoulli in 1695. It was solved by his brother John Bernoulli. (James and John were two of eight mathematicians in the Bernoulli family.) In 1696, Gottfried Leibniz showed that the Bernoulli equation can be reduced to a linear equation by making the substitution $v=y^{1-n}$.

$$y^{-n}\frac{dy}{dx}+P(x)y^{1-n}=Q(x). \tag{2.79}$$

Taking $v=y^{1-n}$, we find via the chain rule that

$$\frac{dv}{dx}=(1-n)y^{-n}\frac{dy}{dx},$$

and so equation (2.79) becomes

$$\frac{1}{1-n}\frac{dv}{dx}+P(x)v=Q(x).$$

Because $1/(1-n)$ is just a constant, the last equation is indeed linear.

Example 2.15 Solve

$$\frac{dy}{dx}-5y=-\frac{5}{2}xy^3. \tag{2.80}$$

Solution This is a Bernoulli equation with $n=3$, $P(x)=-5$, and $Q(x)=-5x/2$. To transform equation (2.80) into a linear equation, we first divide by y^3 to obtain

$$y^{-3}\frac{dy}{dx}-5y^{-2}=-\frac{5}{2}x.$$

Next we make the substitution $v=y^{-2}$. Since $dv/dx=-2y^{-3}\,dy/dx$, the transformed equation is

$$-\frac{1}{2}\frac{dv}{dx}-5v=-\frac{5}{2}x,$$

$$\frac{dv}{dx}+10v=5x. \tag{2.81}$$

Equation (2.81) is linear, so we can solve it for v using the method discussed in Section 2.3. When we do this, it turns out that

$$v=\frac{x}{2}-\frac{1}{20}+Ce^{-10x}.$$

Substituting $v=y^{-2}$ gives the solution

$$y^{-2}=\frac{x}{2}-\frac{1}{20}+Ce^{-10x}.$$

Not included in the last equation is the solution $y\equiv 0$ that was lost in the process of dividing equation (2.80) by y^3. ◆

Equations with Linear Coefficients

We have used various substitutions for y to transform the original equation into a new equation that we could solve. In some cases we must transform both x and y into new variables, say, u and v. This is the situation for **equations with linear** coefficients—that is, equations of the form

$$(a_1x+b_1y+c_1)\mathrm{d}x+(a_2x+b_2y+c_2)\mathrm{d}y=0, \tag{2.82}$$

where the a_i's, b_i's, and c_i's are constants. We leave it as an exercise to show that when $a_1b_2=a_2b_1$, equation (2.82) can be put in the form $\mathrm{d}y/\mathrm{d}x=G(ax+by)$, which we solved via the substitution $z=ax+by$.

Before considering the general case when $a_1b_2\neq a_2b_1$, let's first look at the special situation when $c_1=c_2=0$. Equation (2.82) then becomes

$$(a_1x+b_1y)\mathrm{d}x+(a_2x+b_2y)\mathrm{d}y=0,$$

which can be rewritten in the form

$$\frac{\mathrm{d}y}{\mathrm{d}x}=-\frac{a_1x+b_1y}{a_2x+b_2y}=-\frac{a_1+b_1(y/x)}{a_2+b_2(y/x)}.$$

This equation is homogeneous, so we can solve it using the method discussed earlier in this section.

The above discussion suggests the following procedure for solving equation (2.82). If $a_1b_2\neq a_2b_1$, then we seek a translation of axes of the form

$$x=u+h \quad \text{and} \quad y=v+k,$$

where h and k are constants. That will change $a_1x+b_1y+c_1$ into a_1u+b_1v and change $a_2x+b_2y=c_2$ into a_2u+b_2v. Some elementary algebra shows that such a transformation exists if the system of equations

$$\begin{cases} a_1h+b_1k+c_1=0, \\ a_2h+b_2k+c_2=0 \end{cases} \tag{2.83}$$

has a solution. This is ensured by the assumption $a_1b_2\neq a_2b_1$, which is geometrically equivalent to assuming that the two lines described by the system (2.84) intersect. Now if (h,k) satisfies (2.84), then the substitutions $x=u+h$ and $y=v+k$ transform equation (2.82) into the homogeneous equation

$$\frac{\mathrm{d}v}{\mathrm{d}u}=-\frac{a_1u+b_1v}{a_2u+b_2v}=-\frac{a_1+b_1(v/u)}{a_2+b_2(v/u)}, \tag{2.84}$$

which we know how to solve.

Example 2.16 Solve

$$(-3x+y+6)\mathrm{d}x+(x+y+2)\mathrm{d}y=0. \tag{2.85}$$

Solution Since $a_1b_2=(-3)\times 1\neq 1\times 1=a_2b_1$, we will use the translation of axes $x=u+h$, $y=v+k$, where h and k satisfy the system

$$\begin{cases} -3h+k+6=0, \\ h+k+2=0. \end{cases}$$

Solving the above system for h and k gives $h=1$, $k=-3$. Hence, we let $x=u+1$ and $y=v-3$. Because $\mathrm{d}y=\mathrm{d}v$ and $\mathrm{d}x=\mathrm{d}u$, substituting in equation (2.85) for x and y yields

$$(-3u+v)\mathrm{d}u+(u+v)\mathrm{d}v=0,$$

$$\frac{\mathrm{d}v}{\mathrm{d}u}=\frac{3-(v/u)}{1+(v/u)}.$$

The last equation is homogeneous, so we let $z=v/u$. Then $\mathrm{d}v/\mathrm{d}u=z+u(\mathrm{d}z/\mathrm{d}u)$, and, substituting for v/u, we obtain

$$z+u\frac{\mathrm{d}z}{\mathrm{d}u}=\frac{3-z}{1+z}.$$

Separating variables gives

$$\int \frac{z+1}{z^2+2z-3}\mathrm{d}z=-\int \frac{1}{u}\mathrm{d}u,$$

$$\frac{1}{2}\ln|z^2+2z-3|=-\ln|u|+C_1,$$

from which it follows that

$$z^2+2z-3=Cu^{-2}.$$

When we substitute back in for z, u, and y, we find

$$(v/u)^2+2(v/u)-3=Cu^{-2},$$

$$v^2+2uv-3u^2=C,$$

$$(y+3)^2+2(x-1)(y+3)-3(x-1)^2=C.$$

This last equation gives an implicit solution to equation (2.85). ◆

Exercises 2.5

In Problems 1-8, identify (do not solve) the equation as homogeneous, Bernoulli, linear coefficients, or of the form $y'=G(ax+by)$.

1. $2tx\,\mathrm{d}x+(t^2-x^2)\mathrm{d}t=0$.
2. $(y-4x-1)^2\mathrm{d}x-\mathrm{d}y=0$.
3. $\mathrm{d}y/\mathrm{d}x+y/x=x^3y^2$.
4. $(t+x+2)\mathrm{d}x+(3t-x-6)\mathrm{d}t=0$.
5. $\theta\,\mathrm{d}y-y\,\mathrm{d}\theta=\sqrt{\theta y}\,\mathrm{d}\theta$.

6. $(ye^{-2x}+y^3)dx-e^{-2x}dy=0$.

7. $\cos(x+y)dy=\sin(x+y)dx$.

8. $(y^3-\theta y^2)d\theta+2\theta^2 y\, dy=0$.

Use the method discussed under "Homogeneous Equations" to solve Problems 9-16.

9. $(xy+y^2)dx-x^2dy=0$.

10. $(3x^2-y^2)dx+(xy-x^3y^{-1})dy=0$.

11. $(y^2-xy)dx+x^2\, dy=0$.

12. $(x^2+y^2)dx+2xy\, dy=0$.

13. $\frac{dx}{dt}=\frac{x^2+t\sqrt{t^2+x^2}}{tx}$.

14. $\frac{dy}{d\theta}=\frac{\theta\sec(y/\theta)+y}{\theta}$.

15. $\frac{dy}{dx}=\frac{x^2-y^2}{3xy}$.

16. $\frac{dy}{dx}=\frac{y(\ln y-\ln x+1)}{x}$.

Use the method discussed under "Equations of the Form $dy/dx=G(ax+by)$*" to solve Problems 17-20.*

17. $dy/dx=\sqrt{x+y}-1$.

18. $dy/dx=(x+y+2)^2$.

19. $dy/dx=(x-y+5)^2$.

20. $dy/dx=\sin(x-y)$.

Use the method discussed under "Bernoulli Equations" to solve Problems 21-28.

21. $\frac{dy}{dx}+\frac{y}{x}=x^2y^2$.

22. $\frac{dy}{dx}-y=e^{2x}y^3$.

23. $\frac{dy}{dx}=\frac{2y}{x}-x^2y^2$.

24. $\frac{dy}{dx}+\frac{y}{x-2}=5(x-2)y^{1/2}$.

25. $\frac{dx}{dt}+tx^3+\frac{x}{t}=0$.

26. $\frac{dy}{dx}+y=e^xy^{-2}$.

27. $\frac{dr}{d\theta}=\frac{r^2+2r\theta}{\theta^2}$.

28. $\frac{dy}{dx}+y^3x+y=0$.

Use the method discussed under "Equations with Linear Coefficients" to solve Problems 29-32.

29. $(-3x+y-1)dx+(x+y+3)dy=0$.

30. $(x+y-1)dx+(y-x-5)dy=0$.

31. $(2x-y)dx+(4x+y-3)dy=0$.

32. $(2x+y+4)dx+(x-2y-2)dy=0$.

In Problems 33-40, solve the equation given in:

33. Problem 1.
34. Problem 2.
35. Problem 3.
36. Problem 4.
37. Problem 5.
38. Problem 6.
39. Problem 7.
40. Problem 8.

41. Use the substitution $v=x-y+2$ to solve equation (2.77).

42. Use the substitution $y=vx^2$ to solve

$$\frac{dy}{dx}=\frac{2y}{x}+\cos(y/x^2).$$

43. (1) Show that the equation $dy/dx=f(x,y)$ is homogeneous if and only if $f(tx,ty)=f(x,y)$. [Hint: Let $t=1/x$.]

(2) A function $H(x,y)$ is called **homogeneous of order *n*** if $H(tx,ty)=t^nH(x,y)$. Show that the equation

$$M(x,y)dx+N(x,y)dy=0$$

is homogeneous if $M(x,y)$ and $N(x,y)$ are both homogeneous of the same order.

44. Show that equation (2.82) reduces to an equation of the form

$$\frac{dy}{dx}=G(ax+by),$$

when $a_1b_2=a_2b_1$. [Hint: If $a_1b_2=a_2b_1$, then $a_2/a_1=b_2/b_1=k$, so that $a_2=ka_1$ and $b_2=kb_1$.]

45. **Coupled Equations.** In analyzing coupled equations of the form

$$\begin{cases}\frac{dy}{dt}=ax+by,\\ \frac{dx}{dt}=\alpha x+\beta y,\end{cases}$$

where a, b, α, and β are constants, we may wish to determine the relationship between x and y rather than the individual solutions $x(t)$, $y(t)$. For this purpose, divide the first equation by the second to obtain

$$\frac{dy}{dx}=\frac{ax+by}{\alpha x+\beta y}. \tag{2.86}$$

This new equation is homogeneous, so we can solve it via the substitution $v=y/x$. We refer to the solutions of equation (2.86) as **integral curves.** Determine the integral curves for the system

$$\begin{cases}\dfrac{dy}{dt}=-4x-y,\\ \dfrac{dx}{dt}=2x-y.\end{cases}$$

46. **Magnetic Field Lines.** As described in Problem 20 of Exercises 1.3, the magnetic field lines of a dipole satisfy

$$\frac{dy}{dx}=\frac{3xy}{2x^2-y^2}.$$

Solve this equation and sketch several of these lines.

47. **Riccati Equation.** An equation of the form

$$\frac{dy}{dx}=P(x)y^2+Q(x)y+R(x) \tag{2.87}$$

is called a **generalized Riccati equation.** ①

(1) If one solution—say, $\mu(x)$—of equation (2.87) is known, show that the substitution $y=\mu+1/v$ reduces equation (18) to a linear equation in v.

(2) Given that $\mu(x)=x$ is a solution to

$$\frac{dy}{dx}=x^3(y-x)^2+\frac{y}{x},$$

use the result of part (1) to find all the other solutions to this equation.

Chapter Summary

In this chapter, we have discussed various types of first-order differential

---

① Historical footnote: Count Jacopo Riccati studied a particular case of this equation in 1724 during his investigation of curves whose radii of curvature depend only on the variable y and not the variable x.

equations. The most important are the separable, linear, and exact equations. Their principal features and method of solution are outlined below.

Separable Equations: $dy/dx=g(x)p(y)$. Separate the variables and integrate.

Linear Equations: $dy/dx+P(x)y=Q(x)$. The integrating factor $\mu=e^{\int P(x)dx}$ reduces the equation to $d(\mu y)/dx=\mu Q$, so that $\mu y=\int \mu Q\,dx+C$.

Exact Equations: $dF(x,y)=0$. Solutions are given implicitly by $F(x,y)=C$. If $\partial M/\partial y=\partial N/\partial x$, then $M\,dx+N\,dy=0$ is exact and F is given by

$$F=\int M\,dx+g(y),\ \text{where } g'(y)=N-\frac{\partial}{\partial y}\int M\,dx$$

or

$$F=\int N\,dy+h(x),\ \text{where } h'(x)=M-\frac{\partial}{\partial x}\int N\,dy.$$

When an equation is not separable, linear, or exact, it may be possible to find an integrating factor or perform a substitution that will enable us to solve the equation.

Special Integrating Factors: $\mu M\,dx+\mu N\,dy=0$ is exact. If $(\partial M/\partial y-\partial N/\partial x)/N$ depends only on x, then

$$\mu(x)=\exp\left[\int\left(\frac{\partial M/\partial y-\partial N/\partial x}{N}\right)dx\right]$$

is an integrating factor. If $(\partial N/\partial x-\partial M/\partial y)/M$ depends only on y, then

$$\mu(y)=\exp\left[\int\left(\frac{\partial N/\partial x-\partial M/\partial y}{M}\right)dy\right]$$

is an integrating factor.

Homogeneous Equations: $dy/dx=G(y/x)$. Let $v=y/x$. Then $dy/dx=v+x(dv/dx)$, and the transformed equation in the variables v and x is separable.

Equations of the Form: $dy/dx=G(ax+by)$. Let $z=ax+by$. Then $dz/dx=a+b(dy/dx)$, and the transformed equation in the variables z and x is separable.

Bernoulli Equations: $dy/dx+P(x)y=Q(x)y^n$. For $n\neq 0$ or 1, let $v=y^{1-n}$. Then $dv/dx=(1-n)y^{-n}(dy/dx)$, and the transformed equation in the variables v and x is linear.

Linear Coefficients: $(a_1x+b_1y+c_1)dx+(a_2x+b_2y+c_2)dy=0$. For $a_1b_2\neq a_2b_1$, let $x=u+h$ and $y=v+k$, where h and k satisfy

$$\begin{cases}a_1h+b_1k+c_1=0,\\ a_2h+b_2k+c_2=0.\end{cases}$$

Then the transformed equation in the variables u and v is homogeneous.

Review Problems

In Problems 1-30, solve the equation.

1. $\frac{dy}{dx}=\frac{e^{x+y}}{y-1}$.

2. $\frac{dy}{dx}-4y=32x^2$.

3. $(x^2-2y^{-3})dy+(2xy-3x^2)dx=0$.

4. $\frac{dy}{dx}+\frac{3y}{x}=x^2-4x+3$.

5. $[\sin(xy)+xy\cos(xy)]dx+[1+x^2\cos(xy)]dy=0$.

6. $2xy^3\,dx-(1-x^2)\,dy=0$.

7. $t^3y^2\,dt+t^4y^{-6}\,dy=0$.

8. $\frac{dy}{dx}+\frac{2y}{x}=2x^2y^2$.

9. $(x^2+y^2)dx+3xy\,dy=0$.

10. $[1+(1+x^2+2xy+y^2)^{-1}]dx+[y^{-1/2}+(1+x^2+2xy+y^2)^{-1}]dy=0$.

11. $\frac{dx}{dt}=1+\cos^2(t-x)$.

12. $(y^3+4e^xy)dx+(2e^x+3y^2)dy=0$.

13. $\frac{dy}{dx}-\frac{y}{x}=x^2\sin 2x$.

14. $\frac{dx}{dt}-\frac{x}{t-1}=t^2+2$.

15. $\frac{dy}{dx}=2-\sqrt{2x-y+3}$.

16. $\frac{dy}{dx}+y\tan x+\sin x=0$.

17. $\frac{dy}{d\theta}+2y=y^2$.

18. $\frac{dy}{dx}=(2x+y-1)^2$.

19. $(x^2-3y^2)dx+2xy\,dy=0$.

20. $\frac{dy}{d\theta}+\frac{y}{\theta}=-4\theta y^{-2}$.

21. $(y-2x-1)dx+(x+y-4)dy=0$.

22. $(2x-2y-8)dx+(x-3y-6)dy=0$.

23. $(y-x)dx+(x+y)dy=0$.

24. $(\sqrt{y/x}+\cos x)dx+(\sqrt{x/y}+\sin y)dy=0$.

25. $y(x-y-2)dx+x(y-x-4)dy=0$.

26. $\frac{dy}{dx}+xy=0$.

27. $(3x-y-5)dx+(x-y+1)dy=0$.

28. $\frac{dy}{dx}=\frac{x-y-1}{x+y+5}$.

29. $(4xy^3-9y^2+4xy^2)dx+(3x^2y^2-6xy+2x^2y)dy=0$.

30. $\frac{dy}{dx}=(x+y+1)^2-(x+y-1)^2$.

In Problems 31-40, solve the initial value problem.

31. $(x^3-y)dx+x\,dy=0$, $y(1)=3$.

32. $\frac{dy}{dx}=\left(\frac{x}{y}+\frac{y}{x}\right)$, $y(1)=-4$.

33. $(t+x+3)dt+dx=0$, $x(0)=1$.

34. $\frac{dy}{dx}-\frac{2y}{x}=x^2\cos x$, $y(\pi)=2$.

35. $(2y^2+4x^2)dx-xy\,dy=0$, $y(1)=-2$.

36. $[2\cos(2x+y)-x^2]dx+[\cos(2x+y)+e^y]dy=0$, $y(1)=0$.

37. $(2x-y)dx+(x+y-3)dy=0$, $y(0)=2$.

38. $\sqrt{y}\,dx+(x^2+4)dy=0$, $y(0)=4$.

39. $\frac{dy}{dx}-\frac{2y}{x}=x^{-1}y^{-1}$, $y(1)=3$.

40. $\frac{dy}{dx}-4y=2xy^2$, $y(0)=-4$.

41. Express the solution to the following initial value problem using a definite integral:

$$\frac{dy}{dt}=\frac{1}{1+t^2}-y,\quad y(2)=3.$$

Technical Writing Exercises

1. An instructor at Ivey U. asserted, "All you need to know about first-order differential equations is how to solve those that are exact." Give arguments that support and arguments that refute the instructor's claim.

2. What properties do solutions to linear equations have that are not shared by solutions to either separable or exact equations? Give some specific examples to support your conclusions.

3. Consider the differential equation

$$\frac{dy}{dx}=ay+be^{-x}, \quad y(0)=c,$$

where a, b, and c are constants. Describe what happens to the asymptotic behavior as $x\to+\infty$ of the solution when the constants a, b, and c are varied. Illustrate with figures and/or graphs.

Chapter 3

Existence and Uniqueness Theory

3.1 Picard's Existence and Uniqueness Theorem

In nearly every chapter of this book, the theory, application, or procedure discussed depend on a fundamental theorem concerning the existence and uniqueness of the solution to an initial value problem. In this chapter, we provide the proofs of these basic theorems.

How does one know that a solution exists? When possible, the simplest way is actually to exhibit the solution in an explicit form. While we were able to do this for the entire class of linear first-order differential equations (see Section 2.3) and also for homogeneous equations (or homogeneous normal systems) that have constant coefficients, most differential equations do not fall into these categories.

Another approach to proving the existence of a solution is to construct a sequence of **approximate solutions** to the initial value problem and then show that these approximations converge to the desired solution. This is the strategy we adopt in this chapter.

For the purpose of handling a very general class of differential equations, our procedure will be to construct a sequence of **successive approximations** (**iterations**) and show that these converge to the desired solution.

To describe the procedure, we begin by expressing the initial value problem

$$\frac{\mathrm{d}y}{\mathrm{d}x}=f(x,y), \quad y(x_0)=y_0 \tag{3.1}$$

as an **integral equation.** This is accomplished by integrating both sides of equation (3.1) from x_0 to x:

$$\int_{x_0}^{x} y'(t)\mathrm{d}t=y(x)-y(x_0)=\int_{x_0}^{x} f(t,y(t))\mathrm{d}t.$$

Setting $y(x_0)=y_0$ and solving for $y(x)$, we have

$$y=y_0+\int_{x_0}^{x} f(t,y(t))\mathrm{d}t. \tag{3.2}$$

Our derivation shows that a solution to equation (3.1) must also be a solution to equation (3.2).

Conversely, using the fundamental theorem of calculus, we can show that if f and y are continuous, then a solution to equation (3.2) satisfies the differential equation in (3.1).

Furthermore, such a solution must satisfy the initial condition as can be seen by substituting in equation (3.2). Therefore, with these continuity assumptions, the initial value problem (3.1) and the integral equation (3.2) are equivalent in the sense that they have the same solutions.

Now we are ready to prove the following existence and uniqueness result, utilizing the Picard approximations.

Theorem 3.1 (Existence and uniqueness for the Cauchy problem) Let $A: |x-x_0| \leqslant a, |y-y_0| \leqslant b$ and suppose that $f: A \to \mathbf{R}$ is continuous and satisfies a Lipschitz condition. Then the Cauchy problem (3.1) has a unique solution on the interval $[x_0-\alpha, x_0+\alpha]$, where $\alpha=\min\left(a, \frac{b}{M}\right)$, $M=\max_{(x,y)\in A} |f(x,y)|$.

Remark 3.1 We say that a function $f: A\to\mathbf{R}$ satisfies a Lipschitz condition with respect to y in $\mathbf{R}$ if there exists a constant $L(>0)$ such that

$$|f(x,y_1)-f(x,y_2)| \leqslant L|y_1-y_2|,$$

where $(x,y_1),(x,y_2)\in A$ and L is called a **Lipschitz constant.**

Proof of Theorem 3.1 We will prove the theorem by showing the following four things:

(i) Define the Picard approximations as

$$\begin{cases} y_0(x)=y_0, \\ y_n(x)=y_0+\int_{x_0}^{x} f(t,y_{n-1}(t))\mathrm{d}t, |x-x_0| \leqslant \alpha, \\ n=1,2,\ldots \end{cases} \tag{3.3}$$

Then the sequence $\{y_n(x)\}$ is defined on $[x_0-\alpha, x_0+\alpha]$, and

$$(x, y_n(x))\in A, \quad x_0 \leqslant x \leqslant x_0+\alpha, \quad n=0, 1, \ldots$$

(ii) The sequence $\{y_n(x)\}$ converges uniformly on $[x_0-\alpha, x_0+\alpha]$ to a continuous function, denotes by $y(x)$.

(iii) The function $y(x)$ is a solution of equation (3.2) on $[x_0-\alpha, x_0+\alpha]$.

(iv) The solution is unique.

It is convenient to assume $x \geqslant x_0$; a similar proof will hold for $x \leqslant x_0$.

(i) We first show that

$$|y_n(x)-y_0| \leqslant b$$

holds on $x_0 \leqslant x \leqslant x_0+\alpha$ for $n=1, 2, \ldots$

Clearly the inequality holds for $n=0$.

Assuming that $|y_{n-1}(x)-y_0| \leqslant b$ holds, we have

$$|y_n(x)-y_0| \leqslant \int_{x_0}^{x} |f(s, y_{n-1})| \mathrm{d}s \leqslant (x-x_0)M \leqslant \alpha M \leqslant b.$$

Therefore, by induction, $|y_n(x)-y_0| \leqslant b$ holds on $x_0 \leqslant x \leqslant x_0+\alpha$ for all n.

(ii) We now show that the sequence $\{y_n(x)\}$ converges uniformly on $x_0 \leqslant x \leqslant x_0+\alpha$ to a continuous function $y(x)$ which satisfies the integral equation (3.2).

This is accomplished by writing $y_n(x)$ in the form

$$y_n(x)=y_0+[y_1(x)-y_0]+\ldots+[y_n(x)-y_{n-1}(x)].$$

Clearly, the sequence $\{y_n(x)\}$ converges uniformly if, and only if, the infinite series

$$y_0+[(y_1(x)-y_0]+\ldots+[y_n(x)-y_{n-1}(x)]+\ldots \tag{3.4}$$

converges uniformly. To prove that the infinite series (3.4) converges uniformly on $x_0 \leqslant x \leqslant x_0+\alpha$, it suffices to show that $\sum_{n=1}^{\infty} |y_n(x)-y_{n-1}(x)|$ converges uniformly on $x_0 \leqslant x \leqslant x_0+\alpha$. This is accomplished in the following manner.

Observe that

$$\begin{aligned} |y_{n+1}(x)-y_n(x)| &\leqslant \int_{x_0}^{x} |f(s, y_n(s))-f(s, y_{n-1}(s))| \mathrm{d}s \\ &\leqslant L\int_{x_0}^{x} |y_n(s)-y_{n-1}(s)| \mathrm{d}s, \quad x_0 \leqslant x \leqslant x_0+\alpha. \end{aligned} \tag{3.5}$$

Setting $n=0$ in inequalities (3.5) gives

$$|y_1(x)-y_0| \leqslant \int_{x_0}^{x} |f(s, y_0)| \mathrm{d}s \leqslant M|x-x_0|.$$

This in turn, implies that

$$\begin{aligned} |y_2(x)-y_1(x)| &\leqslant L\int_{x_0}^{x} |y_1(s)-y_0| \mathrm{d}s \\ &\leqslant LM\int_{x_0}^{x} (s-x_0)\mathrm{d}s \leqslant \frac{LM(x-x_0)^2}{2} \leqslant \frac{LM}{2!}\alpha^2. \end{aligned}$$

Proceeding inductively, we see that

$$|y_n(x)-y_{n-1}(x)|\leqslant\frac{ML^{n-1}(x-x_0)^n}{n!}\leqslant\frac{L^{n-1}M}{n!}\alpha^n,\quad x_0\leqslant x\leqslant x_0+\alpha$$

for $n=1, 2, \ldots$

Thanks to the series

$$\sum_{n=1}^{\infty}\frac{L^{n-1}\alpha^n}{n!}$$

converges for α, we obtain a consequence that the sequence $\{y_n(x)\}$ converges uniformly on $x_0\leqslant x\leqslant x_0+\alpha$.

We denote the limit of the sequence $\{y_n(x)\}$ by $y(x)$.

(iii) We will show that $y(x)$ satisfies the integral equation (3.2) and that $y(x)$ is continuous. To this end, recall that the Picard iterations $y_n(x)$ are defined recursively through the equation (3.3). Taking limits of both sides of equation (3.3) gives

$$y(x)=\lim_{n\to\infty}y_n(x)=y_0+\lim_{n\to\infty}\int_{x_0}^{x}f(s,y_{n-1}(s))\mathrm{d}s. \tag{3.6}$$

To show that the right-hand side of (3.6) equals

$$y_0+\int_{x_0}^{x}f(s,y(s))\mathrm{d}s$$

(that is, to justify passing the limit through the integral sign), we must show that $\{f(x,y_{n-1}(x))\}$ converges uniformly $f(x,y(x))$ on $x_0\leqslant x\leqslant x_0+\alpha$ as n approaches infinity. This is accomplished in the following manner. Observe first that the graph of $y(x)$ lies in the rectangle A for $x_0\leqslant x\leqslant x_0+\alpha$, since it is the limit of functions $y_n(x)$ whose graphs lie in A. Hence

$$|f(x,y_{n-1}(x))-f(x,y(x))|\leqslant L|y_{n-1}(x)-y(x)|,\quad x_0\leqslant x\leqslant x_0+\alpha.$$

This shows that $f(x,y_{n-1}(x))$ converges uniformly to $f(x,y(x))$ on $x_0\leqslant x\leqslant x_0+\alpha$. Thus

$$\begin{aligned}y(x)&=\lim_{n\to\infty}y_n(x)=y_0+\lim_{n\to\infty}\int_{x_0}^{x}f(s,y_{n-1}(s))\mathrm{d}s\\&=y_0+\int_{x_0}^{x}f(s,\lim_{n\to\infty}y_{n-1}(s)\mathrm{d}s)\\&=y_0+\int_{x_0}^{x}f(s,y(s))\mathrm{d}s,\end{aligned}$$

that is $y(x)$ is a solution of the integral equation (3.2).

It is clear that $y(x)$ is continuous on $x_0\leqslant x\leqslant x_0+\alpha$, since it is the limit function of the uniform convergence sequence $\{y_n(x)\}$.

(iv) Show that the solution is unique. Let $z(x)$ be also a solution of equation

(3.2). By induction, we show that

$$|z(x)-y_n(x)|\leqslant\frac{ML^n(x-x_0)^{n+1}}{(n+1)!}. \tag{3.7}$$

In fact, we have

$$z(x)-y_0(x)=\int_{x_0}^{x} f(t,y(t))\mathrm{d}t,$$

and therefore $|z(x)-y_0(x)|\leqslant M|x-x_0|$ which inequality (3.7) holds for $n=0$.

If inequality (3.7) holds for $n-1$, we have

$$|z(x)-y_n(x)|=\left|\int_{x_0}^{x}[f(t,z(t))-f(t,y_{n-1}(t))]\mathrm{d}t\right|$$

$$\leqslant L\left|\int_{x_0}^{x}\left|z(t)-y_{n-1}(t)\right|\mathrm{d}t\right|\leqslant L\int_{x_0}^{x}\frac{ML^{n-1}(t-x_0)^n}{n!}\mathrm{d}t$$

$$\leqslant\frac{ML^n(x-x_0)^{n+1}}{(n+1)!},$$

and this proves

$$|z(x)-y_n(x)|\leqslant\frac{ML^n(x-x_0)^{n+1}}{(n+1)!} \quad \text{for } n\in\mathbf{N}.$$

Consequently, $|z(x)-y_n(x)|\leqslant\frac{ML^n\alpha^{n+1}}{(n+1)!}$. $\lim\limits_{n\to\infty}\frac{ML^n\alpha^{n+1}}{(n+1)!}=0$, since $\sum\limits_{n=0}^{\infty}\frac{ML^n\alpha^{n+1}}{(n+1)!}<\infty$. This implies that $\{y_n(x)\}$ converges uniformly to $z(x)$ on $[x_0-\alpha,x_0+\alpha]$. And so $y(x)\equiv z(x), x_0-\alpha\leqslant x\leqslant x_0+\alpha$.

Corollary 3.1 Let f and $\frac{\partial f}{\partial y}$ be continuous on $A=\{(x,y)\,|\,|x-x_0|\leqslant a, |y-y_0|\leqslant b\}$. Compute $M=\max\limits_{(x,y)\in A}|f(x,y)|$, and set $\alpha=\min\left(a,\frac{b}{M}\right)$. Then the initial value problem (3.1) has a unique solution on the interval $[x_0-\alpha,x_0+\alpha]$.

Example 3.1 Show that the solution $y(x)$ of the initial value problem

$$\frac{\mathrm{d}y}{\mathrm{d}x}=x^2+\mathrm{e}^{-y}, \quad y(0)=0$$

exists for $-\frac{1}{2}\leqslant x\leqslant\frac{1}{2}$, and in this interval $|y(x)|\leqslant 1$.

Solution Let A be the rectangle $|t|\leqslant\frac{1}{2}$, $|y|\leqslant 1$. Computing

$$M=\max_{(x,y)\in A}(x^2+\mathrm{e}^{-y^2})=\frac{5}{4},$$

we see that $y(x)$ exists for $|x|\leqslant\min(\frac{1}{2},\frac{1}{5/4})=\frac{1}{2}$, and in this interval $|y(x)|\leqslant 1$. ◆

3.2 Existence of Solutions of Linear Equation

In the preceding section the best we could say was that if f and $\frac{\partial f}{\partial y}$ are continuous in a rectangle $A=\{(x,y)\,|\,|x-x_0|\leqslant a, |y-y_0|\leqslant b\}$, then the solution exists on some subinterval $[x_0-\alpha, x_0+\alpha]\subset[x_0-a, x_0+a]$.

For example, the seemingly simple nonlinear problem

$$y'(x)=y^2(x),\quad y(0)=a(\neq 0)$$

has the solution $y(x)=1/(a^{-1}-x)$, which is undefined at $x=\frac{1}{a}$ even though $f(x,y)=y^2$ is very well behaved.

In this section, we show that for linear equation in normal form, a solution exists over the entire interval. This fact is a consequence of the following theorem.

Theorem 3.2 The initial value problem

$$\frac{\mathrm{d}y}{\mathrm{d}x}=p(x)y+q(x),\quad y(x_0)=y_0, \tag{3.8}$$

with $p,q\in C[a,b]$, $t_0\in[a,b]$ and $y_0\in R$ has a unique solution valid on $[a,b]$.

Proof Define the Picard approximations as

$$\begin{cases} y_0(x)\equiv y_0, \\ y_n(x)=y_0+\int_{x_0}^{x}(p(t)y_{n-1}(t)+q(t))\mathrm{d}t, \\ n=1,2,\ldots \end{cases} \tag{3.9}$$

We show that the Picard iteration converges uniformly to a unique solution on the interval $[a,b]$.

The approximations are well defined. Since $y_0(x)\equiv y_0$ is continuous on $[a,b]$ and f is continuous on the strip $R=\{(x,y)\,|\,a\leqslant x\leqslant b, -\infty<y<+\infty\}$, it follows that

$$y_1(x)=y_0+\int_{x_0}^{x}(p(t)y_0+q(t))\mathrm{d}t$$

is well defined and continuous on $[a,b]$. By induction, we find that $y_n(x)$ is well defined and continuous on $[a,b]$.

The approximations converge uniformly on $[a,b]$. Recall that the sequence of successive approximations $\{y_n(x)\}$ consists of the partial sums of the series

$$y_0+\sum_{n=1}^{\infty}(y_n(x)-y_{n-1}(x)). \tag{3.10}$$

Thus, if this series converges uniformly on $[a,b]$, then so does the sequence.

Since $p(x)$ and $q(x)$ are continuous on $[a,b]$, they are bounded; that is, there are some constants M and L such that $|p(x)y_0+q(x)|\leqslant M$ and $|p(x)|\leqslant L$ on $[a,b]$. To simplify the computations, let's consider x in the interval $[x_0,b]$ (the case when $x\in[a,x_0]$ can be handled in a similar fashion). For such x,

$$|y_1(x)-y_0|\leqslant\int_{x_0}^{x}|p(x)y_0+q(x)|\mathrm{d}s\leqslant M(x-x_0),$$

and

$$\begin{aligned}|y_2(x)-y_1(x)|&\leqslant\int_{x_0}^{x}|p(s)(y_1(s)-y_0(s))|\mathrm{d}s\leqslant L\int_{x_0}^{x}|(y_1(s)-y_0(s))|\mathrm{d}s\\&\leqslant LM\int_{x_0}^{x}(s-x_0)\mathrm{d}s=\frac{LM}{2}(x-x_0)^2\end{aligned}$$

for $x\in[x_0,b]$. Similarly, for $n=1, 2, \ldots$ we find by induction that

$$\begin{aligned}|y_{n+1}(x)-y_n(x)|&\leqslant\int_{x_0}^{x}|p(s)(y_n(s)-y_{n-1}(s))|\mathrm{d}s\leqslant L\int_{x_0}^{x}|(y_n(s)-y_{n-1}(s))|\mathrm{d}s\\&\leqslant\frac{L^nM}{n!}\int_{x_0}^{x}(s-x_0)^n\mathrm{d}s=\frac{L^nM}{(n+1)!}(x-x_0)^{n+1}.\end{aligned}$$

From the last inequalities we see that the terms in series (3. 10) are bounded in absolute value by the terms in the positive series $\sum_{n=0}^{\infty}\frac{L^nM}{(n+1)!}(b-x_0)^{n+1}$ which is convergence.

Therefore, by the Weierstrass M test, the series in (3. 10) converges uniformly on $[x_0,b]$ to a continuous function $y(x)$. Hence, the sequence $\{y_n(x)\}$ converges uniformly on $[x_0,b]$ to $y(x)$.

A similar argument establishes convergence on $[a,x_0]$, so $\{y_n(x)\}$ converges uniformly on $[a,x_0]$ to $y(x)$.

$y(x)$ is a Solution. Since $\{y_n(x)\}$ converges uniformly to $y(x)$, we can take the limit inside the integral. Thus, using equation (3. 9), we obtain

$$\begin{aligned}y(x)&=\lim_{n\to\infty}y_n(x)=y_0+\lim_{n\to\infty}\int_{x_0}^{x}(p(s)y_{n-1}(s)+q(s))\mathrm{d}s\\&=y_0+\int_{x_0}^{x}(p(s)\lim_{n\to\infty}y_{n-1}(s)+q(s))\mathrm{d}s\\&=y_0+\int_{x_0}^{x}(p(s)y(s)+q(s))\mathrm{d}s.\end{aligned}$$

Since $y(x)$ satisfies the equivalent integral equation, it must satisfy the initial value problem (3.8).

Uniqueness. Let z be a solution to initial value problem (3.8) on $[a,b]$. Then z must satisfy the integral equation

$$z(x)=y_0+\int_{x_0}^{x}(p(t)y(t)+q(t))\mathrm{d}t$$

just as $y(x)$ does. Therefore, for x in $[x_0,b]$,

$$|z(x)-y(x)|=\left|\int_{x_0}^{x}[p(t)(z(t)-y(t))]\mathrm{d}t\right|$$

$$\leqslant L\int_{x_0}^{x}|z(t)-y(t)|\mathrm{d}t\leqslant LM(x-x_0), \tag{3.11}$$

where $M=\max\limits_{x_0\leqslant x\leqslant b}|z(x)-y(x)|$. Using inequalities (3.11), we next show by a similar string of inequalities that

$$|z(x)-y(x)|=\left|\int_{x_0}^{x}[p(t)(z(t)-y(t))]\mathrm{d}t\right|\leqslant L\int_{x_0}^{x}|z(t)-y(t)|\mathrm{d}t$$

$$\leqslant L^2M\int_{x_0}^{x}(t-x_0)\mathrm{d}t=\frac{L^2M}{2}(x-x_0)^2.$$

By induction, we find $|z(x)-y(x)|\leqslant\dfrac{ML^n(x-x_0)^n}{n!}$ for x in $[x_0,b]$. Taking the limit as $n\to\infty$, we deduce that $|z(x)-y(x)|\equiv 0$. That is, z and y agree on $[x_0,b]$. A similar argument shows that they agree on $[a,x_0]$, so the initial value problem has a unique solution on the interval $[a,b]$.

3.3 Estimates of Error and Approximate Calculation

The method of Picard iteration not only gave the solution but also found functions which approximate it well, i.e., let $y(t)$ be the exact solution of the initial value problem

$$\frac{\mathrm{d}y}{\mathrm{d}t}=f(t,y),\quad y(t_0)=y_0,$$

and the Picard scheme for solving this is

$$y_n(t)=y_0+\int_{t}^{t_0}f(s,y_{n-1}(s))\mathrm{d}s,\quad y_0(t)=y_0.$$

If we make in approximating the solution $y(t)$ by $y_n(t)$, then the error

$$|y_n(t)-y(t)|\leqslant\frac{ML^nh^{n+1}}{(n+1)!},$$

$y_n(t)$ is called the **nth approximate solution** to the initial value problem $\frac{dy}{dt}=f(t,y)$, $y(t_0)=y_0$.

Example 3.2 Let $\frac{dy}{dx}=x^2+y^2,(x,y)\in R:[-1,1]\times[-1,1]$.

(i) Determine the interval of existence that the existence theorem predicts for the solution $y(x)$ of the initial value problem $\frac{dy}{dx}=x^2+y^2$, $y(0)=0$.

(ii) Determine an approximate solution of the initial value problem above such that the error is less than 0.05.

Solution Let $f(x,y)=x^2+y^2$. Then $f, f_y\in C(R)$, and so the existence and uniqueness theorem holds. From the existence theorem, the initial value problem $\frac{dy}{dx}=x^2+y^2$, $y(0)=0$ has a unique solution on $[-h,h]$. Here

$$M=\max_{(x,y)\in R}\{x^2+y^2\}=2,\quad h=\min\left\{a,\frac{b}{M}\right\}=\min\left\{1,\frac{1}{2}\right\}=\frac{1}{2}.$$

Thus, the interval of solution is $\left[-\frac{1}{2},\frac{1}{2}\right]$.

Since $|f_y|=|2y|\leqslant 2,(x,y)\in R$, we take $L=2$. According to our desire, the error of the solution should satisfy

$$|y_n(x)-y(x)|\leqslant\frac{ML^nh^{n+1}}{(n+1)!}=\frac{1}{(n+1)!}<0.05.$$

Since $\frac{1}{4!}=\frac{1}{24}<\frac{1}{20}=0.05$, we take $n=3$. Therefore, we obtain the successive approximates:

$$y_0(x)=0,$$

$$y_1(x)=\int_0^x[t^2+y_0^2(t)]dt=\int_0^x t^2dt=\frac{x^3}{3},$$

$$y_2(x)=\int_0^x[t^2+y_1^2(t)]dt=\int_0^x\left[t^2+\left(\frac{t^3}{3}\right)^2\right]dt=\frac{x^3}{3}+\frac{x^7}{63},$$

$$y_3(x)=\int_0^x[t^2+y_2^2(t)]dt=\int_0^x\left[t^2+\left(\frac{t^3}{3}+\frac{t^7}{63}\right)^2\right]dt$$

$$=\int_0^x\left[t^2+\frac{t^6}{9}+\frac{2t^{10}}{189}+\frac{t^{14}}{3969}\right]dt=\frac{x^3}{3}+\frac{x^7}{63}+\frac{x^{11}}{2079}+\frac{x^{15}}{59535}.$$

$y_3(x)$ is the approximate solution desired, $|y_3(x)-y(x)|\leqslant 0.05$, $x\in\left[-\frac{1}{2},\frac{1}{2}\right]$.

Example 3. 3 Use the method of successive approximation to the solution to initial value problem

$$y'=-y, \quad y(0)=1.$$

Solution Rewriting the differential equation as an integral equation, we obtain

$$y(x)=1-\int_0^x y(t)\mathrm{d}t.$$

Setting $y_0(x)=1$, we obtain

$$y_1(x)=1-\int_0^x \mathrm{d}t=1-x,$$

$$y_2(x)=1-\int_0^x (1-t)\mathrm{d}t=1-x+\frac{x^2}{2!},$$

$$y_3(x)=1-\int_0^x \left(1-t+\frac{t^2}{2!}\right)\mathrm{d}t=1-x+\frac{x^2}{2!}-\frac{x^3}{3!}.$$

It is easy to prove by induction that

$$y_n(x)=1-x+\frac{x^2}{2!}-\ldots+(-1)^n\frac{x^n}{n!}.$$

Recalling the Taylor series expansion of e^{-x}, we see that

$$\lim_{n\to\infty} y_n(x)=\mathrm{e}^{-x}.$$

The approximate solution approaches the true solution.

3.4 Continuation of Solutions

So far we only considered local solutions, i. e. , solutions which are defined in a neighborhood of (x_0, y_0). Simple example shows that the solution $x(t)$ may not exist for all t, for example the equation $y'=1+y^2$ has solution $y(x)=\tan(x-c)$ and this solution does not exist beyond the interval $\left(c-\frac{\pi}{2}, c+\frac{\pi}{2}\right)$, and we have $y(x)\to\pm\infty$ as $x\to\pm\frac{\pi}{2}$. This is typical.

If $y(t)$ is a solution of

$$\frac{\mathrm{d}y}{\mathrm{d}x}=f(x,y) \tag{3.12}$$

defined on an interval I, we say that $z(t)$ is a continuation of or extension of $y(t)$ if $z(t)$ is itself a solution of equation (3. 12) defined on an interval J which properly contains I and z restricted to I equals y. A solution is non-continuable or saturated if no such extension exists; i. e. , I is the maximal interval on which a solution to equation (3. 12) exists.

To extend the solution we solve the Cauchy problem locally, say from x_0 to $x_0 + h$ and then we can try to continue the solution by solving the Cauchy problem $y' = f(x, y)$ with new initial condition $y(x_0 + h)$. In order to do this, we should be able to solve it locally everywhere and we will assume that f satisfies a local Lipschitz condition.

Definition 3. 1 A function $f: U \to R$ (where U is an open set of $R \times R$) satisfies a local Lipschitz condition if for any $(x_0, y_0) \in U$ there exist a neighborhood $V \subset U$ such that f satisfies a Lipschitz condition on V.

If the function f is of class C^1 in U, then it satisfies a local Lipschitz condition.

Lemma 3. 1 Let $U \subset R \times R$ be an open set and assume that f: $U \to R$ is continuous and satisfies a local Lipschitz condition. Then for any $(x_0, y_0) \in U$ there exists an open interval $I_{\max} = (\omega_-, \omega_+)$ with $-\infty \leqslant \omega_- < t_0 < \omega_+ \leqslant +\infty$ such that

- The Cauchy problem

$$y' = f(x, y), \quad y(x_0) = y_0 \tag{3. 13}$$

has a unique solution on $I_{\max}$.

- If $z: I \to R$ is a solution of Cauchy problem (3. 13), then $I \subset I_{\max}$, and $z = y|_I$.

Proof (i) Let $y: I \to R$ and $z: J \to R$ be two solutions of the Cauchy problem with $t_0 \in I, J$. Then $y(x) \equiv z(x)$ on $I \cap J$. Suppose it is not true, there is point t_1 such that $y(t_1) \neq z(t_1)$. Consider the first point where the solutions separate. The local existence and uniqueness theorem shows that it is impossible.

(ii) Let us define the interval: $I_{\max} = U\{I$; I is an open interval, $t_0 \in I$, there exists a solution of Cauchy problem (3. 13) on $I\}$.

This interval is open and we can define the solution on $I_{\max}$ as follows: If $x \in I_{\max}$, then there exists I where the Cauchy problem has a solution and we can define $y(x)$. The part (i) shows that $y(x)$ is uniquely defined on $I_{\max}$.

Theorem 3. 3 Let $U \subset R \times R$ be an open set and let us assume that f: $U \to R$ is continuous and satisfies a local Lipschitz condition. Then every solution of $y' = f(x, y)$ has a continuation up to the boundary of U. More precisely, if $y: I_{\max} \to R$ is the solution passing through $(x_0, y_0) \in U$, then for any compact $K \subset U$ there exists $x_1, x_2 \in I_{\max}$ with $x_1 < x_0 < x_2$ such that $(x_1, y(x_1)) \notin K$, $(x_2, y(x_2)) \notin K$.

Remark 3. 2 If $U = R \times R$, Theorem 3. 3 means that either

- $y(x)$ exists for all x,
- There exists x^* such that $\lim\limits_{x \to x^*} |y(x)| = \infty$,

that is, the solution exists globally or the solution "blows up" at a certain point.

Proof Let $I_{\max} = (\omega_-, \omega_+)$. If $\omega_+ = +\infty$, clearly there exists a point x_2 such that $x_2 > x_0$ and $(x_2, y(x_2)) \notin K$ because K is bounded. If $\omega_+ < +\infty$, let us assume that there exists a compact K such that $(x, y(x)) \in K$ for any $x \in (x_0, \omega_+)$. Since $f(x,y)$ is bounded on the compact set K, we have, for x, x' sufficiently close to ω_+,

$$|y(x) - y(x')| = \left| \int_x^{x'} f(t, y(t)) \mathrm{d}t \right| \leqslant M |x' - x| < \varepsilon. \tag{3.14}$$

This shows that $\lim\limits_{x \to \omega_+'} y(x) = y_+$ exists and $(\omega_+, y_+) \in K$, since K is closed.

Theorem 3.3 for the Cauchy Problem with $y(\omega_+) = y_+$ implies that there exists a solution in a neighborhood of ω_+. This contradicts the maximality of the interval $I_{\max}$. For x_1, the argument is similar.

Theorem 3.4 Let $U \subset R \times R$ be an open set and assume that $f: U \to R$ is continuous and satisfies a local Lipschitz condition. Then every solution of $y' = f(x, y)$ can be extended to the left and to the right up to the boundary of U.

Remark 3.3 Solution φ can be extended to the right up to the boundary of U means that φ exists to the right in an interval $x_0 \leqslant x < b$ ($b = +\infty$ is allowed), and one of the following cases applies:

(i) $b = +\infty$, the solution exists for all $x \geqslant x_0$.

(ii) $b < +\infty$ and $\limsup\limits_{x \to b_-} |\varphi(t)| = \infty$.

(iii) $b < +\infty$ and $\liminf\limits_{t \to b_-} \rho(t, \varphi(t)) = 0$, where $\rho(x, y)$ denotes the distance from the point (t, y) to the boundary of U; the solution comes arbitrarily close to the boundary of U.

If U is a bounded region, then (iii) is the only case.

Remark 3.4 Solution ψ can be extended to the right up to the boundary of U means that ψ exists to the right in an interval $a < x \leqslant x_0$ ($a = -\infty$ is allowed), and one of the following cases applies:

(i) $a = -\infty$, the solution exists for all $x \leqslant x_0$.

(ii) $a > -\infty$ and $\limsup\limits_{x \to a_+} |\varphi(x)| = \infty$; the solution "becomes infinite".

(iii) $a > -\infty$ and $\liminf\limits_{x \to a_+} \rho(x, \psi(x)) = 0$.

If U is a bounded region, then (iii) is the only case.

Example 3.4 Determine the maximal intervals on which the solutions to the differential equation $\frac{dy}{dx}=\frac{y^2-1}{2}$ passing through $(0,0)$ and $(\ln 2,-3)$ exist.

Solution The function $f(x,y)=\frac{y^2-1}{2}$ is defined on the whole xy-plane and satisfies the conditions of existence theorem and continuation theorem. The general solution to the given differential equation is $y=\frac{1+ce^t}{1-ce^t}$. In particular, the solution passing through $(0,0)$ is $y=\frac{1-e^x}{1+e^x}$. So the maximal interval is $-\infty<x<+\infty$.

The solution passing through $(\ln 2,-3)$ is $y=\frac{1+e^x}{1-e^x}$. The domain of this solution is $(-\infty,0)\cup(0,+\infty)$. Note that $\ln 2>0$ and $y\to-\infty$, as $t\to 0_+$. Hence, the maximal interval of this solution is $0<x<+\infty$.

3.5 Continuous Dependence of Solutions

In the preceding sections, we discussed the existence and uniqueness of solutions to the initial value problem

$$y'=f(x,y),\quad y(x_0)=y_0. \tag{3.15}$$

In this section, we investigate properties of the solution. Namely, we give conditions for the solution to be a continuous function of the initial value or of the function f.

To show that the solution to equation (3.15) is a continuous function of the initial value, we need the following inequality due to T. H. Gronwall.

Gronwall's Inequality

Lemma 3.2 Let u and y be continuous functions satisfying $u(x)>0$ and $v(x)\geqslant 0$ on $[a,b]$. Let $c\geqslant 0$ be a constant. If

$$v(x)\leqslant c+\int_a^x u(t)v(t)\,dt,\quad a\leqslant x\leqslant b, \tag{3.16}$$

then

$$v(x)\leqslant c\exp\left\{\int_a^x u(t)\,dt\right\},\quad a\leqslant x\leqslant b. \tag{3.17}$$

Proof First assume that $c>0$ and define $V(x)=c+\int_a^x u(t)v(t)\,dt$, $a\leqslant x\leqslant b$. Then $v(x)\leqslant V(x)$ by inequality (3.16), and since u and y are nonnegative, $V(x)\geqslant V(a)=c>0$ on $[a,b]$. Moreover, $V'(x)=u(x)v(x)\leqslant u(x)V(x)$. Dividing by $V(x)$, we get

$$\frac{V'(x)}{V(x)} \leqslant u(x). \qquad (3.18)$$

When we integrate both sides of inequality (3.18) from a to x, we find

$$\ln V(x) - \ln V(a) = \ln \frac{V(X)}{c} \leqslant \int_a^x u(t)\,dt.$$

Solving for $V(x)$ (recall that the exponential function is increasing), we obtain

$$V(x) \leqslant c \exp\left[\int_a^x u(t)\,dt\right], \quad a \leqslant x \leqslant b. \qquad (3.19)$$

Since $v(x) \leqslant V(x)$, inequality (3.17) follows from inequality (3.19).

If inequality (3.17) holds with $c=0$, then it certainly holds for all $c>0$. Hence, as we have shown in the preceding discussion, inequality (3.18) is valid for all $c>0$. Taking the limit as $c \to 0_+$, we get $v(x) \leqslant 0$, which is the same as inequality (3.18) with $c=0$.

We first investigate the continuous dependence of the solution $y(x, x_0, y_0)$ on (x_0, y_0).

Lemma 3.3 Let f: $U \to R \times R$ (U is an open set of $R \times R$) be continuous and satisfy a local Lipschitz condition. Then for any compact $K \subset U$ there exists $L \geqslant 0$ such that

$$|f(x,y) - f(x,z)| \leqslant L|y-z| \text{ for all } (x,y), (x,z) \in K.$$

Proof Let us assume the contrary. There exists sequences (x_n, y_n) and (x_n, z_n) such that

$$|f(x_n, y_n) - f(x_n, z_n)| > n|y_n - z_n|. \qquad (3.20)$$

By Bolzano-Weierstrass, the sequence (x_n, y_n) has an accumulation point (x, y). By assumption the function $f(x,y)$ satisfies a Lipschitz condition in a neighborhood V of (x, y). Since f is bounded on K with $M = \max\limits_{(x,y)\in K} |f(x, y)|$, it follows from inequality (3.20) that $|y_n - z_n| \leqslant \frac{2K}{n}$. Therefore there exists infinitely many indices n such that (x_n, y_n), $(x_n, z_n) \in V$. Then inequality (3.20) contradicts the Lipschitz condition on V.

Theorem 3.5 Let f:$U \to R \times R$ (U is an open set of $R \times R$) be continuous and satisfy a local Lipschitz condition. Then the solution $y(x, x_0, y_0)$ of the Cauchy problem

$$y' = f(x,y), \quad y(x_0) = y_0$$

is a continuous function of (x_0, y_0).

Proof We choose a closed interval $[a,b]\subset I_{max}(x_0, y_0)$ such that $(x_0, y_0)\in[a,b]$. We choose ε small enough such that the tubular neighborhood K of $y(x,x_0,y_0)$,

$$K=\{(x,y):\ x\in[a,b],\quad |y-y(x,x_0,y_0)|\leqslant\varepsilon\}$$

is contained in the open set U. By Lemma 3. 3, $f(x,y)$ satisfies a Lipschitz condition on K with a Lipschitz constant L. The set

$$V=\{(x_1,y_1):\ x_1\in[a,b],\quad |y_1-y(x_1,x_0,y_0)|\leqslant\varepsilon e^{-L(b-a)}\}$$

is a neighborhood of (x_0, y_0) which satisfies $V\subset K\subset U$. If $(x_1,y_1)\in V$, we have

$$|y(x,x_1,y_1)-y(x,x_0,y_0)|=|y(x,x_1,y_1)-y(x,x_1,y(x_1,x_0,y_0))|$$

$$=\left|y_1+\int_{x_1}^{x} f(s,y(s,x_1,y_1))ds-[y(x_1,x_0,y_0)+\int_{x_1}^{x} f(s,y(s,x_1,y(x_1,x_0,y_0))ds]\right|$$

$$\leqslant|y_1-y(x_1,x_0,y_0)|+L\int_{x_1}^{x}|y(s,x_1,y_1)-y(s,x_1,y(x_1,x_0,y_0))|.$$

From Gronwall lemma, we conclude that

$$|y(x,x_1,y_1)-y(x,x_0,y_0)|\leqslant e^{L|x-x_1|}|y_1-y(x_1,x_0,y_0)|\leqslant\varepsilon$$

and this concludes the proof.

Theorem 3. 6 (Differentiability of the solution on initial value) If $f(x,y)$ and f'_y are continuous in some region G, then the solution $y=\varphi(x,x_0,y_0)$ of the equation $y'=f(x,y)$ is continuously differentiable and we have formula

$$\frac{\partial\varphi}{\partial x_0}=-f(x_0,y_0)\exp\left[\int_{x_0}^{x}\frac{\partial f(t,\varphi)}{\partial y}dt\right],\quad \frac{\partial\varphi}{\partial y_0}=\exp\left[\int_{x_0}^{x}\frac{\partial f(t,\varphi)}{\partial y}dt\right].$$

Proof Continuity of the solution is obvious. Now we prove differentiability of the solution. We mainly prove that φ'_{x_0} exists and continues.

Take two initial points (x_0, y_0) and $(x_0+\Delta x_0, y_0)$, where $|\Delta x_0|\leqslant\alpha$, α is an enough small and positive constant. Let two solutions are φ and ψ, namely,

$$\varphi=y_0+\int_{x_0}^{x} f(t,\varphi)dt,\quad \psi=y_0+\int_{x_0+\Delta x_0}^{x} f(t,\psi)dt.$$

Then, we have

$$\psi-\varphi=\int_{x_0+\Delta x_0}^{x} f(t,\psi)dt-\int_{x_0}^{x} f(t,\varphi)dt$$

$$=-\int_{x_0}^{x_0+\Delta x_0} f(t,\psi)dt+\int_{x_0}^{x}\frac{\partial f[t,\varphi+\theta(\psi-\varphi)]}{\partial y}(\psi-\varphi)dt,$$

where $0<\theta<1$. In consideration of continuity of f'_y, φ and ψ, we have

$$\frac{\partial f[t,\varphi+\theta(\psi-\varphi)]}{\partial y}=\frac{\partial f(t,\varphi)}{\partial y}+r_1-\frac{1}{\Delta x_0}\int_{x_0}^{x_0+\Delta x_0} f(t,\psi)\mathrm{d}t$$
$$=-f(x_0,y_0)+r_2,$$

where $r_i\to 0$ if $\Delta x_0\to 0$ and $r_i=0$ if $\Delta x_0=0$ $(i=1,2)$. Therefore, we have

$$\frac{\psi-\varphi}{\Delta x_0}=-f(x_0,y_0)+r_2+\int_{x_0}^{x}\left[\frac{\partial f(t,\varphi)}{\partial y}+r_1\right]\frac{\psi-\varphi}{\Delta x_0}\mathrm{d}t.$$

Letting $\Delta x_0\to 0$, we have

$$\varphi'_{x_0}=-f(x_0,y_0)+\int_{x_0}^{x}\frac{\partial f(t,\varphi)}{\partial y}\varphi'_{x_0}\,\mathrm{d}t.$$

It implies that φ'_{x_0} is the solution of initial value problem

$$\text{DE}: \frac{\mathrm{d}z}{\mathrm{d}x}=\frac{\partial f(x,\varphi)}{\partial y}z;$$
$$\text{IC}: z(x_0)=-f(x_0,y_0).$$

Solving this initial value problem, we obtain

$$\frac{\partial\varphi}{\partial x_0}=-f(x_0,y_0)\ \exp\left[\int_{x_0}^{x}\frac{\partial f(t,\varphi)}{\partial y}\mathrm{d}t\right].$$

It is in the same way to obtain

$$\frac{\partial\varphi}{\partial y_0}=\exp\left[\int_{x_0}^{x}\frac{\partial f(t,\varphi)}{\partial y}\mathrm{d}t\right].$$

Example 3.5 Given first-order differential equation

$$\frac{\mathrm{d}y}{\mathrm{d}x}=\tan\frac{y}{x},$$

find the values of

$$\frac{\partial\varphi}{\partial x_0},\quad \frac{\partial\varphi}{\partial y_0},\quad \frac{\partial\varphi}{\partial x}$$

where $x_0=1$, $y_0=0$, $y=\varphi(x,x_0,y_0)$ is a solution of the initial value problem.

Solution Note that

$$\frac{\partial}{\partial y}\left(\tan\frac{y}{x}\right)=\frac{1}{x}\sec^2\frac{y}{x}$$

is continuous $(x\neq 0)$. Therefore, the initial value problem

$$\frac{\mathrm{d}y}{\mathrm{d}x}=\tan\frac{y}{x},\qquad y(1)=0$$

possesses a unique solution and this solution obviously is $y=0$. Hence, we have

$$\left.\frac{\partial\varphi}{\partial x_0}\right|_{(1,0)}=-f(x_0,y_0)\ \exp\left.\left[\int_{x_0}^{x}\frac{\partial f(t,\varphi)}{\partial y}\mathrm{d}t\right]\right|_{(1,0)}$$
$$=-\tan\frac{0}{1}\exp\left[\int_1^x\frac{1}{t}\sec^2\frac{0}{t}\mathrm{d}t\right]=0\cdot|x|=0,$$

$$\frac{\partial\varphi}{\partial y_0}=\exp\left[\int_{x_0}^{x}\frac{\partial f(t,\varphi)}{\partial y}\mathrm{d}t\right]=|x|,\qquad \frac{\partial\varphi}{\partial x}=\tan\frac{0}{x}=0.$$ ◆

Chapter 4

Theory of Higher-Order Linear Differential Equations

In this chapter, we will discuss the basic theory of linear higher-order differential equations. In the statements and proofs of results, we use concepts usually covered in an elementary linear algebra course—namely, linear dependence, determinants, and methods for solving systems of linear equations. These concepts also arise in the matrix approach for solving systems of differential equations and will be discussed in Chapter 5.

4.1 Basic Theory of Linear Differential Equations

A linear differential equation of order n is an equation that can be written in the form

$$a_n(x)y^{(n)}(x)+a_{n-1}(x)y^{(n-1)}(x)+\dots+a_0(x)y(x)=b(x), \tag{4.1}$$

where $a_0(x)$, $a_1(x)$, ..., $a_n(x)$ and $b(x)$ depend only on x, not y. When a_0, a_1, ..., a_n are all constants, we say equation (4.1) has **constant coefficients**; otherwise it has **variable coefficients.** If $b(x)\equiv 0$, equation (4.1) is called **homogeneous**; otherwise it is **nonhomogeneous.**

In developing a basic theory, we assume that $a_0(x)$, $a_1(x)$, ..., $a_n(x)$ and $b(x)$ are all continuous on an interval I and $a_n(x)\neq 0$ on I. Then, on dividing by $a_n(x)$, we can rewrite equation (4.1) in the **standard form**

$$y^{(n)}(x)+p_1(x)y^{(n-1)}(x)+\dots+p_n(x)y(x)=g(x), \tag{4.2}$$

where the functions $p_1(x)$, $p_2(x)$, ..., $p_n(x)$, and $g(x)$ are continuous on I.

For a linear higher-order differential equation, the initial value problem always has a unique solution.

Existence and Uniqueness

Theorem 4.1 Suppose $p_1(x)$, ..., $p_n(x)$ and $g(x)$ are each continuous on an interval (a,b) that contains the point x_0. Then, for any choice of the initial values γ_0,

$\gamma_1, \ldots, \gamma_{n-1}$, there exists a unique solution $y(x)$ on the whole interval (a,b) to the initial value problem

$$y^{(n)}(x)+p_1(x)y^{(n-1)}(x)+\ldots+p_n(x)y(x)=g(x), \tag{4.3}$$

$$y(x_0)=\gamma_0,\quad y'(x_0)=\gamma_1,\ \ldots,\ y^{(n-1)}(x_0)=\gamma_{n-1}. \tag{4.4}$$

The proof of Theorem 4.1 can be found in Chapter 5.

Example 4.1 For the initial value problem

$$x(x-1)y'''-3xy''+6x^2y'-(\cos x)y=\sqrt{x+5}, \tag{4.5}$$

$$y(x_0)=1,\quad y'(x_0)=0,\quad y''(x_0)=7, \tag{4.6}$$

determine the values of x_0 and the intervals (a,b) containing x_0 for which Theorem 4.1 guarantees the existence of a unique solution on (a,b).

Solution Putting equation (4.5) in standard form, we find that $p_1(x)=-3/(x-1)$, $p_2(x)=6x/(x-1)$, $p_3(x)=-(\cos x)/[x(x-1)]$, and $g(x)=\sqrt{x+5}/[x(x-1)]$. Now $p_1(x)$ and $p_2(x)$ are continuous on every interval not containing $x=1$, while $p_3(x)$ is continuous on every interval not containing $x=0$ or $x=1$. The function $g(x)$ is not defined for $x<-5$, $x=0$, and $x=1$, but is continuous on $(-5, 0)$, $(0, 1)$, and $(1, +\infty)$. Hence, the functions p_1, p_2, p_3, and g are simultaneously continuous on the intervals $(-5, 0)$, $(0, 1)$, and $(1, +\infty)$. From Theorem 4.1 it follows that if we choose $x_0\in(-5, 0)$, then there exists a unique solution to the initial value problem (4.5)-(4.6) on the whole interval $(-5, 0)$. Similarly, for $x_0\in(0, 1)$, there is a unique solution on $(0, 1)$ and, for $x_0\in(1, +\infty)$, a unique solution on $(1, +\infty)$. ◆

If we let the left-hand side of equation (4.3) define the differential operator L,

$$L[y]:=\frac{d^n y}{dx^n}+p_1\frac{d^{n-1}y}{dx^{n-1}}+\ldots+p_n y=(D^n+p_1D^{n-1}+\ldots+p_n)[y], \tag{4.7}$$

then we can express equation (4.3) in the operator form

$$L[y](x)=g(x). \tag{4.8}$$

It is essential to keep in mind that L is a linear operator—that is, it satisfies

$$L[y_1+y_2+\ldots+y_m]=L[y_1]+L[y_2]+\ldots+L[y_m], \tag{4.9}$$

$$L[cy]=cL[y]\ (c \text{ is any constant}). \tag{4.10}$$

These are familiar properties for the differentiation operator D, from which equations (4.9) and (4.10) follow (see Problem 25 in Exercises 4.1).

As a consequence of this linearity, if $y_1, y_2, \ldots, y_m$ are solutions to the homogeneous equation

$$L[y](x)=0, \tag{4.11}$$

then any linear combination of these functions, $C_1 y_1+C_2 y_2+\ldots+C_m y_m$, is also a solution, because

$$L[C_1 y_1+C_2 y_2+\ldots+C_m y_m]=C_1\cdot 0+C_2\cdot 0+\ldots+C_m\cdot 0=0.$$

Imagine now that we have found n solutions $y_1, y_2, \ldots, y_n$ to the nth-order linear equation (4. 11). Is it true that every solution to equation (4. 11) can be represented by

$$C_1 y_1+C_2 y_2+\ldots+C_n y_n \tag{4.12}$$

for appropriate choices of the constants $C_1, C_2, \ldots, C_n$? The answer is yes, provided the solutions $y_1, y_2, \ldots, y_n$ satisfy a certain property that we now derive.

Let $\varphi(x)$ be a solution to equation (4. 11) on the interval (a,b) and let x_0 be a fixed number in (a,b). If it is possible to choose the constants $C_1, C_2, \ldots, C_n$ so that

$$\begin{cases} C_1 y_1(x_0)+C_2 y_2(x_0)+\ldots+C_n y_n(x_0)=\varphi(x_0), \\ C_1 y_1'(x_0)+C_2 y_2'(x_0)+\ldots+C_n y_n'(x_0)=\varphi'(x_0), \\ \cdots\cdots \\ C_1 y_1^{(n-1)}(x_0)+C_2 y_2^{(n-1)}(x_0)+\ldots+C_n y_n^{(n-1)}(x_0)=\varphi^{(n-1)}(x_0), \end{cases} \tag{4.13}$$

then, since $\varphi(x)$ and $C_1 y_1(x)+C_2 y_2(x)+\ldots+C_n y_n(x)$ are two solutions satisfying the same initial conditions at x_0, the uniqueness conclusion of Theorem 4. 1 gives

$$\varphi(x)=C_1 y_1(x)+C_2 y_2(x)+\ldots+C_n y_n(x) \tag{4.14}$$

for all x in (a,b).

The system (4. 13) consists of n linear equations in the n unknowns $C_1, C_2, \ldots, C_n$. It has a unique solution for all possible values of $\varphi(x_0), \varphi'(x_0), \ldots, \varphi^{(n-1)}(x_0)$ if and only if the determinant of the coefficients is different from zero; that is, if and only if

$$\begin{vmatrix} y_1(x_0) & y_2(x_0) & \cdots & y_n(x_0) \\ y_1'(x_0) & y_2'(x_0) & \cdots & y_n'(x_0) \\ \vdots & \vdots & & \vdots \\ y_1^{(n-1)}(x_0) & y_2^{(n-1)}(x_0) & \cdots & y_n^{(n-1)}(x_0) \end{vmatrix} \neq 0. \tag{4.15}$$

Hence, if $y_1, y_2, \ldots, y_n$ are solutions to equation (4. 11) and there is some point x_0 in (a,b) such that (4. 15) holds, then every solution $\varphi(x)$ to equation (4. 11) is a linear combination of $y_1, y_2, \ldots, y_n$. Before formulating this fact as a theorem, it is convenient to identify the determinant by name.

Wronskian

Definition 4.1 Let $f_1, f_2, \ldots, f_n$ be any n functions that are $(n-1)$ times differentiable. The function

$$W[f_1, f_2, \ldots, f_n](x) := \begin{vmatrix} f_1(x) & f_2(x) & \ldots & f_n(x) \\ f_1'(x) & f_2'(x) & \ldots & f_n'(x) \\ \vdots & \vdots & & \vdots \\ f_1^{(n-1)}(x) & f_2^{(n-1)}(x) & \ldots & f_n^{(n-1)}(x) \end{vmatrix} \tag{4.16}$$

is called the **Wronskian** of $f_1, f_2, \ldots, f_n$.

We now state the representation theorem that we proved above for solutions to homogeneous linear differential equations.

Representation of Solutions (Homogeneous Case)

Theorem 4.2 Let $y_1, y_2, \ldots, y_n$ be n solutions on (a,b) of

$$y^{(n)}(x) + p_1(x)y^{(n-1)}(x) + \ldots + p_n(x)y(x) = 0, \tag{4.17}$$

where $p_1, p_2, \ldots, p_n$ are continuous on (a,b). If at some point x_0 in (a,b) these solutions satisfy

$$W[y_1, y_2, \ldots, y_n](x_0) \neq 0, \tag{4.18}$$

then every solution of equation (17) on (a,b) can be expressed in the form

$$y(x) = C_1 y_1(x) + C_2 y_2(x) + \ldots + C_n y_n(x), \tag{4.19}$$

where $C_1, C_2, \ldots, C_n$ are constants.

The linear combination of $y_1, y_2, \ldots, y_n$ in equation (4.19), written with arbitrary constants $C_1, C_2, \ldots, C_n$, is referred to as a **general solution** to equation (4.17).

In linear algebra, a set of m column vectors $\{\boldsymbol{v}_1, \boldsymbol{v}_2, \ldots, \boldsymbol{v}_m\}$, each having m components, is said to be linearly dependent if and only if at least one of them can be expressed as a linear combination of the others. ① A basic theorem then states that if a determinant is zero, its column vectors are linearly dependent, and conversely. So if a Wronskian of solutions to equation (4.17) is zero at a point x_0, one of its columns (the final column, say; we can always renumber!) equals a linear combination of the others:

---

① This is equivalent to saying there exist constants $C_1, C_2, \ldots, C_m$ **not all zero**, such that $C_1\boldsymbol{v}_1 + C_2\boldsymbol{v}_2 + \ldots + C_m\boldsymbol{v}_m$ equals the zero vector.

$$\begin{bmatrix} y_n(x_0) \\ y_n'(x_0) \\ \vdots \\ y_n^{(n-1)}(x_0) \end{bmatrix} = d_1 \begin{bmatrix} y_1(x_0) \\ y_1'(x_0) \\ \vdots \\ y_1^{(n-1)}(x_0) \end{bmatrix} + d_2 \begin{bmatrix} y_2(x_0) \\ y_2'(x_0) \\ \vdots \\ y_2^{(n-1)}(x_0) \end{bmatrix} + \ldots + d_{n-1} \begin{bmatrix} y_{n-1}(x_0) \\ y_{n-1}'(x_0) \\ \vdots \\ y_{n-1}^{(n-1)}(x_0) \end{bmatrix}. \tag{4.20}$$

Now consider the two functions $y_n(x)$ and $[d_1 y_1(x)+d_2 y_2(x)+\ldots+d_{n-1} y_{n-1}(x)]$. They are both solutions to equation (4.17), and we can interpret (4.20) as stating that they satisfy the same initial conditions at $x=x_0$. By the uniqueness theorem, then, they are one and the same function:

$$y_n(x)=d_1 y_1(x)+d_2 y_2(x)+\ldots+d_{n-1} y_{n-1}(x) \tag{4.21}$$

for all x in the interval I. Consequently, their derivatives are the same also, and so

$$\begin{bmatrix} y_n(x) \\ y_n'(x) \\ \vdots \\ y_n^{(n-1)}(x) \end{bmatrix} = d_1 \begin{bmatrix} y_1(x) \\ y_1'(x) \\ \vdots \\ y_1^{(n-1)}(x) \end{bmatrix} + d_2 \begin{bmatrix} y_2(x) \\ y_2'(x) \\ \vdots \\ y_2^{(n-1)}(x) \end{bmatrix} + \ldots + d_{n-1} \begin{bmatrix} y_{n-1}(x) \\ y_{n-1}'(x) \\ \vdots \\ y_{n-1}^{(n-1)}(x) \end{bmatrix} \tag{4.22}$$

for all x in I. Hence, the final column of the Wronskian $W[y_1, y_2, \ldots, y_n]$ is always a linear combination of the other columns, and consequently the Wronskian is always zero.

In summary, the Wronskian of n solutions to the homogeneous equation (4.17) is either identically zero, or never zero, on the interval (a,b). We have also shown that, in the former case, equation (4.21) holds throughout (a,b). Such a relationship among functions is an extension of the notion of linear dependence introduced in Section 4.2. We employ the same nomenclature for the general case.

Linear Dependence of Functions

Definition 4.2 The m functions $f_1, f_2, \ldots, f_m$ are said to be **linearly dependent on an interval *I*** if at least one of them can be expressed as a linear combination of the others on I; equivalently, they are linearly dependent if there exist constants C_1, $C_2, \ldots, C_m$, not all zero, such that

$$C_1 f_1(x)+C_2 f_2(x)+\ldots+C_m f_m(x)=0 \tag{4.23}$$

for all x in I. Otherwise, they are said to be **linearly independent on *I*.**

Example 4.2 Show that the functions $f_1(x)=\mathrm{e}^x$, $f_2(x)=\mathrm{e}^{-2x}$, and $f_3(x)=3\mathrm{e}^x-2\mathrm{e}^{-2x}$ are linearly dependent on $(-\infty, +\infty)$.

Solution Obviously, f_3 is a linear combination of f_1 and f_2:

$$f_3(x)=3e^x-2e^{-2x}=3f_1(x)-2f_2(x).$$

Note further that the corresponding identity $3f_1(x)-2f_2(x)-f_3(x)=0$ matches the pattern (4.23). Moreover, observe that f_1, f_2, and f_3 are pairwise linearly independent on $(-\infty, +\infty)$, but this does not suffice to make the triplet independent. ◆

To prove that functions f_1, f_2, ..., f_m are linearly independent on the interval (a,b), a convenient approach is the following: Assume that equation (4.23) holds on (a,b) and show that this forces $C_1=C_2=\ldots=C_m=0$.

Example 4.3 Show that the functions $f_1(x)=x$, $f_2(x)=x^2$, and $f_3(x)=1-2x^2$ are linearly independent on $(-\infty, +\infty)$.

Solution Assume C_1, C_2, and C_3 are constants for which

$$C_1x+C_2x^2+C_3(1-2x^2)=0 \tag{4.24}$$

holds at every x. If we can prove that equation (4.24) implies $C_1=C_2=C_3=0$, then linear independence follows. Let's set $x=0$, 1, and -1 in equation (4.24). Substituting in (4.24) gives

$$\begin{aligned} C_3=0 \quad (x=0), \\ C_1+C_2-C_3=0 \quad (x=1), \\ -C_1+C_2-C_3=0 \quad (x=-1). \end{aligned} \tag{4.25}$$

When we solve this system (or compute the determinant of the coefficients), we find that the only possible solution is $C_1=C_2=C_3=0$. Consequently, the functions f_1, f_2, and f_3 are linearly independent on $(-\infty, +\infty)$.

A neater solution is to note that if equation (4.24) holds for all x, so do its first and second derivatives. At $x=0$ these conditions are $C_3=0$, $C_1=0$, and $2C_2-4C_3=0$. Obviously, each coefficient must be zero. ◆

Linear dependence of functions is different from linear dependence of vectors in the Euclidean space $\mathbf{R}^n$, because (4.23) is a functional equation that imposes a condition every point of an interval. However, we have seen in equation (4.21) that when the functions are all solutions to the same homogeneous differential equation, linear dependence of the column vectors of the Wronskian (at any point x_0) implies linear dependence of the functions. The converse is also true, as demonstrated by equations (4.21) and (4.22). Theorem 4.3 summarizes our deliberations.

Linear Dependence and the Wronskian

Theorem 4.3 If $y_1, y_2, \ldots, y_n$ are n solutions to $y^{(n)}+p_1 y^{(n-1)}+\ldots+p_n y=0$ on the interval (a,b), with $p_1, p_2, \ldots, p_n$ continuous on (a,b), then the following statements are equivalent:

(i) $y_1, y_2, \ldots, y_n$ are linearly dependent on (a,b).

(ii) The Wronskian $W[y_1, y_2, \ldots, y_n](x_0)$ is zero at some point x_0 in (a,b).

(iii) The Wronskian $W[y_1, y_2, \ldots, y_n](x)$ is identically zero on (a,b).

The contrapositives of these statements are also equivalent:

(iv) $y_1, y_2, \ldots, y_n$ are linearly independent on (a,b).

(v) The Wronskian $W[y_1, y_2, \ldots, y_n](x_0)$ is nonzero at some point x_0 in (a,b).

(vi) The Wronskian $W[y_1, y_2, \ldots, y_n](x)$ is never zero on (a,b).

Whenever (iv), (v), or (vi) is met, $\{y_1, y_2, \ldots, y_n\}$ is called a **fundamental solution set** for equation (4.17) on (a,b).

The Wronskian is a curious function. If we take $W[f_1, f_2, \ldots, f_n](x)$ for n arbitrary functions, we simply get a function of x with no particularly interesting properties. But if the n functions are all solutions to the same homogeneous differential equation, then either it is identically zero or never zero. In fact, one can prove **Abel's identity** when the functions are all solutions to equation (4.17):

$$W[y_1, y_2, \ldots, y_n](x)=W[y_1, y_2, \ldots, y_n](x_0)\exp\left[\int_{x_0}^{x} p_1(t)\mathrm{d}t\right], \tag{4.26}$$

which clearly exhibits this property. Problem 30 in Exercises 4.1 outlines a proof of equation (4.26) for $n=3$.

It is useful to keep in mind that the following sets consist of functions that are linearly independent on every open interval (a,b):

$$\{1, x, x^2, \ldots, x^n\}$$

$$\{1, \cos x, \sin x, \cos 2x, \sin 2x, \ldots, \cos nx, \sin nx\},$$

$$\{e^{\alpha_1 x}, e^{\alpha_2 x}, \ldots, e^{\alpha_n x}\} \quad (\alpha_i\text{'s are distinct constants}).$$

If we combine the linearity (superposition) properties (4.9) and (4.10) with the representation theorem for solutions of the homogeneous equation, we obtain the following representation theorem for nonhomogeneous equations.

Representation of Solutions (Nonhomogeneous Case)

Theorem 4.4 Let $y_p(x)$ be a particular solution to the nonhomogeneous equation

$$y^{(n)}(x)+p_1(x)y^{(n-1)}(x)+\ldots+p_n(x)y(x)=g(x) \tag{4.27}$$

on the interval (a,b) with $p_1, p_2, \ldots, p_n$ continuous on (a,b), and let $\{y_1, y_2, \ldots, y_n\}$ be a fundamental solution set for the corresponding homogeneous equation

$$y^{(n)}(x)+p_1(x)y^{(n-1)}(x)+\ldots+p_n(x)y(x)=0. \tag{4.28}$$

Then every solution of equation (27) on the interval (a,b) can be expressed in the form

$$y(x)=y_p(x)+C_1y_1(x)+\ldots+C_ny_n(x). \tag{4.29}$$

Proof Let $\varphi(x)$ be any solution to equation (4.27). Because both $\varphi(x)$ and $y_p(x)$ are solutions to equation (4.27), by linearity the difference $\varphi(x)-y_p(x)$ is a solution to the homogeneous equation (4.28). It then follows from Theorem 4.2 that

$$\varphi(x)-y_p(x)=C_1y_1(x)+C_2y_2(x)+\ldots+C_ny_n(x)$$

for suitable constants $C_1, C_2, \ldots, C_n$. The last equation is equivalent to equation (4.29) [with $\varphi(x)$ in place of $y(x)$], so the theorem is proved. ◆

The linear combination of $y_p, y_1, \ldots, y_n$ in equation (4.29) written with arbitrary constants $C_1, C_2, \ldots, C_n$ is, for obvious reasons, referred to as a **general solution** to equation (4.27). Theorem 4.4 can be easily generalized. For example, if L denotes the operator appearing as the left-hand side in equation (4.27) and if $L[y_{p1}]=g_1$ and $L[y_{p2}]=g_2$, then any solution of $L[y]=c_1g_1+c_2g_2$ can be expressed as

$$y(x)=c_1y_{p1}(x)+c_2y_{p2}(x)+C_1y_1(x)+C_2y_2(x)+\ldots+C_ny_n(x),$$

for a suitable choice of the constants $C_1, C_2, \ldots, C_n$.

Example 4.4 Find a general solution on the interval $(-\infty, +\infty)$ to

$$L[y]:=y'''-2y''-y'+2y=2x^2-2x-4-24e^{-2x}, \tag{4.30}$$

given that $y_{p1}(x)=x^2$ is a particular solution to $L[y]=2x^2-2x-4$, that $y_{p2}(x)=e^{-2x}$ is a particular solution to $L[y]=-12e^{-2x}$, and that $y_1(x)=e^{-x}$, $y_2(x)=e^x$, and $y_3(x)=e^{2x}$ are solutions to the corresponding homogeneous equation.

Solution We previously remarked that the functions e^{-x}, e^x, e^{2x} are linearly independent because the exponents -1, 1, and 2 are distinct. Since each of these functions is a solution to the corresponding homogeneous equation, then $\{e^{-x}, e^x, e^{2x}\}$ is a fundamental solution set. It now follows from the remarks above for nonhomogeneous equations that a general solution to equation (4.30) is

$$\begin{aligned} y(x) &= y_{p1}+2y_{p2}+C_1y_1+C_2y_2+C_3y_3 \\ &= x^2+2e^{-2x}+C_1e^{-x}+C_2e^x+C_3e^{2x}. \quad ◆ \end{aligned} \tag{4.31}$$

Exercises 4.1

In Problems 1-6, determine the largest interval (a,b) for which Theorem 4.1 *guarantees the existence of a unique solution on* (a,b) *to the given initial value problem.*

1. $xy''' - 3y' + e^x y = x^2 - 1$; $y(-2)=1$, $y'(-2)=0$, $y''(-2)=2$.
2. $y''' - \sqrt{x}y = \sin x$; $y(\pi)=0$, $y'(\pi)=11$, $y''(\pi)=3$.
3. $y''' - y'' + \sqrt{x-1}y = \tan x$; $y(5)=y'(5)=y''(5)=1$.
4. $x(x+1)y''' - 3xy' + y = 0$; $y(-1/2)=1$, $y'(-1/2)=y''(-1/2)=0$.
5. $x\sqrt{x+1}y''' - y' + xy = 0$; $y(1/2)=y'(1/2)=-1$, $y''(1/2)=1$.
6. $(x^2-1)y''' + e^x y = \ln x$; $y(3/4)=1$, $y'(3/4)=y''(3/4)=0$.

In Problems 7-14, determine whether the given functions are linearly dependent or linearly independent on the specified interval. Justify your decisions.

7. $\{e^{3x}, e^{5x}, e^{-x}\}$ on $(-\infty,+\infty)$.
8. $\{x^2, x^2-1, 5\}$ on $(-\infty,+\infty)$.
9. $\{\sin^2 x, \cos^2 x, 1\}$ on $(-\infty,+\infty)$.
10. $\{\sin x, \cos x, \tan x\}$ on $(-\pi/2, \pi/2)$.
11. $\{x^{-1}, x^{1/2}, x\}$ on $(0,+\infty)$.
12. $\{\cos 2x, \cos^2 x, \sin^2 x\}$ on $(-\infty,+\infty)$.
13. $\{x, x^2, x^3, x^4\}$ on $(-\infty,+\infty)$.
14. $\{x, xe^x, 1\}$ on $(-\infty,+\infty)$.

Using the Wronskian in Problems 15-18, verify that the given functions form a fundamental solution set for the given differential equation and find a general solution.

15. $y''' + 2y'' - 11y' - 12y = 0$; $\{e^{3x}, e^{-x}, e^{-4x}\}$.
16. $y''' - y'' + 4y' - 4y = 0$; $\{e^x, \cos 2x, \sin 2x\}$.
17. $x^3y''' - 3x^2y'' + 6xy' - 6y = 0$, $x>0$; $\{x, x^2, x^3\}$.
18. $y^{(4)} - y = 0$; $\{e^x, e^{-x}, \cos x, \sin x\}$.

In Problems 19-22, a particular solution and a fundamental solution set are given for a nonhomogeneous equation and its corresponding homogeneous equation. (i) *Find a general solution to the nonhomogeneous equation.* (ii) *Find the solution that satisfies the specified initial conditions.*

19. $y''' + y'' + 3y' - 5y = 2 + 6x - 5x^2$;

    $y(0)=-1$, $y'(0)=1$, $y''(0)=-3$;

    $y_p = x^2$; $\{e^x, e^{-x}\cos 2x, e^{-x}\sin 2x\}$.

20. $xy''' - y'' = -2$; $y(1)=2$, $y'(1)=-1$, $y''(1)=-4$;

$y_p=x^2$; $\{1, x, x^3\}$.

21. $x^3y'''+xy'-y=3-\ln x$, $x>0$;

$y(1)=3$, $y'(1)=3$, $y''(1)=0$;

$y_p=\ln x$; $\{x, x\ln x, x(\ln x)^2\}$.

22. $y^{(4)}+4y=5\cos x$;

$y(0)=2$, $y'(0)=1$, $y''(0)=-1$, $y'''(0)=-2$;

$y_p=\cos x$; $\{e^x\cos x, e^x\sin x, e^{-x}\cos x, e^{-x}\sin x\}$.

23. Let $L[y]:=y'''+y'+xy$, $y_1(x):=\sin x$, and $y_2(x):=x$. Verify that $L[y_1](x)=x\sin x$ and $L[y_2](x)=x^2+1$. Then use the superposition principle (linearity) to find a solution to the differential equation:

(1) $L[y]=2x\sin x-x^2-1$.

(2) $L[y]=4x^2+4-6x\sin x$.

24. Let $L[y]:=y'''-xy''+4y'-3xy$, $y_1(x):=\cos 2x$, and $y_2(x):=-1/3$. Verify that $L[y_1](x)=x\cos 2x$ and $L[y_2](x)=x$. Then use the superposition principle (linearity) to find a solution to the differential equation:

(1) $L[y]=7x\cos 2x-3x$.

(2) $L[y]=-6x\cos 2x+11x$.

25. Prove that L defined in equation (4.7) is a linear operator by verifying that properties (4.9) and (4.10) hold for any n-times differentiable functions $y, y_1, \ldots, y_m$ on (a,b).

26. **Existence of Fundamental Solution Sets.** By Theorem 4.1, for each $j=1, 2, \ldots, n$, there is a unique solution $y_j(x)$ to equation (4.17) satisfying the initial conditions

$$y_j^{(k)}(x_0)=\begin{cases}1, & \text{for } k=j-1,\\ 0, & \text{for } k\neq j-1,\ 0\leqslant k\leqslant n-1.\end{cases}$$

(1) Show that $\{y_1, y_2, \ldots, y_n\}$ is a fundamental solution set for equation (4.17). [Hint: Write out the Wronskian at x_0.]

(2) For given initial values $\gamma_0, \gamma_1, \ldots, \gamma_{n-1}$, express the solution $y(x)$ to equation (4.17) satisfying $y^{(k)}(x_0)=\gamma_k$, $k=0, 1, \ldots, n-1$ [as in equation (4.4)], in terms of this fundamental solution set.

27. Show that the set of functions $\{1, x, x^2, \ldots, x^n\}$, where n is a positive integer, is linearly independent on every open interval (a,b). [Hint: Use the fact that

a polynomial of degree at most n has no more than n zeros unless it is identically zero.]

28. The set of functions

$$\{1, \cos x, \sin x, \ldots, \cos nx, \sin nx\},$$

where n is a positive integer, is linearly independent on every interval (a,b). Prove this in the special case $n=2$ and $(a,b)=(-\infty, +\infty)$.

29. (1) Show that if $f_1, f_2, \ldots, f_m$ are linearly independent on $(-1, 1)$, then they are linearly independent on $(-\infty, +\infty)$.

(2) Give an example to show that if $f_1, f_2, \ldots, f_m$ are linearly independent on $(-\infty, +\infty)$, then they need not be linearly independent on $(-1, 1)$.

30. To prove Abel's identity (4.26) for $n=3$, proceed as follows:

(1) Let $W(x) := W[y_1, y_2, y_3](x)$. Use the product rule for differentiation to show

$$W'(x)=\begin{vmatrix} y_1' & y_2' & y_3' \\ y_1' & y_2' & y_3' \\ y_1'' & y_2'' & y_3'' \end{vmatrix}+\begin{vmatrix} y_1 & y_2 & y_3 \\ y_1'' & y_2'' & y_3'' \\ y_1'' & y_2'' & y_3'' \end{vmatrix}+\begin{vmatrix} y_1 & y_2 & y_3 \\ y_1' & y_2' & y_3' \\ y_1''' & y_2''' & y_3''' \end{vmatrix}.$$

(2) Show that the above expression reduces to

$$W'(x)=\begin{vmatrix} y_1 & y_2 & y_3 \\ y_1' & y_2' & y_3' \\ y_1''' & y_2''' & y_3''' \end{vmatrix}. \tag{4.32}$$

(3) Since each y_i satisfies equation (4.17), show that

$$y_i^{(3)}(x)=-\sum_{k=1}^{3} p_k(x)y_i^{(3-k)}(x) \quad (i=1, 2, 3) \tag{4.33}$$

(4) Substituting the expressions in equation (4.33) into (4.32), show that

$$W'(x)=-p_1(x)W(x). \tag{4.34}$$

(5) Deduce Abel's identity by solving the first-order differential equation (4.34).

31. **Reduction of Order.** If a nontrivial solution $f(x)$ is known for the homogeneous equation

$$y^{(n)}+p_1(x)y^{(n-1)}+\ldots+p_n(x)y=0,$$

the substitution $y(x)=v(x)f(x)$ can be used to reduce the order of the equation, as was shown in Section 4.7 for second-order equations. By completing the following steps, demonstrate the method for the third-order equation

$$y'''-2y''-5y'+6y=0, \tag{4.35}$$

given that $f(x)=e^x$ is a solution.

(1) Set $y(x)=v(x)e^x$ and compute y', y'', and y'''.

(2) Substitute your expressions from part (1) into equation (4.35) to obtain a second-order equation in $w:=v'$.

(3) Solve the second-order equation in part (2) for w and integrate to find v. Determine two linearly independent choices for v, say, v_1 and v_2.

(4) By part (3), the functions $y_1(x)=v_1(x)e^x$ and $y_2(x)=v_2(x)e^x$ are two solutions to equation (4.35). Verify that the three solutions e^x, $y_1(x)$, and $y_2(x)$ are linearly independent on $(-\infty, +\infty)$.

32. Given that the function $f(x)=x$ is a solution to $y'''-x^2y'+xy=0$, show that the substitution $y(x)=v(x)f(x)=v(x)x$ reduces this equation to $xw''+3w'-x^3w=0$, where $w=v'$.

33. Use the reduction of order method described in Problem 31 to find three linearly independent solutions to $y'''-2y''+y'-2y=0$, given that $f(x)=e^2x$ is a solution.

34. **Constructing Differential Equations.** Given three functions $f_1(x)$, $f_2(x)$, $f_3(x)$ that are each three times differentiable and whose Wronskian is never zero on (a,b), show that the equation

$$\begin{vmatrix} f_1(x) & f_2(x) & f_3(x) & y \\ f_1'(x) & f_2'(x) & f_3'(x) & y' \\ f_1''(x) & f_2''(x) & f_3''(x) & y'' \\ f_1'''(x) & f_2'''(x) & f_3'''(x) & y''' \end{vmatrix}=0$$

is a third-order linear differential equation for which $\{f_1, f_2, f_3\}$ is a fundamental solution set. What is the coefficient of y''' in this equation?

35. Use the result of Problem 34 to construct a third-order differential equation for which $\{x, \sin x, \cos x\}$ is a fundamental solution set.

4.2 Homogeneous Linear Equations with Constant Coefficients

Our goal in this section is to obtain a general solution to an nth-order linear differential equation with constant coefficients.

Let's consider the homogeneous linear nth-order differential equation

$$a_ny^{(n)}(x)+a_{n-1}y^{(n-1)}(x)+\ldots+a_1y'(x)+a_0y(x)=0, \tag{4.36}$$

where $a_n(\neq 0)$, a_{n-1}, …, a_0 are real constants. ① Since constant functions are everywhere continuous, equation (4.36) has solutions defined for all x in $(-\infty, +\infty)$. If we can find n linearly independent solutions to equation (4.36) on $(-\infty, +\infty)$, say, y_1, y_2, …, y_n, then we can express a general solution to equation (4.36) in the form

$$y(x)=C_1y_1(x)+C_2y_2(x)+\ldots+C_ny_n(x), \tag{4.37}$$

with C_1, C_2, …, C_n as arbitrary constants.

To find these n linearly independent solutions, we begin by trying a function of the form $y=e^{rx}$.

If we let L be the differential operator defined by the left-hand side of equation (4.36), that is,

$$L[y]:=a_ny^{(n)}+a_{n-1}y^{(n-1)}+\ldots+a_1y'+a_0y, \tag{4.38}$$

then we can write equation (4.36) in the operator form

$$L[y](x)=0. \tag{4.39}$$

For $y=e^{rx}$, we find

$$\begin{aligned} L[e^{rx}](x) &= a_nr^ne^{rx}+a_{n-1}r^{n-1}e^{rx}+\ldots+a_0e^{rx} \\ &= e^{rx}(a_nr^n+a_{n-1}r^{n-1}+\ldots+a_0)=e^{rx}P(r), \end{aligned} \tag{4.40}$$

where $P(r)$ is the polynomial $a_nr^n+a_{n-1}r^{n-1}+\ldots+a_0$. Thus, e^{rx} is a solution to equation (4.39), provided r is a root of the **auxiliary** (or **characteristic**) **equation**

$$P(r)=a_nr^n+a_{n-1}r^{n-1}+\ldots+a_0=0. \tag{4.41}$$

According to the fundamental theorem of algebra, the auxiliary equation has n roots (counting multiplicities), which may be either real or complex.

We proceed to discuss the various possibilities.

Distinct Real Roots

If the roots r_1, r_2, …, r_n of the auxiliary equation (4.41) are real and distinct, then n solutions to equation (4.36) are

$$y_1(x)=e^{r_1x}, \quad y_2(x)=e^{r_2x}, \quad \ldots, \quad y_n(x)=e^{r_nx}. \tag{4.42}$$

As stated in the previous section, these functions are linearly independent on $(-\infty, +\infty)$, a fact that we now officially verify. Let's assume that C_1, C_2, …, C_n are constants such that

---

① Historical footnote: In a letter to John Bernoulli dated September 15, 1739, Leonhard Euler claimed to have solved the general case of the homogeneous linear nth-order equation with constant coefficients.

$$C_1 e^{r_1 x}+C_2 e^{r_2 x}+\ldots+C_n e^{r_n x}=0 \tag{4.43}$$

for all x in $(-\infty, +\infty)$. Our goal is to prove that $C_1=C_2=\ldots=C_n=0$.

One way to show this is to construct a linear operator L_k that annihilates (maps to zero) everything on the left-hand side of equation (4.43) except the kth term. For this purpose, we note that since $r_1, r_2, \ldots, r_n$ are the zeros of the auxiliary polynomial $P(r)$, then $P(r)$ can be factored as

$$P(r)=a_n(r-r_1)\ldots(r-r_n). \tag{4.44}$$

Consequently, the operator $L[y]=a_n y^{(n)}+a_{n-1}y^{(n-1)}+\ldots+a_0 y$ can be expressed in terms of the differentiation operator D as the following composition:

$$L=P(D)^{①}=a_n(D-r_1)\ldots(D-r_n). \tag{4.45}$$

We now construct the polynomial $P_k(r)$ by deleting the factor $(r-r_k)$ from $P(r)$. Then we set $L_k:=P_k(D)$; that is,

$$L_k:=P_k(D)=a_n(D-r_1)\ldots(D-r_{k-1})(D-r_{k+1})\ldots(D-r_n). \tag{4.46}$$

Applying L_k to both sides of equation (4.43), we get, via linearity,

$$c_1 L_k[e^{r_1 x}]+c_2 L_k[e^{r_2 x}]+\ldots+c_n L_k[e^{r_n x}]=0. \tag{4.47}$$

Also, since $L_k=P_k(D)$, we find [just as in equation (4.40)] that $L_k[e^{rx}](x)=e^{rx}P_k{}^{(r)}$ for all r. Thus, equation (4.47) can be written as

$$c_1 e^{r_1 x}P_k(r_1)+c_2 e^{r_2 x}P_k(r_2)+\ldots+c_n e^{r_n x}P_k(r_n)=0,$$

which simplifies to

$$c_k e^{r_k x}P_k(r_k)=0, \tag{4.48}$$

because $P_k(r_i)=0$ for $i\neq k$. Since r_k is not a root of $P_k(r)$, then $P_k(r_k)\neq 0$. It now follows from equation (4.48) that $c_k=0$. But as k is arbitrary, all the constants c_1, $c_2, \ldots, c_n$ must be zero. Thus, $y_1(x), y_2(x), \ldots, y_n(x)$ as given in equation (4.42) are linearly independent (see Problem 26 in Exercises 4.2 for an alternative proof).

We have proved that, in the case of n distinct real roots, a general solution to equation (4.36) is

$$y(x)=C_1 e^{r_1 x}+C_2 e^{r_2 x}+\ldots+C_n e^{r_n x}, \tag{4.49}$$

where $C_1, C_2, \ldots, C_n$ are arbitrary constants.

Example 4.5 Find a general solution to

$$y'''-2y''-5y'+6y=0. \tag{4.50}$$

---

① Historical footnote: The symbolic notation $P(D)$ was introduced by Augustin Cauchy in 1827.

Solution The auxiliary equation is

$$r^3-2r^2-5r+6=0. \tag{4.51}$$

By inspection, we find that $r=1$ is a root. Then, using polynomial division, we get

$$r^3-2r^2-5r+6=(r-1)(r^2-r-6),$$

which further factors into $(r-1)(r+2)(r-3)$. Hence, the roots of equation (4.51) are $r_1=1$, $r_2=-2$, $r_3=3$. Since these roots are real and distinct, a general solution to equation (4.50) is

$$y(x)=C_1\mathrm{e}^{x}+C_2\mathrm{e}^{-2x}+C_3\mathrm{e}^{3x}. \blacklozenge$$

Complex Roots

If $\alpha+\mathrm{i}\beta$ (α, β are real) is a complex root of the auxiliary equation (4.41), then so is its complex conjugate $\alpha-\mathrm{i}\beta$, since the coefficients of $P(r)$ are real-valued (see Problem 24 in Exercises 4.2). If we accept complex-valued functions as solutions, then both $\mathrm{e}^{(\alpha+\mathrm{i}\beta)x}$ and $\mathrm{e}^{(\alpha-\mathrm{i}\beta)x}$ are solutions to equation (4.36). Moreover, if there are no repeated roots, then a general solution to equation (4.36) is again given by equation (4.49). To find two real-valued solutions corresponding to the roots $\alpha\pm\mathrm{i}\beta$, we can just take the real and imaginary parts of $\mathrm{e}^{(\alpha+\mathrm{i}\beta)x}$. That is, since

$$\mathrm{e}^{(\alpha+\mathrm{i}\beta)x}=\mathrm{e}^{\alpha x}\cos\beta x+\mathrm{i}\mathrm{e}^{\alpha x}\sin\beta x, \tag{4.52}$$

then two linearly independent solutions to equation (4.36) are

$$\mathrm{e}^{\alpha x}\cos\beta x, \quad \mathrm{e}^{\alpha x}\sin\beta x. \tag{4.53}$$

In fact, using these solutions in place of $\mathrm{e}^{(\alpha+\mathrm{i}\beta)x}$ and $\mathrm{e}^{(\alpha-\mathrm{i}\beta)x}$ in equation (4.49) preserves the linear independence of the set of n solutions. Thus, treating each of the conjugate pairs of roots in this manner, we obtain a real-valued general solution to equation (4.36).

Example 4.6 Find a general solution to

$$y'''+y''+3y'-5y=0. \tag{4.54}$$

Solution The auxiliary equation is

$$r^3+r^2+3r-5=(r-1)(r^2+2r+5)=0, \tag{4.55}$$

which has distinct roots $r_1=1$, $r_2=-1+2\mathrm{i}$, $r_3=-1-2\mathrm{i}$. Thus, a general solution is

$$y(x)=C_1\mathrm{e}^{x}+C_2\mathrm{e}^{-x}\cos 2x+C_3\mathrm{e}^{-x}\sin 2x. \blacklozenge \tag{4.56}$$

Repeated Roots

If r_1 is a root of multiplicity m, then the n solutions given in equation (4.42) are not even distinct, let alone linearly independent. Recall that for a second-order equation, when we had a repeated root r_1 to the auxiliary equation, we obtained two linearly independent solutions by taking e^{r_1x} and xe^{r_1x}. So if r_1 is a root of equation (4.41) of multiplicity m, we might expect that m linearly independent solutions are

$$e^{r_1x},\quad xe^{r_1x},\quad x^2e^{r_1x},\quad \dots,\quad x^{m-1}e^{r_1x}. \tag{4.57}$$

To see that this is the case, observe that if r_1 is a root of multiplicity m, then the auxiliary equation can be written in the form

$$a_n(r-r_1)^m(r-r_{m+1})\dots(r-r_n)=(r-r_1)^m\widetilde{P}(r)=0, \tag{4.58}$$

where $\widetilde{P}(r):=a_n(r-r_{m+1})\dots(r-r_n)$ and $\widetilde{P}(r_1)\neq 0$. With this notation, we have the identity

$$L[e^{rx}](x)=e^{rx}(r-r_1)^m\widetilde{P}(r) \tag{4.59}$$

[see equation (4.40)]. Setting $r=r_1$ in equation (4.59), we again see that e^{r_1x} is a solution to $L[y]=0$.

To find other solutions, we take the kth partial derivative with respect to r of both sides of equation (4.59):

$$\frac{\partial^k}{\partial r^k}L[e^{rx}](x)=\frac{\partial^k}{\partial r^k}[e^{rx}(r-r_1)^m\widetilde{P}(r)]. \tag{4.60}$$

Carrying out the differentiation on the right-hand side of equation (4.60), we find that the resulting expression will still have $(r-r_1)$ as a factor, provided $k\leqslant m-1$. Thus, setting $r=r_1$ in equation (4.60) gives

$$\left.\frac{\partial^k}{\partial r^k}L[e^{rx}](x)\right|_{r=r_1}=0,\ \text{if}\quad k\leqslant m-1. \tag{4.61}$$

Now notice that the function e^{rx} has continuous partial derivatives of all orders with respect to r and x. Hence, for mixed partial derivatives of e^{rx}, it makes no difference whether the differentiation is done first with respect to x, then with respect to r, or vice versa. Since L involves derivatives with respect to x, this means we can interchange the order of differentiation in equation (4.61) to obtain

$$L\left[\left.\frac{\partial^k}{\partial r^k}(e^{rx})\right|_{r=r_1}\right](x)=0.$$

Thus,

$$\left.\frac{\partial^k}{\partial r^k}(e^{rx})\right|_{r=r_1}=x^ke^{r_1x} \tag{4.62}$$

will be a solution to equation (4.36) for $k=0, 1, \dots, m-1$. So m distinct solutions to equation (4.36), due to the root $r=r_1$ of multiplicity m, are indeed given by (4.57). We leave it as an exercise to show that the m functions in (4.57) are linearly independent on $(-\infty,+\infty)$ (see Problem 25 in Exercises 4.2).

If $\alpha+i\beta$ is a repeated complex root of multiplicity m, then we can replace the $2m$ complex-valued functions

$$e^{(\alpha+i\beta)x}, \quad xe^{(\alpha+i\beta)x}, \quad \ldots, \quad x^{m-1}e^{(\alpha+i\beta)x},$$
$$e^{(\alpha-i\beta)x}, \quad xe^{(\alpha-i\beta)x}, \quad \ldots, \quad x^{m-1}e^{(\alpha-i\beta)x}$$

by the $2m$ linearly independent real-valued functions

$$\begin{aligned} &e^{\alpha x}\cos\beta x, \quad xe^{\alpha x}\cos\beta x, \quad \ldots, \quad x^{m-1}e^{\alpha x}\cos\beta x, \\ &e^{\alpha x}\sin\beta x, \quad xe^{\alpha x}\sin\beta x, \quad \ldots, \quad x^{m-1}e^{\alpha x}\sin\beta x. \end{aligned} \tag{4.63}$$

Using the results of the three cases discussed above, we can obtain a set of n linearly independent solutions that yield a real-valued general solution for equation (4.36).

Example 4.7 Find a general solution to

$$y^{(4)}-y'''-3y''+5y'-2y=0. \tag{4.64}$$

Solution The auxiliary equation is

$$r^4-r^3-3r^2+5r-2=(r-1)^3(r+2)=0,$$

which has roots $r_1=1$, $r_2=1$, $r_3=1$, and $r_4=-2$. Because the root at 1 has multiplicity 3, a general solution is

$$y(x)=C_1e^x+C_2xe^x+C_3x^2e^x+C_4e^{-2x}. \ \blacklozenge \tag{4.65}$$

Example 4.8 Find a general solution to

$$y^{(4)}-8y'''+26y''-40y'+25y=0, \tag{4.66}$$

whose auxiliary equation can be factored as

$$r^4-8r^3+26r^2-40r+25=(r^2-4r+5)^2=0. \tag{4.67}$$

Solution The auxiliary equation (4.67) has repeated complex roots: $r_1=2+i$, $r_2=2+i$, $r_3=2-i$, and $r_4=2-i$. Hence, a general solution is

$$y(x)=C_1e^{2x}\cos x+C_2xe^{2x}\cos x+C_3e^{2x}\sin x+C_4xe^{2x}\sin x. \ \blacklozenge$$

Exercises 4.2

In Problems 1-14, find a general solution for the differential equation with x as the independent variable.

1. $y'''+2y''-8y'=0$.
2. $y'''-3y''-y'+3y=0$.
3. $6z'''+7z''-z'-2z=0$.
4. $y'''+2y''-19y'-20y=0$.
5. $y'''+3y''+28y'+26y=0$.
6. $y'''-y''+2y=0$.

7. $2y''' - y'' - 10y' - 7y = 0$.

8. $y''' + 5y'' - 13y' + 7y = 0$.

9. $u''' - 9u'' + 27u' - 27u = 0$.

10. $y''' + 3y'' - 4y' - 6y = 0$.

11. $y^{(4)} + 4y''' + 6y'' + 4y' + y = 0$.

12. $y''' + 5y'' + 3y' - 9y = 0$.

13. $y^{(4)} + 4y'' + 4y = 0$.

14. $y^{(4)} + 2y''' + 10y'' + 18y' + 9y = 0$.

[Hint: $y(x)\sin 3x$ is a solution.]

In Problems 15-18, find a general solution to the given homogeneous equation.

15. $(D-1)^2(D+3)(D^2+2D+5)^2[y]=0$.

16. $(D+1)^2(D-6)^3(D+5)(D^2+1)(D^2+4)[y]=0$.

17. $(D+4)(D-3)(D+2)^3(D^2+4D+5)^2D^5[y]=0$.

18. $(D-1)^3(D-2)(D^2+D+1)(D^2+6D+10)^3[y]=0$.

In Problems 19-21, solve the given initial value problem.

19. $y''' - y'' - 4y' + 4y = 0$; $y(0) = -4$, $y'(0) = -1$, $y''(0) = -19$.

20. $y''' + 7y'' + 14y' + 8y = 0$; $y(0) = 1$, $y'(0) = -3$, $y''(0) = 13$.

21. $y''' - 4y'' + 7y' - 6y = 0$; $y(0) = 1$, $y'(0) = 0$, $y''(0) = 0$.

In Problems 22 and 23, find a general solution for the given linear system using the elimination method.

22. $d^2x/dt^2 - x + 5y = 0$, $2x + d^2y/dt^2 + 2y = 0$.

23. $d^3x/dt^3 - x + dy/dt + y = 0$, $dx/dt - x + y = 0$.

24. Let $P(r) = a_n r^n + \ldots + a_1 r + a_0$ be a polynomial with real coefficients $a_n, \ldots, a_1, a_0$. Prove that if r_1 is a zero of $P(r)$, then so is its complex conjugate $\bar{r}_1$. [Hint: Show that $\overline{P(r)} = P(\bar{r})$, where the bar denotes complex conjugation.]

25. Show that the m functions e^{rx}, xe^{rx}, $\ldots$, $x^{m-1}e^{rx}$ are linearly independent on $(-\infty, +\infty)$. [Hint: Show that these functions are linearly independent if and only if $1, x, \ldots, x^{m-1}$ are linearly independent.]

26. As an alternative proof that the functions e^{r_1x}, e^{r_2x}, $\ldots$, e^{r_nx} are linearly independent on $(-\infty, +\infty)$ when $r_1, r_2, \ldots, r_n$ are distinct, assume

$$C_1e^{r_1x} + C_2e^{r_2x} + \ldots + C_ne^{r_nx} = 0 \qquad (4.68)$$

holds for all x in $(-\infty, +\infty)$ and proceed as follows:

(1) Because the r_i's are distinct, we can (if necessary) relabel them so that

$$r_1>r_2>\ldots>r_n.$$

Divide equation (4.68) by e^{r_1x} to obtain

$$C_1+C_2\frac{e^{r_2x}}{e^{r_1x}}+\ldots+C_n\frac{e^{r_nx}}{e^{r_1x}}=0.$$

Now let $x\to+\infty$ on the left-hand side to obtain $C_1=0$.

(2) Since $C_1=0$, equation (4.68) becomes

$$C_2e^{r_2x}+C_3e^{r_3x}+\ldots+C_ne^{r_nx}=0$$

for all x in $(-\infty,+\infty)$. Divide this equation by e^{r_2x} and let $x\to+\infty$ to conclude that $C_2=0$.

(3) Continuing in the manner of (2), argue that all the coefficients, C_1, C_2, ..., C_n are zero and hence e^{r_1x}, e^{r_2x}, ..., e^{r_nx} are linearly independent on $(-\infty,+\infty)$.

27. (1) Derive the form

$$y(x)=A_1e^x+A_2e^{-x}+A_3\cos x+A_4\sin x$$

for the general solution to the equation $y^{(4)}=y$, from the observation that the fourth roots of unity are 1, −1, i, and −i.

(2) Derive the form

$$y(x)=A_1e^x+A_2e^{-x/2}\cos(\sqrt{3}x/2)+A_3e^{-x/2}\sin(\sqrt{3}x/2)$$

for the general solution to the equation $y^{(3)}=y$, from the observation that the cube roots of unity are 1, $e^{i2\pi/3}$, and $e^{-i2\pi/3}$.

28. **Higher-Order Cauchy-Euler Equations.** A differential equation that can be expressed in the form

$$a_nx^ny^{(n)}(x)+a_{n-1}x^{n-1}y^{(n-1)}(x)+\ldots+a_0y(x)=0,$$

where a_n, a_{n-1}, ..., a_0 are constants, is called a homogeneous **Cauchy-Euler** equation. (The second-order case will be discussed in Section 4.7.) Use the substitution $y=x^r$ to help determine a fundamental solution set for the following Cauchy-Euler equations:

(1) $x^3y'''+x^2y''-2xy'+2y=0$, $x>0$.

(2) $x^4y^{(4)}+6x^3y'''+2x^2y''-4xy'+4y=0$, $x>0$.

(3) $x^3y'''-2x^2y''+13xy'-13y=0$, $x>0$

[Hint: $x^{\alpha+i\beta}=e^{(\alpha+i\beta)\ln x}=x^\alpha\{\cos(\beta\ln x)+i\sin(\beta\ln x)\}$.]

29. Let $y(x)=Ce^{rx}$, where C ($\neq0$) and r are real numbers, be a solution to a

differential equation. Suppose we cannot determine r exactly but can only approximate it by $\tilde{r}$. Let $\tilde{y}(x):=Ce^{\tilde{r}x}$ and consider the error $|y(x)-\tilde{y}(x)|$.

(1) If r and $\tilde{r}$ are positive, $r\neq\tilde{r}$, show that the error grows exponentially large as x approaches $+\infty$.

(2) If r and $\tilde{r}$ are negative, $r\neq\tilde{r}$, show that the error goes to zero exponentially as x approaches $+\infty$.

30. **Vibrating Beam.** In studying the transverse vibrations of a beam, one encounters the homogeneous equation

$$EI\frac{d^4y}{dx^4}-ky=0,$$

where $y(x)$ is related to the displacement of the beam at position x, the constant E is Young's modulus, I is the area moment of inertia, and k is a parameter. Assuming E, I, and k are positive constants, find a general solution in terms of sines, cosines, hyperbolic sines, and hyperbolic cosines.

4.3 Undetermined Coefficients and the Annihilator Method

In this section, we are going to introduce the method of undetermined coefficients to nonhomogeneous linear equation with constant coefficients.

At the outset, we'll describe the new point of view that will be adopted for the analysis. Then we illustrate its implications and ultimately derive a simplified set of rules for its implemention. The rigorous approach is known as the **annihilator method.**

The first premise of the annihilator method is the observation, gleaned from the analysis of the previous section, that all of the "suitable types" of nonhomogeneities $f(x)$ (products of polynomials times exponentials times sinusoids) are themselves solutions to **homogeneous** differential equations with constant coefficients. Observe the following:

(i) Any nonhomogeneous term of the form $f(x)=e^{rx}$ satisfies $(D-r)[f]=0$.

(ii) Any nonhomogeneous term of the form $f(x)=x^k e^{rx}$ satisfies $(D-r)^m[f]=0$ for $k=0, 1, \ldots, m-1$.

(iii) Any nonhomogeneous term of the form $f(x)=\cos\beta x$ or $\sin\beta x$ satisfies $(D^2+\beta^2)[f]=0$.

(iv) Any nonhomogeneous term of the form $f(x)=x^k e^{\alpha x}\cos\beta x$ or $x^k e^{\alpha x}\sin\beta x$

satisfies $[(D-\alpha)^2+\beta^2]^m[f]=0$ for $k=0, 1, \ldots, m-1$.

In other words, each of these nonhomogeneities is annihilated by a differential operator with constant coefficients.

Annihilator

Definition 4.3 A linear differential operator A is said to **annihilate** a function f if

$$A[f](x)=0, \tag{4.69}$$

for all x. That is, A annihilates f if f is a solution to the homogeneous linear differential equation (4.69) on $(-\infty,+\infty)$.

Example 4.9 Find a differential operator that annihilates

$$6xe^{-4x}+5e^x \sin 2x. \tag{4.70}$$

Solution Consider the two functions whose sum appears in (4.70). Observe that $(D+4)^2$ annihilates the function $f_1(x):=6xe^{-4x}$. Further, $f_2(x):=5e^x \sin 2x$ is annihilated by the operator $(D-1)^2+4$. Hence, the composite operator

$$A:=(D+4)^2[(D-1)^2+4],$$

which is the same as the operator

$$[(D-1)^2+4](D+4)^2,$$

annihilates both f_1 and f_2. But then, by linearity, A also annihilates the sum f_1+f_2. ◆

We now show how annihilators can be used to determine particular solutions to certain nonhomogeneous equations. Consider the nth-order differential equation with constant coefficients

$$a_ny^{(n)}(x)+a_{n-1}y^{(n-1)}(x)+\ldots+a_0y(x)=f(x), \tag{4.71}$$

which can be written in the operator form

$$L[y](x)=f(x), \tag{4.72}$$

where

$$L=a_nD^n+a_{n-1}D^{n-1}+\ldots+a_0.$$

Assume that A is a linear differential operator with constant coefficients that annihilates $f(x)$. Then

$$AL[y](x)=A[f](x)=0,$$

so any solution to equation (4.72) is also a solution to the homogeneous equation

$$AL[y](x)=0, \tag{4.73}$$

involving the composition of the operators A and L. But equation (4.73) has constant coefficients, and we are experts on differential equations with constant coefficients.

From this we can deduce the form of a particular solution to equation (4.72). Let's look at some examples and then summarize our findings.

Example 4.10 Find a general solution to

$$y''-y=xe^x+\sin x. \tag{4.74}$$

Solution The homogeneous equation corresponding to equation (4.74) is $y''-y=0$, with the general solution $C_1e^{-x}+C_2e^x$. Since e^x is a solution of the homogeneous equation, the nonhomogeneity xe^x demands a solution form $x(C_3+C_4x)e^x$. To accommodate the nonhomogeneity $\sin x$, we need an undetermined coefficient form $C_5\sin x+C_6\cos x$. Values for C_3 through C_6 in the particular solution are determined by substitution:

$$y_p''-y_p=(C_3xe^x+C_4x^2e^x+C_5\sin x+C_6\cos x)''$$
$$-(C_3xe^x+C_4x^2e^x+C_5\sin x+C_6\cos x)=\sin x+xe^x,$$

eventually leading to the conclusion $C_3=-1/4$, $C_4=1/4$, $C_5=-1/2$, and $C_6=0$. Thus (for future reference), a general solution to equation (4.74) is

$$y(x)=C_1e^{-x}+C_2e^x+x\left(-\frac{1}{4}+\frac{1}{4}x\right)e^x-\frac{1}{2}\sin x. \tag{4.75}$$

For the annihilator method, observe that (D^2+1) annihilates $\sin x$ and $(D-1)^2$ annihilates xe^x. Therefore, any solution to equation (4.74), expressed for convenience in operator form as $(D^2-1)[y](x)=xe^x+\sin x$, is annihilated by the composition $(D^2+1)(D-1)^2(D^2-1)$; that is, it satisfies the constant coefficient homogeneous equation

$$(D^2+1)(D-1)^2(D^2-1)[y]=(D+1)(D-1)^3(D^2+1)[y]=0. \tag{4.76}$$

We deduce that the general solution to equation (4.76) is given by

$$y=C_1e^{-x}+C_2e^x+C_3xe^x+C_4x^2e^x+C_5\sin x+C_6\cos x. \tag{4.77}$$

This is precisely the solution form generated by the methods of Chapter 4; the first two terms are the general solution to the associated homogeneous equation, and the remaining four terms express the particular solution to the nonhomogeneous equation with undetermined coefficients. Substitution of equation (4.77) into equation (4.74) will lead to the quoted values for C_3 through C_6, and indeterminant values for C_1 and C_2; the latter are available to fit initial conditions.

Note how the annihilator method automatically accounts for the fact that the nonhomogeneity xe^x requires the form $C_3xe^x+C_4x^2e^x$ in the particular solution, by counting the total number of factors of $(D-1)$ in the annihilator and the original differential operator. ◆

Example 4. 11 Find a general solution, using the annihilator method, to

$$y'''-3y''+4y=xe^{2x}. \tag{4.78}$$

Solution The associated homogeneous equation takes the operator form

$$(D^3-3D^2+4)[y]=(D+1)(D-2)^2[y]=0. \tag{4.79}$$

The nonhomogeneity xe^{2x} is annihilated by $(D-2)^2$. Therefore, every solution of equation (4. 78) also satisfies

$$(D-2)^2(D^3-3D^2+4)[y]=(D+1)(D-2)^4[y]=0. \tag{4.80}$$

A general solution to equation (4. 80) is

$$y(x)=C_1e^{-x}+C_2e^{2x}+C_3xe^{2x}+C_4x^2e^{2x}+C_5x^3e^{2x}. \tag{4.81}$$

Comparison with equation (4. 79) shows that the first three terms in right-hand side of equation (4. 81) give a general solution to the associated homogeneous equation and the last two terms constitute a particular solution form with undetermined coefficients. Direct substitution reveals $C_4=-1/18$ and $C_5=1/18$ and so a general solution to equation (4. 78) is

$$y(x)=C_1e^{-x}+C_2e^{2x}+C_3xe^{2x}-\frac{1}{18}x^2e^{2x}+\frac{1}{18}x^3e^{2x}.$$ ◆

Note that we don't have to implement the annihilator method directly; we simply need to introduce the following method of undetermined coefficients.

Method of Undetermined Coefficients

To find a particular solution to the constant-coefficient differential equation $L[y]=Cx^me^{rx}$, where m is a nonnegative integer, use the form

$$y_p(x)=x^s(A_mx^m+\ldots+A_1x+A_0)e^{rx}, \tag{4.82}$$

with $s=0$ if r is not a root of the associated auxiliary equation; otherwise, take s equal to the multiplicity of this root.

To find a particular solution to the constant-coefficient differential equation $L[y]=Cx^me^{\alpha x}\cos\beta x$ or $L[y]=Cx^me^{\alpha x}\sin\beta x$, where $\beta\neq 0$, use the form

$$y_p(x)=x^s(A_mx^m+\ldots+A_1x+A_0)e^{\alpha x}\cos\beta x$$
$$+x^s(B_mx^m+\ldots+B_1x+B_0)e^{\alpha x}\sin\beta x, \tag{4.83}$$

with $s=0$ if $\alpha+i\beta$ is not a root of the associated auxiliary equation; otherwise, take s equal to the multiplicity of this root.

Exercises 4. 3

In Problems 1-4, use the method of undetermined coefficients to determine the form of a particular solution for the given equation.

1. $y'''-2y''-5y'+6y=e^x+x^2$.

2. $y'''+y''-5y'+3y=e^{-x}+\sin x$.

3. $y'''+3y''-4y=e^{-2x}$.

4. $y'''+y''-2y=xe^x+1$.

In Problems 5-10, find a general solution to the given equation.

5. $y'''-2y''-5y'+6y=e^x+x^2$.

6. $y'''+y''-5y'+3y=e^{-x}\sin x$.

7. $y'''+3y''-4y=e^{-2x}$.

8. $y'''+y''-2y=xe^x+1$.

9. $y'''-3y''+3y'-y=e^x$.

10. $y'''+4y''+y'-26y=e^{-3x}\sin 2x+x$.

In Problems 11-20, find a differential operator that annihilates the given function.

11. x^4-x^2+11.　　12. $3x^2-6x+1$.

13. e^{-7x}.　　14. e^{5x}.

15. $e^{2x}-6e^x$.　　16. x^2-e^x.

17. $x^2e^{-x}\sin 2x$.　　18. $xe^{3x}\cos 5x$.

19. $xe^{-2x}+xe^{-5x}\sin 3x$.　　20. $x^2e^x-x\sin 4x+x^3$.

In Problems 21-30, use the annihilator method to determine the form of a particular solution for the given equation.

21. $u''-5u'+6u=\cos 2x+1$.

22. $y''+6y'+8y=e^{3x}-\sin x$.

23. $y''-5y'+6y=e^{3x}-x^2$.

24. $\theta''-\theta=xe^x$.

25. $y''-6y'+9y=\sin 2x+x$.

26. $y''+2y'+y=x^2-x+1$.

27. $y''+2y'+2y=e^{-x}\cos x+x^2$.

28. $y''-6y'+10y=e^{3x}-x$.

29. $z'''-2z''+z'=x-e^x$.

30. $y'''+2y''-y'-2y=e^x-1$.

In Problems 31-33, solve the given initial value problem.

31. $y'''+2y''-9y'-18y=-18x^2-18x+22$;

$y(0)=-2$, $y'(0)=-8$, $y''(0)=-12$.

32. $y'''-2y''+5y'=-24e^{3x}$; $y(0)=4$, $y'(0)=-1$, $y''(0)=-5$.

33. $y'''-2y''-3y'+10y=34xe^{-2x}-16e^{-2x}-10x^2+6x+34$;

$y(0)=3$, $y'(0)=0$, $y''(0)=0$.

34. Use the annihilator method to show that if $a_0\neq 0$ in equation (4.71) and $f(x)$ has the form

$$f(x)=b_m x^m+b_{m-1}x^{m-1}+\ldots+b_1 x+b_0, \tag{4.84}$$

then

$$y_p(x)=B_m x^m+B_{m-1}x^{m-1}+\ldots+B_1 x+B_0$$

is the form of a particular solution to equation (4.71).

35. Use the annihilator method to show that if $a_0=0$ and $a_1\neq 0$ in equation (4.71) and $f(x)$ has the form given in equation (4.84), then equation (4.71) has a particular solution of the form

$$y_p(x)=x(B_m x^m+B_{m-1}x^{m-1}+\ldots+B_1 x+B_0).$$

36. Use the annihilator method to show that if $f(x)$ in equation (4.71) has the form $f(x)=Be^{\alpha x}$, then equation (4.71) has a particular solution of the form $y_p(x)=x^s Be^{\alpha x}$, where s is chosen to be the smallest nonnegative integer such that $x^s e^{\alpha x}$ is not a solution to the corresponding homogeneous equation.

37. Use the annihilator method to show that if $f(x)$ in equation (4.71) has the form

$$f(x)=a\cos\beta x+b\sin\beta x,$$

then equation (4.71) has a particular solution of the form

$$y_p(x)=x^s(A\cos\beta x+B\sin\beta x), \tag{4.85}$$

where s is chosen to be the smallest nonnegative integer such that $x^s\cos\beta x$ and $x^s\sin\beta x$ are not solutions to the corresponding homogeneous equation.

In Problems 38 and 39, use the elimination method to find a general solution to the given system.

38. $x-d^2y/dt^2=t+1$, $dx/dt+dy/dt-2y=e^t$.

39. $d^2x/dt^2-x+y=0$, $x+d^2y/dt^2-y=e^{3t}$.

4.4 Method of Variation of Parameters

In the previous section, we discussed the method of undetermined coefficients and the annihilator method. These methods work only for linear equations with constant coefficients and when the nonhomogeneous term is a solution to some homogeneous linear equation with constant coefficients. In this section, we show how the **method of variation of parameters** works for higher-order linear equations with variable coefficients.

Our goal is to find a particular solution to the standard form equation

$$L[y](x)=g(x), \tag{4.86}$$

where $L[y]:=y^{(n)}+p_1y^{(n-1)}+\ldots+p_ny$ and the coefficient functions $p_1, p_2, \ldots, p_n$, as well as g, are continuous on (a,b). The method to be described requires that we already know a fundamental solution set $\{y_1, y_2, \ldots, y_n\}$ for the corresponding homogeneous equation

$$L[y](x)=0. \tag{4.87}$$

A general solution to equation (4.87) is then

$$y_h(x)=C_1y_1(x)+C_2y_2(x)+\ldots+C_ny_n(x), \tag{4.88}$$

where C_1, C_2, …, C_n are arbitrary constants. In the method of variation of parameters, we assume there exists a particular solution to equation (4.86) of the form

$$y_p(x)=v_1(x)y_1(x)+v_2(x)y_2(x)+\ldots+v_n(x)y_n(x) \tag{4.89}$$

and try to determine the functions $v_1, v_2, \ldots, v_n$.

There are n unknown functions, so we will need n conditions (equations) to determine them. These conditions are obtained as follows. Differentiating y_p in equation (4.89) gives

$$y_p'=(v_1y_1'+v_2y_2'+\ldots+v_ny_n')+(v_1'y_1+v_2y_2'+\ldots+v_n'y_n). \tag{4.90}$$

To prevent second derivatives of the unknowns $v_1, v_2, \ldots, v_n$ from entering the formula for y_p'', we impose the condition

$$v_1'y_1+v_2'y_2+\ldots+v_n'y_n=0.$$

In a like manner, as we compute y_p'', y_p''', …, $y_p^{(n-1)}$, we impose $(n-2)$ additional conditions involving $v_1', v_2', \ldots, v_n'$; namely,

$$v_1'y_1'+\ldots+v_n'y_n'=0, \ldots, v_1'y_1^{(n-2)}+\ldots+v_n'y_n^{(n-2)}=0.$$

Finally, the nth condition that we impose is that y_p satisfy the given equation (4.86). Using the previous conditions and the fact that y_1, y_2, …, y_n are solutions to the homogeneous equation, then $L[y_p]=g$ reduces to

$$v_1' y_1^{(n-1)} + v_2' y_2^{(n-1)} + \dots + v_n' y_n^{(n-1)} = g \qquad (4.91)$$

(see Problem 12 in Exercises 4. 4). We therefore seek n functions $v_1', v_2', \dots, v_n'$ that satisfy the system

$$\begin{cases} y_1 v_1' + y_2 v_2' + \dots + y_n v_n' = 0, \\ \cdots\cdots \\ y_1^{(n-2)} v_1' + y_1^{(n-2)} v_2' + \dots + y_n^{(n-2)} v_n' = 0, \\ y_1^{(n-1)} v_1' + y_1^{(n-1)} v_2' + \dots + y_n^{(n-1)} v_n' = g. \end{cases} \qquad (4.92)$$

Caution. This system was derived under the assumption that the coefficient of the highest derivative $y^{(n)}$ in equation (4. 86) is one. If, instead, the coefficient of this term is the constant a, then in the last equation in system (4. 92) the right-hand side becomes g/a.

A sufficient condition for the existence of a solution to system (4. 92) for x in (a,b) is that the determinant of the matrix made up of the coefficients of $v_1', v_2', \dots, v_n'$ be different from zero for all x in (a,b). But this determinant is just the Wronskian:

$$\begin{vmatrix} y_1 & y_2 & \cdots & y_n \\ \vdots & \vdots & & \vdots \\ y_1^{(n-2)} & y_2^{(n-2)} & \cdots & y_n^{(n-2)} \\ y_1^{(n-1)} & y_2^{(n-2)} & \cdots & y_n^{(n-1)} \end{vmatrix} = W[y_1, y_2, \dots, y_n](x), \qquad (4.93)$$

which is never zero on (a,b) because $\{y_1, y_2, \dots, y_n\}$ is a fundamental solution set. Solving system (4. 92), we find

$$v_k'(x) = \frac{g(x) W_k(x)}{W[y_1, y_2, \dots, y_n](x)}, \quad k = 1, 2, \dots, n, \qquad (4.94)$$

where $W_k(x)$ is the determinant of the matrix obtained from the Wronskian $W[y_1, y_2, \dots, y_n](x)$ by replacing the kth column by col $[0, \dots, 0, 1]$. Using a cofactor expansion about this column, we can express $W_k(x)$ in terms of an $(n-1)$th-order Wronskian:

$$W_k(x) = (-1)^{n-k} W[y_1, \dots, y_{k-1}, y_{k+1}, \dots, y_n](x), \quad k = 1, 2, \dots, n. \qquad (4.95)$$

Integrating $v_k'(x)$ in equation (4. 94) gives

$$v_k(x) = \int \frac{g(x) W_k(x)}{W[y_1, y_2, \dots, y_n](x)} \mathrm{d}x, \quad k = 1, 2, \dots, n. \qquad (4.96)$$

Finally, substituting the v_k's back into equation (4. 89), we obtain a particular solution to equation (4. 86):

$$y_p(x)=\sum_{k=1}^{n} y_k(x)=\int \frac{g(x)W_k(x)}{W[y_1, y_2, \ldots, y_n](x)}dx. \tag{4.97}$$

Note that in equation (4.86) we presumed that the coefficient of the leading term, $y^{(n)}$, was unity. If, instead, it is $p_0(x)$, we must replace $g(x)$ by $g(x)/p_0(x)$ in equation (4.97).

Although equation (4.97) gives a neat formula for a particular solution to equation (4.86), its implementation requires one to evaluate $(n+1)$ determinants and then perform n integrations. This may entail several tedious computations. However, the method does work in cases when the technique of undetermined coefficients does not apply (provided, of course, we know a fundamental solution set).

Example 4.12 Find a general solution to the Cauchy-Euler equation

$$x^3y'''+x^2y''-2xy'+2y=x^3\sin x, \quad x>0, \tag{4.98}$$

given that $\{x, x^{-1}, x^2\}$ is a fundamental solution set to the corresponding homogeneous equation.

Solution An important first step is to divide equation (4.98) by x^3 to obtain the standard form

$$y'''+\frac{1}{x}y''-\frac{2}{x^2}y'+\frac{2}{x^3}y=\sin x, \quad x>0, \tag{4.99}$$

from which we see that $g(x)=\sin x$. Since $\{x, x^{-1}, x^2\}$ is a fundamental solution set, we can obtain a particular solution of the form

$$y_p(x)=v_1(x)x+v_2(x)x^{-1}+v_3(x)x^2. \tag{4.100}$$

To use formula (4.97), we must first evaluate the four determinants:

$$W[x,x^{-1},x^2](x)=\begin{vmatrix} x & x^{-1} & x^2 \\ 1 & -x^{-2} & 2x \\ 0 & 2x^{-3} & 2 \end{vmatrix}=-6x^{-1},$$

$$W_1(x)=(-1)^{(3-1)}W[x^{-1},x^2](x)=(-1)^2\begin{vmatrix} x^{-1} & x^2 \\ -x^{-2} & 2x \end{vmatrix}=3,$$

$$W_2(x)=(-1)^{(3-2)}\begin{vmatrix} x & x^2 \\ 1 & 2x \end{vmatrix}=-x^2,$$

$$W_3(x)=(-1)^{(3-3)}\begin{vmatrix} x & x^{-1} \\ 1 & -x^{-2} \end{vmatrix}=-2x^{-1}.$$

Substituting the above expressions into formula (4.97), we find

$$y_p(x)=x\int \frac{(\sin x)3}{-6x^{-1}}dx+x^{-1}\int \frac{(\sin x)(-x^2)}{-6x^{-1}}dx+x^2\int \frac{(\sin x)(-2x^{-1})}{-6x^{-1}}dx$$

$$=x\int\left(-\frac{1}{2}x\sin x\right)dx+x^{-1}\int\frac{1}{6}x^3\sin x dx+x^2\int\frac{1}{3}\sin x dx,$$

which after some labor simplifies to

$$y_p(x)=\cos x-x^{-1}\sin x+C_1x+C_2x^{-1}+C_3x^2, \qquad (4.101)$$

where C_1, C_2, and C_3 denote the constants of integration. Since $\{x, x^{-1}, x^2\}$ is a fundamental solution set for the homogeneous equation, we can take C_1, C_2, and C_3 to be arbitrary constants. The right-hand side of equation (4.101) then gives the desired general solution. ◆

In the preceding example, the fundamental solution set $\{x, x^{-1}, x^2\}$ can be derived by substituting $y=x^r$ into the homogeneous equation corresponding to equation (4.98). However, in dealing with other equations that have variable coefficients, the determination of a fundamental set may be extremely difficult. In Section 4.6, we will tackle this problem using power series methods.

Exercises 4.4

In Problems 1-6, use the method of variation of parameters to determine a particular solution to the given equation.

1. $y'''-3y''+4y=e^{2x}$.
2. $y'''-2y''+y'=x$.
3. $z'''+3z''-4z=e^{2x}$.
4. $y'''-3y''+3y'-y=e^x$.
5. $y'''+y'=\tan x$, $\quad 0<x<\pi/2$.
6. $y'''+y'=\sec\theta\tan\theta$, $\quad 0<\theta<\pi/2$.
7. Find a general solution to the Cauchy-Euler equation

$$x^3y'''-3x^2y''+6xy'-6y=x^{-1}, \quad x>0,$$

given that $\{x, x^2, x^3\}$ is a fundamental solution set for the corresponding homogeneous equation.

8. Find a general solution to the Cauchy-Euler equation

$$x^3y'''-2x^2y''+3xy'-3y=x^2, \quad x>0,$$

given that $\{x, x\ln x, x^3\}$ is a fundamental solution set for the corresponding homogeneous equation.

9. Given that $\{e^x, e^{-x}, e^{2x}\}$ is a fundamental solution set for the homogeneous equation corresponding to the equation

$$y'''-2y''-y'+2y=g(x),$$

determine a formula involving integrals for a particular solution.

10. Given that $\{x, x^{-1}, x^4\}$ is a fundamental solution set for the homogeneous equation corresponding to the equation

$$x^3 y''' - x^2 y'' - 4xy' + 4y = g(x), \quad x > 0,$$

determine a formula involving integrals for a particular solution.

11. Find a general solution to the Cauchy-Euler equation

$$x^3 y''' - 3xy' + 3y = x^4 \cos x, \quad x > 0.$$

12. Derive the system (4. 92) in the special case when $n=3$. [Hint: To determine the last equation, require that $L[y_p]=g$ and use the fact that y_1, y_2, and y_3 satisfy the corresponding homogeneous equation.]

13. Show that

$$W_k(x) = (-1)^{(n-k)} W[y_1, \ldots, y_{k-1}, y_{k+1}, \ldots, y_n](x).$$

14. **Deflection of a Beam under Axial Force.** A uniform beam under a load and subject to a constant axial force is governed by the differential equation

$$y^{(4)}(x) - k^2 y''(x) = q(x), \quad 0 < x < L,$$

where $y(x)$ is the deflection of the beam, L is the length of the beam, k^2 is proportional to the axial force, and $q(x)$ is proportional to the load (see Figure 4. 1).

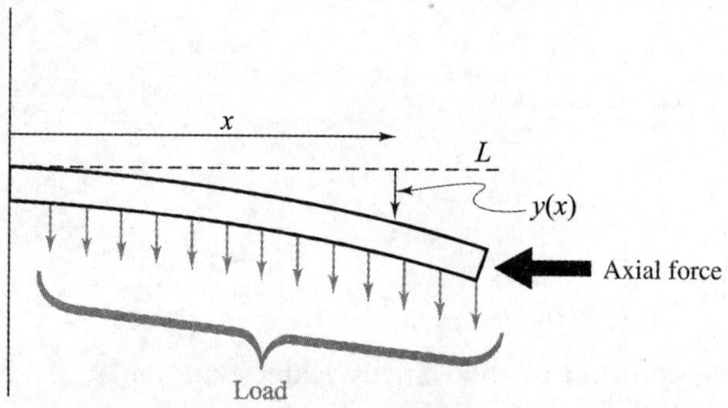

Figure 4. 1 Deformation of a beam under axial force and load

(1) Show that a general solution can be written in the form

$$y(x) = C_1 + C_2 x + C_3 e^{kx} + C_4 e^{-kx} + \frac{1}{k^2}\int q(x) x \, dx - \frac{x}{k^2}\int q(x) \, dx + \frac{e^{kx}}{2k^3}\int q(x) e^{-kx} \, dx - \frac{e^{-kx}}{2k^3}\int q(x) e^{kx} \, dx.$$

(2) Show that the general solution in part (1) can be rewritten in the form

$$y(x) = c_1 + c_2 x + c_3 e^{kx} + c_4 e^{-kx} + \int_0^x q(s) G(s, x) \, ds,$$

where

$$G(s,x):=\frac{s-x}{k^2}-\frac{\sin h[k(s-x)]}{k^3}.$$

(3) Let $q(x)\equiv 1$. First compute the general solution using the formula in part (1) and then using the formula in part (2). Compare these two general solutions with the general solution

$$y(x)=B_1+B_2x+B_3\mathrm{e}^{kx}+B_4\mathrm{e}^{-kx}-\frac{1}{2k^2}x^2,$$

which one would obtain using the method of undetermined coefficients.

(4) What are some advantages of the formula in part (2)?

4.5 Cauchy-Euler Equation

The differential equation

$$ax^2y''(x)+bxy'(x)+cy(x)=h(x), \tag{4.102}$$

where a, b and c are constants and $y(x)$ is the unknown function. Such equations are called **second-order Cauchy-Euler equations.**

We consider the homogeneous equation corresponding to equation (4.102), namely,

$$ax^2y''(x)+bxy'(x)+cy(x)=0. \tag{4.103}$$

Once the general solution of equation (4.103) has been found, equation (4.102) can be solved by variation of parameters. We shall restrict ourselves to the interval $x>0$ so that the coefficient of y'' does not vanish and the existence theorem will apply. The interval $x<0$ could also be used.

We can show that the substitution $x=\mathrm{e}^t$ transforms equation (4.103) into a constant coefficient equation in the new independent variable t.

(i) Using the chain rule, we have $\frac{\mathrm{d}y}{\mathrm{d}t}=x\frac{\mathrm{d}y}{\mathrm{d}x}$.

(ii) Differentiate the equation in part (i) with respect to t to obtain $\frac{\mathrm{d}^2y}{\mathrm{d}t^2}=\frac{\mathrm{d}y}{\mathrm{d}t}+x^2\frac{\mathrm{d}^2y}{\mathrm{d}x^2}$.

(iii) Substituting the formulas in parts (i) and (ii) into equation (4.103), we derive an equation with constant coefficient $a\frac{\mathrm{d}^2y}{\mathrm{d}t^2}+(b-a)\frac{\mathrm{d}y}{\mathrm{d}t}+cy=0$.

A second-order homogenous Cauchy-Euler equation has the form

$$ax^2y''(x)+bxy'(x)+cy(x)=0,\quad x>0, \tag{4.104}$$

where $a(\neq 0)$, b and c are (real) constants. Equation (4.104) has solutions of the form $y=x^r$. To determine the values for r, we can proceed as follows. Let L be the differential operator defined by the left-hand side of equation (4.104), that is

$$L[y]=ax^2y''(x)+bxy'(x)+cy(x), \tag{4.105}$$

and set $w(x,r)=x^r$, $x>0$. When we substitute $w(x,r)$ for $y(x)$ in equation (4.104), we find

$$L[w](x)=ax^2r(r-1)x^{r-2}+bxrx^{r-1}+cx^r=\{ar^2+(b-a)r+c\}x^r.$$

From this, we see that $w=x^r$ is a solution to equation (4.103) if and only if r satisfies

$$ar^2+(b-a)r+c=0. \tag{4.106}$$

This equation is referred to as the auxiliary, or indicial equation for equation (4.104). When the indicial equation has two distinct roots, we have $L[w](x)=a(r-r_1)(r-r_2)x^r$, from which it follows that equation (4.104) has the two solutions:

$$y_1(x)=w(x,r_1)=x^{r_1},\quad x>0, \tag{4.107}$$

$$y_2(x)=w(x,r_2)=x^{r_2},\quad x>0. \tag{4.108}$$

When r_1, r_2 are complex conjugates, $\alpha\pm i\beta$, we can use Euler's formula to express

$$x^{\alpha+i\beta}=e^{(\alpha+i\beta)\ln x}=x^\alpha\cos(\beta\ln x)+ix^\alpha\sin(\beta\ln x).$$

Since the real and imaginary parts of $x^{\alpha+i\beta}$ must also be solutions to equation (4.104), we can replace equations (4.107) and (4.108) by the two linearly independent real-values solutions:

$$y_1(x)=x^\alpha\cos(\beta\ln x),\quad y_2(x)=x^\alpha\sin(\beta\ln x).$$

If the indicial equation (4.106) has a repeated real root r_0, then it turns out that x^{r_0}, $x^{r_0}\ln x$ are two linearly independent solutions.

Example 4.13 Find a general solution to

$$4x^2y(x)+y(x)=0,\quad x>0. \tag{4.109}$$

Solution Let $w(x,r)=x^r$ and let L denote the left of equation (4.109). A short calculation gives

$$L[w](x)=(4r^2-4r+1)x^r.$$

Solving the indicial equation $(4r^2-4r+1)=0$ yields the repeated root $r_0=\frac{1}{2}$. Thus, a general solution to equation (4.109) is $y(x)=c_1\sqrt{x}+c_2\sqrt{x}\ln x$.

Example 4.14 Find a general solution to

$$x^2y''(x)+3xy'(x)+5y(x)=0, \quad x>0. \tag{4.110}$$

Solution Let $w(x,r)=x^r$ and let L denote the left of equation (4.110). A short calculation gives $L[w](x)=(r^2+2r+5)x^r$. Solving the indicial equation $r^2+2r+5=0$ yields a pair of complex conjugates root $r_{1,2}=-1\pm 2\mathrm{i}$. Thus, a general solution to equation (4.110) is

$$y(x)=c_1\sqrt{x}+c_2\sqrt{x}\ln x.$$

The equation in the form

$$x^n y^{(n)}+a_1x^{n-1}y^{(n-1)}+\ldots+a_{n-1}xy'+a_ny=0$$

is called ***n*th-order Euler's equation**, where $a_i\ (i=1, 2, \ldots, n)$ are constants.

As mentioned in second-order case, we assume a solution of the form $y=x^m$.

Substituting into the differential equation, we find that index m must satisfy the indicial equation

$$m(m-1)\ldots(m-n+1)+a_1m(m-1)\ldots(m-n+2)+\ldots+a_n=0.$$

If m is a real root, then $y=x^m$ is a solution.

If $m=\alpha+\mathrm{i}\beta$ is a complex root, then $\overline{m}=\alpha-\mathrm{i}\beta$ is also a root and

$$x^\alpha(\cos\beta\ln|x|+\mathrm{i}\sin\beta\ln|x|), \quad x^\alpha(\cos\beta\ln|x|-\mathrm{i}\sin\beta\ln|x|)$$

are two complex-value solutions, and consequently we obtain the real solutions

$$x^\alpha\cos(\beta\ln|x|), \quad x^\alpha\sin(\beta\ln|x|).$$

If m is a root of multiplicity k, then it can be shown that functions

$$x^m, \quad x^m\ln|x|, \quad x^m\ln^2|x|, \quad \ldots, \quad x^m\ln^{k-1}|x|$$

are k linearly independent solutions. ◆

Exercises 4.5

In Problems 1-10, use the substitution $y=x^r$ to find a general solution to the given equation for $x>0$.

1. $x^2y''(x)+6xy'(x)+6y(x)=0.$
2. $2x^2y''(x)+13xy'(x)+15y(x)=0.$
3. $x^2y''(x)-xy'(x)+17y(x)=0.$
4. $x^2y''(x)+2xy'(x)-3y(x)=0.$
5. $\dfrac{\mathrm{d}^2y}{\mathrm{d}x^2}=\dfrac{5}{x}\dfrac{\mathrm{d}y}{\mathrm{d}x}-\dfrac{13}{x^2}y.$
6. $\dfrac{\mathrm{d}^2y}{\mathrm{d}x^2}=\dfrac{1}{x}\dfrac{\mathrm{d}y}{\mathrm{d}x}-\dfrac{4}{x^2}y.$

7. $x^3y'''(x)+4x^2y''(x)+10xy'(x)-10y(x)=0$.

8. $x^3y'''(x)+4x^2y''(x)+xy'(x)=0$.

9. $x^3y'''(x)+3x^2y''(x)+5xy'(x)-5y(x)=0$.

10. $x^3y'''(x)+9x^2y''(x)+19xy'(x)+8y(x)=0$.

In Problems 11 and 12, use a substitution of the form $y=(x-c)^r$ to find a general solution to the given equation for $x>c$.

11. $2(x-3)^2y''(x)+5(x-3)y'(x)-2y(x)=0$.

12. $4(x+2)^2y''(x)+5y(x)=0$.

In Problems 13 and 14, use variation of parameters to find a general solution to the given equation for $x>0$.

13. $x^2y''(x)-2xy'(x)+2y(x)=x^{-1/2}$.

14. $x^2y''(x)+2xy'(x)-2y(x)=6x^{-2}+3x$.

In Problems 15-17, solve the given initial value problem.

15. $t^2x''(t)-12x(t)=0$; $x(1)=3$, $x'(1)=5$.

16. $x^2y''(x)+5xy'(x)+4y(x)=0$; $y(1)=3$, $y'(1)=7$.

17. $x^3y'''(x)+6x^2y''(x)+29xy'(x)-29y(x)=0$;
$y(1)=2$, $y'(1)=-3$, $y''(1)=19$.

18. Let $L[y](x):=x^3y'''(x)+xy'(x)-y(x)$.

(1) Show that $L[x^r](x)=(r-1)^3x^r$.

(2) Using an extension of the argument given in this section for the case when the indicial equation has a double root, show that $L[y]=0$ has the general solution

$$y(x)=C_1x+C_2x\ln x+C_3x(\ln x)^2.$$

4.6 Solution by Power Series

One of the most powerful methods of solving the second-order homogenous linear equation

$$y''+p(x)y'+q(x)y=0 \tag{4.111}$$

is the method of power series. This technique has the advantage of simplicity and is especially useful when one desires only to approximate a solution by the first few terms.

Noticeably, although we consider only second-order linear equations, this method can often be used to approximate the solution of higher-order linear equations even non-linear ones.

Before stating these conditions, we need several definitions.

Definition 4.4 The point $x=a$ is called an **ordinary point** of the equation (4.111) if both $p(x)$ and $q(x)$ are analytic at $x=a$.

If either $p(x)$ or $q(x)$ is not analytic at $x=a$, then the point is called a **singular point** or a **singularity** of the equation.

For example, second-order Euler's differential equation has a singularity at $x=0$ but all other points are ordinary points.

We shall now state an important theorem which gives the conditions which ensure the existence of a power-series solution.

Theorem 4.5 If $x=a$ is an ordinary point of the equation (4.111), then there exist two linearly independent power-series solution of the form

$$y=\sum_{n=0}^{\infty} a_n (x-a)^n.$$

These series will converge at least for all values of x in the interval $|x-a|<R$, where R is the distance from the point $x=a$ to the nearest singular point of the differential equation in the complex plane.

Example 4.15 Consider the equation

$$y''+xy'+2y=0. \tag{4.112}$$

Here $p(x)=x$ and $q(x)=2$, both of which are analytic for all x, and by Theorem 4.5 there exist solutions of the form $y=\sum_{n=0}^{\infty} a_n x^n$. The coefficients a_n can be determined by substituting the series into the differential equation (4.112). We first compute the derivative y' and y'':

$$y'=\sum_{n=1}^{\infty} n a_n x^{n-1}, \quad y''=\sum_{n=2}^{\infty} n(n-1)a_n x^{n-2}=\sum_{n=0}^{\infty} (n+2)(n+1)a_{n+2}x^n.$$

Substituting into equation (4.112) and collecting terms, we have

$$\sum_{n=0}^{\infty} [(n+2)(n+1)a_{n+2}+(n+2)a_n]x^n=0.$$

Now, in order for a power series to vanish for every value of x in some interval, it is necessary that the coefficient of each power of x must vanish. Hence we have

$$a_{n+2}=-\frac{1}{n+1}a_n, \quad n\geqslant 0.$$

From these equations, we see that any coefficients a_n can be expressed ultimately in terms of a_0 or a_1. Thus, we have

$$y=a_0\left[1-x^2+\frac{1}{3}x^4-\frac{1}{3\times5}x^6+\ldots+(-1)^n\frac{1}{3\times5\times\ldots\times(2n-1)}x^{2n}+\ldots\right]$$
$$+a_1\left[x-\frac{1}{2}x^3+\frac{1}{2\times4}x^5-\ldots+(-1)^{n-1}\frac{1}{2\times4\times\ldots\times(2n-2)}x^{2n-1}+\ldots\right].$$
$$(4.113)$$

This is the general solution of equation (4.112) with a_0 and a_1 as arbitrary constants. In order to determine the interval of validity of the solution, we must determine the interval of convergence of the series. From equation (4.113), we know that

$$\lim_{n\to\infty}\left|\frac{a_nx^n}{a_{n-2}x^{n-2}}\right|=\lim_{n\to\infty}\frac{1}{n-1}x^2=0,$$

so the series converges for all x.

If initial conditions are given, for example, $y(0)=2$, $y'(0)=3$, then we obtain immediately $a_0=2$, $a_1=3$.

In solving an initial value problem, that is one in which $y(a)$ and $y'(a)$ are given, the following method is often more convenient than the general series method developed above.

The reader will recall that if a function $y(x)$ can be represented by a power series near $x=a$, then this series is the Taylor series:

$$y(x)=y(a)+y'(a)(x-a)+\frac{y''(a)}{2!}(x-a)^2+\ldots+\frac{y^{(n)}(a)}{n!}(x-a)^n+\ldots$$

In particular, if $x=0$ is an ordinary point of the equation:

$$y''+p(x)y'+q(x)y=0,$$

then the solution near $x=0$ is

$$y(x)=y(0)+y'(0)x+\frac{y''(0)}{2!}x^2+\ldots+\frac{y^{(n)}(0)}{n!}x^n+\ldots$$

Example 4.16 Find the solution of

$$y''+(x+3)y'-2y=0,$$

which satisfies the initial conditions:

$$y(0)=1,\quad y'(0)=2.$$

Solution Since $x=0$ is an ordinary point of the equation, the solution will be of the form

$$y(x)=y(0)+y'(0)x+\frac{y''(0)}{2!}x^2+\ldots+\frac{y^{(n)}(0)}{n!}x^n+\ldots$$

The value of $y(0)$ and $y'(0)$ are given. The quantity $y''(0)$ can be computed from the equation itself, since

$$y''=-(x+3)y'+2y,$$

and hence for $x=0$ we have

$$y''(0)=-3y'(0)+2y(0)=-4.$$

In order to obtain $y'''(0)$, we differentiate the equation to get

$$y'''+(x+3)y''+y'-2y'=0,$$

and hence

$$y'''(0)=-3y''(0)+y'(0)=14.$$

For $y^{(4)}$, we have

$$y^{(4)}+(x+3)y'''+3y''-y''=0,$$

and

$$y^{(4)}(0)=-3y'''(0)-2y''(0)=-34.$$

Therefore, the solution is

$$y(x)=1+2x-2x^2+\frac{7}{3}x^3-\frac{17}{12}x^4+\ldots$$

From Theorem 4.5, we know that this series converges for all x, since the differential equation has no singularities. ◆

Definition 4.5 A point x_0 is said to be a **regular singular point** or **regular singularity** of the differential equation (4.111) if x_0 is a singular point, and $(x-x_0)\ p(x)$ and $(x-x_0)^2\ q(x)$ are analytic at x_0. Otherwise, x_0 is called an **irregular singular point.**

Definition 4.6 (Indicial Equation) If x_0 is a regular singular point of equation (4.111), then the indicial equation for this point is

$$r(r-1)+p_0r+q_0=0, \tag{4.114}$$

where $p_0=\lim\limits_{x\to x_0}(x-x_0)p(x)$, $q_0=\lim\limits_{x\to x_0}(x-x_0)^2\ q(x)$. The roots of the indicial equation are called the **exponents (indices)** of the singularity x_0.

To derive a series solution about the singular point x_0 of

$$a_2(x)y''(x)+a_1(x)p(x)y'(x)+a_0(x)q(x)y(x)=0,\quad x>0, \tag{4.115}$$

we have two methods:

(i) Set $p(x)=\dfrac{a_1(x)}{a_2(x)}$, $q(x)=\dfrac{a_0(x)}{a_2(x)}$. If both $(x-x_0)p(x)$ and $(x-x_0)^2\ q(x)$ are analytic at x_0, then x_0 is a regular singular point and the remaining steps apply.

(ii) Let $w(r,x)=(x-x_0)^{\alpha}\sum\limits_{n=0}^{\infty}a_n(x-x_0)^n=\sum\limits_{n=0}^{\infty}a_n(x-x_0)^{n+r}$, and, using termwise differentiation, substitute $w(r,x)$ into equation (4. 115) to obtain an equation of the form.

Theorem 4. 6 (Frobenius Theorem) If equation (4. 111) has a regular singularity at $x=a$, then the equation has the power series solutions in the form

$$y=(x-a)^{\alpha}\sum_{n=0}^{\infty}a_n(x-a)^n$$

where α is an undetermined constant.

Since $x=a$ is a regular singular point, we know that $(x-a)p(x)$ and $(x-a)^2q(x)$ are analytic. It is no harm to set

$$(x-a)p(x)=\sum_{n=0}^{\infty}p_n(x-a)^n,$$

$$(x-a)^2q(x)=\sum_{n=0}^{\infty}q_n(x-a)^n.$$

Substituting y, $p(x)$ and $q(x)$ into the equation and equating the coefficients of the same time power of x on both sides of the equation, we have

$$\alpha^2+(p_0-1)\alpha+q_0=0.$$

This equation is called the **indicial equation** since it is used to determine the index α.

Example 4. 17 Solve

$$4xy''+2y'-y=0.$$

Solution From Theorem 4. 6, the equation has the power-series solutions in the form

$$y=x^{\alpha}\sum_{n=0}^{\infty}a_nx^n.$$

In consideration of

$$xp(x)=\frac{1}{2},\quad x^2q(x)=0-\frac{1}{4}x,$$

we have the indicial equation

$$\alpha^2-\frac{1}{2}\alpha=0.$$

It yields $\alpha_1=0$, $\alpha_2=1/2$.

Substituting $y=x^{\alpha_1}\sum_{n=0}^{\infty}a_nx^n$ into the equation, we obtain

$$2n(2n-1)a_n-a_{n-1}=0,\quad n=1,2,\ldots$$

Therefore, the solution is

$$y=a_0\left[1+\frac{x}{2!}+\ldots+\frac{x^n}{(2n)!}+\ldots\right]$$

The solution for $\alpha_2=1/2$ is found in the same way, and the general solution is

$$y=a_0\left[1+\frac{x}{2!}+\ldots+\frac{x^n}{(2n)!}+\ldots\right]+b_0x^{\frac{1}{2}}\left[1+\frac{x}{3!}+\ldots+\frac{x^n}{(2n+1)!}+\ldots\right].◆$$

Example 4.18 Solve

$$x^2y''+xy'+(x^2-\beta^2)y=0 \quad (\beta\geqslant 0).$$

Solution This equation is called **β-order Bessel equation.**

From Theorem 4.6, the equation has the power series solutions in the form

$$y=x^{\alpha}\sum_{n=0}^{\infty}a_nx^n.$$

In consideration of

$$xp(x)=1, \quad x^2q(x)=-\beta^2+x^2,$$

we have the indicial equation

$$\alpha^2-\beta^2=0.$$

It yields $\alpha_1=\beta$, $\alpha_2=-\beta$.

Substituting $y=x^{\alpha_1}\sum_{n=0}^{\infty}a_nx^n$ into the equation, we obtain

$$(2\beta+1)a_1=0,$$

$$(2n\beta+n^2)a_n+a_{n-2}=0, \quad n\geqslant 2.$$

From these equations, we obtain

$$a_{2k-1}=0, \quad k=1, 2, \ldots$$

$$a_{2k}=(-1)^k\frac{1}{k!\cdot 4^k(\beta+k)(\beta+k-1)\ldots(\beta+1)}a_0, \quad k=1, 2, \ldots$$

If we take

$$a_0=\frac{1}{2^{\beta}\Gamma(\beta+1)},$$

where $\Gamma(\beta+1)$ is the gamma function, then we have

$$y_1=J_{\beta}(x)=\sum_{n=0}^{\infty}\frac{(-1)^k}{k!\,\Gamma(\beta+k+1)}\left(\frac{x}{2}\right)^{\beta+2k}.$$

If β is not equal to zero or a positive integer, we can obtain a second, linearly independent solution by substituting

$$y=x^{\alpha_2}\sum_{n=0}^{\infty}a_nx^n$$

into Bessel equation. In this case, the general solution of Bessel equation is

$$y=c_1\mathrm{J}_\beta(x)+c_2\mathrm{J}_{-\beta}(x).\ ◆$$

If β is equal to zero or a positive integer, then a second solution (which is linearly independent with y_1) can not be found by the above method. At the time, we have to use Liouville's formula.

$\mathrm{J}_\beta(x)$ is called **β-order Bessel function**, and $\mathrm{J}_{-\beta}(x)$ is called **$-\beta$-order Bessel function.** Utilizing the solution of Bessel equation, we can solve some equations.

Example 4.19 Solve

$$x^2y''+xy'+\left(4x^2-\frac{9}{25}\right)y=0.$$

Solution Letting $t=2x$, we can rewrite the equation as

$$t^2\frac{\mathrm{d}^2y}{\mathrm{d}t^2}+t\frac{\mathrm{d}y}{\mathrm{d}t}+\left(t^2-\frac{9}{25}\right)y=0.$$

This is a Bessel equation and $\beta=3/5$. Therefore, we have

$$y=c_1\mathrm{J}_{\frac{3}{5}}(t)+c_2\mathrm{J}_{-\frac{3}{5}}(t),$$

and the solution of original equation is

$$y=c_1\mathrm{J}_{\frac{3}{5}}(2x)+c_2\mathrm{J}_{-\frac{3}{5}}(2x).\ ◆$$

Exercises 4.6

In Problems 1-10, determine all the singular points of the given differential equation.

1. $(x+1)y''-x^2y'+3y=0$.
2. $x^2y''+3y'-xy=0$.
3. $(\theta^2-2)y''+2y'+(\sin\theta)y=0$.
4. $(x^2+x)y''+3y'-6xy=0$.
5. $(t^2-t-2)x''+(t+1)x'-(t-2)x=0$.
6. $(x^2-1)y''+(1-x)y'+(x^2-2x+1)y=0$.
7. $(\sin x)y''+(\cos x)y=0$.
8. $\mathrm{e}^xy''-(x^2-1)y'+2xy=0$.
9. $(\sin\theta)y''-(\ln\theta)y=0$.
10. $[\ln(x-1)]y''+(\sin 2x)y'-\mathrm{e}^xy=0$.

In Problems 11-18, find at least the first four nonzero terms in a power series expansion about $x=0$ for a general solution to the given differential equation.

11. $y'+(x+2)y=0$.

12. $y'-y=0$.

13. $z''-x^2z=0$.

14. $(x^2+1)y''+y=0$.

15. $y''+(x-1)y'+y=0$.

16. $y''-2y'+y=0$.

17. $w''-x^2w'+w=0$.

18. $(2x-3)y''-xy'+y=0$.

In Problems 19-24, find a power series expansion about $x=0$ for a general solution to the given differential equation. Your answer should include a general formula for the coefficients.

19. $y'-2xy=0$.

20. $y''+y=0$.

21. $y''-xy'+4y=0$.

22. $y''-xy=0$.

23. $z''-x^2z'-xz=0$.

24. $(x^2+1)y''-xy'+y=0$.

In Problems 25-28, find at least the first four nonzero terms in a power series expansion about $x=0$ for the solution to the given initial value problem.

25. $w'''+3xw'-w=0$; $w(0)=2$, $w'(0)=0$.

26. $(x^2-x+1)y''-y'-y=0$; $y(0)=0$, $y'(0)=1$.

27. $(x+1)y''-y=0$; $y(0)=0$, $y'(0)=1$.

28. $y''+(x-2)y'-y=0$; $y(0)=-1$, $y'(0)=0$.

In Problems 29-31, use the first few terms of the power series expansion to find a cubic polynomial approximation for the solution to the given initial value problem. Graph the linear, quadratic, and cubic polynomial approximations for $-5\leqslant x\leqslant 5$.

29. $y''+y'-xy=0$; $y(0)=1$, $y'(0)=-2$.

30. $y''-4xy'+5y=0$; $y(0)=-1$, $y'(0)=1$.

31. $(x^2+2)y''+2xy'+3y=0$; $y(0)=1$, $y'(0)=2$.

32. Consider the initial value problem

$$y''-2xy'-2y=0;\quad y(0)=a_0,\ y'(0)=a_1,$$

where a_0 and a_1 are constants.

(1) Show that if $a_0=0$, then the solution will be an odd function [that is, $y(-x)$

$=-y(x)$ for all x]. What happens when $a_1=0$?

(2) Show that if a_0 and a_1 are positive, then the solution is increasing on $(0, +\infty)$.

(3) Show that if a_0 is negative and a_1 is positive, then the solution is increasing on $(-\infty, 0)$.

(4) What conditions on a_0 and a_1 would guarantee that the solution is increasing on $(-\infty, +\infty)$?

In Problems 33-38, express a general solution to the given equation using Bessel functions of either the first or second kind.

33. $4x^2y''+4xy'+(4x^2-1)y=0$.

34. $9x^2y''+9xy'+(9x^2-16)y=0$.

35. $x^2y''+xy'+(x^2-1)y=0$.

36. $x^2y''+xy'+x^2y=0$.

37. $9t^2x''+9tx'+(9t^2-4)x=0$.

38. $x^2z''+xz'+(x^2-16)z=0$.

In Problems 39-44, express a general solution to the given equation using Bessel functions of either the first or second kind.

39. $4x^2y''+4xy'+(4x^2-1)y=0$.

40. $9x^2y''+9xy'+(9x^2-16)y=0$.

41. $x^2y''+xy'+(x^2-1)y=0$.

42. $x^2y''+xy'+x^2y=0$.

43. $9t^2x''+9tx'+(9t^2-4)x=0$.

44. $x^2z''+xz'+(x^2-16)z=0$.

4.7 Laplace Transforms

Definition 4.7 Let f be a function on $[0, +\infty)$. The **Laplace transform** of f is the function F defined by the integral

$$F(s):=\int_0^{+\infty} e^{-st}f(t)\mathrm{d}t. \tag{4.116}$$

The domain is all the values of s for which the integral in (4.116) exists. The Laplace transform of f is denoted by both $F(s)$ and $\mathscr{L}\{f\}(s)$.

Existence of the Transform

Theorem 4.7 If f is piecewise continuous on $[0, +\infty)$ and of exponential order

α, that is there exist positive constants T and M such that $|f(t)| \leqslant Me^{\alpha t}$, for all $t \geqslant T$. Then $\mathscr{L}\{f\}(s)$ exists for $s > \alpha$.

Important Property of the Laplace Transform

Theorem 4.8 (Linearity of the Transform) Let f, g, and h be functions whose Laplace transforms exist for $s > \alpha$ and let c be a constant. Then, for $s > \alpha$,

$$\mathscr{L}\{f(x)+g(x)\}=\mathscr{L}\{f(x)\}+\mathscr{L}\{g(x)\},$$

$$\mathscr{L}\{\alpha f(x)\}=\alpha\mathscr{L}\{f(x)\}.$$

Theorem 4.9 (Laplace Transform of the Derivative) Let f be continuous on $[0, +\infty)$ and $f'(t)$ be piecewise continuous on $[0, +\infty)$, with both of exponential order α. Then, for $s > \alpha$,

$$\mathscr{L}\{f'\}(s)=s\mathscr{L}\{f\}(s)-f(0).$$

Proof $\mathscr{L}\{f'(x)\}=\int_0^{+\infty} e^{-sx} f'(x)\mathrm{d}x = e^{-sx} f(x)\Big|_0^{+\infty} + s\int_0^{+\infty} e^{-sx} f(x)\mathrm{d}x$

$$=s\mathscr{L}\{f(x)\}-f(0).$$

Theorem 4.10 (Laplace Transform of Higher-Order Derivatives) Let f, f', f'', …, $f^{(n-1)}$ be continuous on $[0, +\infty)$ and let $f^{(n)}$ be piecewise continuous on $[0, +\infty)$, with all these functions of exponential order α. Then, for $s > \alpha$,

$$\mathscr{L}\{f^{(n)}(x)\}=s^n\mathscr{L}\{f\}(s)-s^{n-1}f(0)-s^{n-2}f'(0)-\dots-f^{(n-1)}(0).$$

A table of Laplace transforms is as follows:

Laplace transforms table

Object function $f(x)$	Transform function $F(s)$	Conditions
1	$\frac{1}{s}$	$s>0$
x	$\frac{1}{s^2}$	$s>0$
x^n	$\frac{n!}{s^{n+1}}$	$s>0$
e^{ax}	$\frac{1}{s-a}$	$s>a$
xe^{ax}	$\frac{1}{(s-a)^2}$	$s>a$
$x^n e^{ax}$	$\frac{n!}{(s-a)^{n+1}}$	$s>a$
$\sin \varpi x$	$\frac{\varpi}{s^2+\varpi^2}$	$s>0$

Continued

Object function $f(x)$	Transform function $F(s)$	Conditions
$\cos \overline{\omega}x$	$\frac{s}{s^2+\overline{\omega}^2}$	$s>0$
$x\sin \overline{\omega}x$	$\frac{2s\overline{\omega}}{(s^2+\overline{\omega}^2)^2}$	$s>0$
$x\cos \overline{\omega}x$	$\frac{(s^2-\overline{\omega}^2)}{(s^2+\overline{\omega}^2)^2}$	$s>0$
$e^{ax}\sin \overline{\omega}x$	$\frac{\overline{\omega}}{[(s-a)^2+\overline{\omega}^2]}$	$s>a$
$e^{ax}\cos \overline{\omega}x$	$\frac{(s-a)}{[(s-a)^2+\overline{\omega}^2]^2}$	$s>a$
$xe^{ax}\sin \overline{\omega}x$	$\frac{2\overline{\omega}(s-a)}{[(s-a)^2+\overline{\omega}^2]^2}$	$s>a$
$xe^{ax}\cos \overline{\omega}x$	$\frac{[(s-a)^2-\overline{\omega}^2]}{[(s-a)^2+\overline{\omega}^2]^2}$	$s>a$
$\mathrm{sh}\,\overline{\omega}x$	$\frac{\overline{\omega}}{s^2-\overline{\omega}^2}$	$s>\lvert\overline{\omega}\rvert$
$\mathrm{ch}\,\overline{\omega}x$	$\frac{s}{s^2-\overline{\omega}^2}$	$s<\lvert\overline{\omega}\rvert$

Definition 4.8 (Inverse Laplace Transform) Given a function $F(s)$, if there is a function f that is continuous on $[0,+\infty)$ and satisfies $\mathscr{L}\{f\}=F$, then we say that $f\{t\}$ is the **inverse Laplace transform** of $F(s)$ and employ the notation $\mathscr{L}^{-1}\{f\}=f$.

Theorem 4.11 (Linearity of the Inverse Transform) Assume that $\mathscr{L}^{-1}\{F\}$, $\mathscr{L}^{-1}\{F_1\}$, and $\mathscr{L}^{-1}\{F_2\}$ exist and are continuous on $[0,+\infty)$ and let α be any constant. Then

$$\mathscr{L}^{-1}\{F_1+F_2\}=\mathscr{L}^{-1}\{F_1\}+\mathscr{L}^{-1}\{F_2\},$$
$$\mathscr{L}^{-1}\{\alpha F\}=\alpha\mathscr{L}^{-1}\{F\}.$$

Our goal is to show how Laplace transforms can be used to solve initial value problems for linear differential equations. Recall that we have already studied ways of solving such initial value problems in Chapter 4. These previous methods required that we first find a general solution of the differential equation and then use the initial conditions to determine the desired solution. As we will see, the method of Laplace transforms leads to the solution of the initial value problem without first finding a general solution.

Method of Laplace Transforms

To solve an initial value problem

$$a_n y^n + a_{n-1} y^{(n-1)} + \ldots + a_1 y' + a_0 y = f(x), \tag{4.117}$$

$$y(0) = y_0, \quad y'(0) = y_0', \quad \ldots, \quad y^{(n-1)}(0) = y_0^{(n-1)}, \tag{4.118}$$

we take the following three steps:

(i) Take the Laplace transform of both sides of the equation.

(ii) Use the properties of the Laplace transform and the initial conditions to obtain an equation for the Laplace transform of the solution and then solve this equation for the transform.

(iii) Determine the inverse Laplace transform of the solution by looking it up in a table or by using a suitable method (such as partial fractions) in combination with the table.

Remark 4.1 If $f(x)$ is of exponential order and y is a solution of equation (4.117), where a_n, a_{n-1}, ..., a_1, and a_0 are constants, then y, y', y'', ..., $y^{(n)}$ are of exponential order.

Example 4.20 Solve the initial value problem

$$y'' + y = \sin 2x, \tag{4.119}$$

$$y(0) = 0, \quad y'(0) = 1. \tag{4.120}$$

Solution The differential equation (4.119) is an identity between two functions of x. Hence equality holds for the Laplace transforms of these functions:

$$\mathscr{L}\{y'' + y\} = \mathscr{L}\{\sin 2x\}.$$

Using the linearity property of $\mathscr{L}$ and the previously computed transform of the exponential function, we can write

$$\mathscr{L}\{y''\} + \mathscr{L}\{y\} = \frac{2}{s^2+4}. \tag{4.121}$$

From the formulas for the Laplace transform of higher-order derivatives and the initial conditions in (4.120), we find

$$\mathscr{L}\{y''\}(s) = s^2 \mathscr{L}\{y\}(s) - sy(0) - y'(0).$$

Substituting these expressions into equation (4.121) and solving for $\mathscr{L}\{y\}(s)$ yields

$$\mathscr{L}\{y\}(s) = \frac{s^2+6}{(s^2+1)(s^2+4)}.$$

Using partial fractions decomposition, we obtain

$$\mathscr{L}\{y\}(s)=\frac{5}{3(s^2+1)}-\frac{2}{3(s^2+4)}.$$

Taking inverse Laplace transform, we get

$$y(x)=\frac{5}{3}\sin x-\frac{1}{3}\sin 2x.$$

Example 4.21 Find the general solution to the differential equation $y''-y'-2y=0$ using Laplace transform method.

Solution Taking the Laplace transform of both sides of the equation, we have

$$[s^2\mathscr{L}\{y\}(s)-sy(0)-y'(0)]-[s\mathscr{L}\{y\}(s)-y(0)]-2\mathscr{L}\{y\}(s)=0.$$

By writing $A=y(0)$ and $B=y'(0)$ and solving for $\mathscr{L}\{y\}(s)$, we have

$$(s^2-s-2)\mathscr{L}\{y\}(s)=As+B-A.$$

Hence

$$\mathscr{L}\{y\}(s)=\frac{As+B-A}{s^2-s-2}=\frac{1}{3}\frac{2A-B}{s+1}+\frac{1}{3}\frac{A+B}{s-2}$$

Taking inverse Laplace transform, we obtain

$$\begin{aligned}y(x)&=\frac{1}{3}(2A-B)\mathrm{e}^{-x}+\frac{1}{3}(A+B)\mathrm{e}^{2x}\\&=C_1\mathrm{e}^{-x}+C_2\mathrm{e}^{2x}\end{aligned}$$

as the general solution to the given equation, where C_1 and C_2 are arbitrary constants. ◆

Exercises 4.7

In Problems 1-14, *solve the given initial value problem using the method of Laplace transforms.*

1. $y''-2y'+5y=0$; $y(0)=2$, $y'(0)=4$.
2. $y''-y'-2y=0$; $y(0)=-2$, $y'(0)=5$.
3. $y''+6y'+9y=0$; $y(0)=-1$, $y'(0)=6$.
4. $y''+6y'+5y=12\mathrm{e}^t$; $y(0)=-1$, $y'(0)=7$.
5. $w''+w=t^2+2$; $w(0)=1$, $w'(0)=-1$.
6. $y''-4y'+5y=4\mathrm{e}^{3t}$; $y(0)=2$, $y'(0)=7$.
7. $y''-7y'+10y=9\cos t+7\sin t$; $y(0)=5$, $y'(0)=-4$.
8. $y''+4y=4t^2-4t+10$; $y(0)=0$, $y'(0)=3$.
9. $z''+5z'-6z=21\mathrm{e}^{t-1}$; $z(1)=-1$, $z'(1)=9$.
10. $y''-4y=4t-8\mathrm{e}^{-2t}$; $y(0)=0$, $y'(0)=5$.

11. $y''-y=t-2$; $y(2)=3$, $y'(2)=0$.

12. $w''-2w'+w=6t-2$; $w(-1)=3$, $w'(-1)=7$.

13. $y''-y'-2y=-8\cos t-2\sin t$; $y(\pi/2)=1$, $y'(\pi/2)=0$.

14. $y''+y=t$; $y(\pi)=0$, $y'(\pi)=0$.

In Problems 15-24, solve for $Y(s)$, the Laplace transform of the solution $y(t)$ to the given initial value problem.

15. $y''-3y'+2y=\cos t$; $y(0)=0$, $y'(0)=-1$.

16. $y''+6y=t^2-1$; $y(0)=0$, $y'(0)=-1$.

17. $y''+y-y=t^3$; $y(0)=1$, $y'(0)=0$.

18. $y''-2y'-y=e^{2t}-e^t$; $y(0)=1$, $y'(0)=3$.

19. $y''+5y'-y=e^t-1$; $y(0)=1$, $y'(0)=1$.

20. $y''+3y=t^3$; $y(0)=0$, $y'(0)=0$.

21. $y''-2y'+y=\cos t-\sin t$; $y(0)=1$, $y'(0)=3$.

22. $y''-6y'+5y=te^t$; $y(0)=2$, $y'(0)=-1$.

23. $y''+4y=g(t)$; $y(0)=-1$, $y'(0)=0$, where $g(t)=\begin{cases} t, & t<2, \\ 5, & t>2. \end{cases}$

24. $y''-y=g(t)$; $y(0)=1$, $y'(0)=2$, where $g(t)=\begin{cases} 1, & t<3, \\ t, & t>3. \end{cases}$

In Problems 25-28, solve the given third-order initial value problem for $y(t)$ using the method of Laplace transforms.

25. $y'''-y''+y'-y=0$; $y(0)=1$, $y'(0)=1$, $y''(0)=3$.

26. $y'''+4y''+y'-6y=-12$; $y(0)=1$, $y'(0)=4$, $y''(0)=-2$.

27. $y'''+3y''+3y'+y=0$; $y(0)=-4$, $y'(0)=4$, $y''(0)=-2$.

28. $y'''+y''+3y'-5y=16e^{-t}$; $y(0)=0$, $y'(0)=2$, $y''(0)=-4$.

In Problems 29-32, use the method of Laplace transforms to find a general solution to the given differential equation by assuming $y(0)=a$ and $y'(0)=b$, where a and b are arbitrary constants.

29. $y''-4y'+3y=0$.

30. $y''+6y'+5y=t$.

31. $y''+2y'+2y=5$.

32. $y''-5y'+6y=-6te^{2t}$.

Chapter Summary

The theory and techniques for solving an nth-order linear differential equation

$$y^{(n)}+p_1(x)y^{(n-1)}+\ldots+p_n(x)y=g(x) \tag{4.122}$$

are natural extensions of the development for linear second-order equations given in Chapter 4. Assuming that $p_1, p_2, \ldots, p_n$ and $g(x)$ are continuous functions on an open interval I, there is a unique solution to equation (4.122) on I that satisfies the n initial conditions: $y(x_0)=\gamma_0$, $y'(x_0)=\gamma_1$, $\ldots$, $y^{(n-1)}(x_0)=\gamma_{n-1}$, where $x_0 \in I$.

For the corresponding homogeneous equation

$$y^{(n)}+p_1(x)y^{(n-1)}+\ldots+p_n(x)y=0, \tag{4.123}$$

there exists a set of n **linearly independent** solutions $\{y_1, y_2, \ldots, y_n\}$ on I. Such functions are said to form a **fundamental solution** set, and every solution to equation (4.123) can be written as a linear combination of these functions:

$$y(x)=C_1y_1(x)+C_2y_2(x)+\ldots+C_ny_n(x).$$

The linear independence of solutions to equation (4.123) is equivalent to the nonvanishing on I of the **Wronskian**

$$W[y_1, y_2, \ldots, y_n](x):=\begin{bmatrix} y_1(x) & y_2(x) & \ldots & y_n(x) \\ y_1'(x) & y_2'(x) & \ldots & y_n'(x) \\ y_1^{(n-1)}(x) & y_2^{(n-1)}(x) & \ldots & y_n^{(n-1)}(x) \end{bmatrix}$$

When equation (4.123) has (real) constant coefficients so that it is of the form

$$a_ny^{(n)}+a_{n-1}y^{(n-1)}+\ldots+a_0y=0, \quad a_n\neq 0, \tag{4.124}$$

then the problem of determining a fundamental solution set can be reduced to the algebraic problem of solving the **auxiliary** or **characteristic equation**

$$a_nr^n+a_{n-1}r^{n-1}+\ldots+a_0=0. \tag{4.125}$$

If the n roots of equation (4.125)—say, $r_1, r_2, \ldots, r_n$—are all distinct, then

$$\{e^{r_1x}, e^{r_2x}, \ldots, e^{r_nx}\} \tag{4.126}$$

is a fundamental solution set for equation (4.124). If some real root—say, r_1—occurs with multiplicity m (e.g., $r_1=r_2=\ldots=r_m$), then m of the functions in set (4.126) are replaced by

$$e^{r_1x}, \quad xe^{r_1x}, \quad \ldots, \quad x^{m-1}e^{r_1x}.$$

When a complex root $\alpha+i\beta$ to equation (4.125) occurs with multiplicity m, then so does its conjugate and $2m$ members of the set equation (4.126) are replaced by the real-valued functions

$$e^{\alpha x}\sin\beta x,\quad xe^{\alpha x}\sin\beta x,\quad \ldots,\quad x^{m-1}e^{\alpha x}\sin\beta x,$$
$$e^{\alpha x}\cos\beta x,\quad xe^{\alpha x}\cos\beta x,\quad \ldots,\quad x^{m-1}e^{\alpha x}\cos\beta x.$$

A general solution to the nonhomogeneous equation (4.122) can be written as

$$y(x)=y_p(x)+y_h(x),$$

where y_p is some particular solution to equation (4.122) and y_h is a general solution to the corresponding homogeneous equation. Two useful techniques for finding particular solutions are the **annihilator method** (undetermined coefficients) and the method of **variation of parameters.**

The annihilator method applies to equations of the form

$$L[y]=g(x), \tag{4.127}$$

where L is a linear differential operator with constant coefficients and the forcing term $g(x)$ is a polynomial, exponential, sine, or cosine, or a linear combination of products of these. Such a function $g(x)$ is annihilated (mapped to zero) by a linear differential operator A that also has constant coefficients. Every solution to the nonhomogeneous equation (4.127) is then a solution to the homogeneous equation $AL[y]=0$, and, by comparing the solutions of the latter equation with a general solution to $L[y]=0$, we can obtain the form of a particular solution to equation (4.127). These forms have previously been studied in Section 4.4 for the method of undetermined coefficients.

The method of variation of parameters is more general in that it applies to arbitrary equations of the form (4.122). The idea is, starting with a fundamental solution set $\{y_1, y_2, \ldots, y_n\}$ for equation (4.123), to determine functions v_1, v_2, $\ldots$, v_n such that

$$y_p(x)=v_1(x)y_1(x)+v_2(x)y_2(x)+\ldots+v_n(x)y_n(x) \tag{4.128}$$

satisfies equation (4.122). This method leads to the formula

$$y_p(x)=\sum_{k=1}^{n} y_k(x)\int\frac{g(x)W_k(x)}{W[y_1, y_2, \ldots, y_n](x)}\mathrm{d}x, \tag{4.129}$$

where

$$W_k(x)=(-1)^{n-k}W[y_1, \ldots, y_{k-1}, y_{k+1}, \ldots, y_n](x),\quad k=1, 2, \ldots, n.$$

Power Series Method for an Ordinary Point

In the case of a linear equation of the form

$$y''+p(x)y'+q(x)y=0, \tag{4.130}$$

where $p(x)$ and $q(x)$ are analytic at x_0, the point x_0 is called an **ordinary point**, and the equation has a pair of linearly independent solutions expressible as power series about x_0. The radii of convergence of these series solutions are at least as large as the distance from x_0 to the nearest singularity (real or complex) of the equation. To find power series solutions to equation (4.130), we substitute $y(x)=\sum_{n=0}^{\infty} a_n(x-x_0)^n$ into equation (4.130), group like terms, and set the coefficients of the resulting power series equal to zero. This leads to a recurrence relation for the coefficients a_n, which, in some cases, may even yield a general formula for the a_n. The same method applies to the nonhomogeneous version of equation (4.130), provided the forcing function is also analytic at x_0.

Regular Singular Points

If, in equation (4.130), either $p(x)$ or $q(x)$ fails to be analytic at x_0, then x_0 is a **singular point** of (4.130). If x_0 is a singular point for which $(x-x_0)p(x)$ and $(x-x_0)^2q(x)$ are both analytic at x_0, then x_0 is a **regular singular point.** The Cauchy-Euler equation

$$ax^2\frac{d^2y}{dx^2}+bx\frac{dy}{dx}+cy=0,\quad x>0 \tag{4.131}$$

has a regular singular point at $x=0$, and a general solution to equation (4.131) can be obtained by substituting $y=x^r$ and examining the roots of the resulting indicial equation $ar^2+(b-a)r+c=0$.

Method of Frobenius

For an equation of the form (4.130) with a regular singular point at x_0, a series solution can be found by the **method of Frobenius.** This is obtained by substituting

$$w(r,x)=(x-x_0)^r\sum_{n=0}^{\infty} a_n(x-x_0)^n$$

into equation (4.130), finding a recurrence relation for the coefficient and choosing $r=r_1$, the larger root of the **indicial equation**

$$r(r-1)+p_0r+q_0=0, \tag{4.132}$$

where $p_0:=\lim\limits_{x\to x_0}(x-x_0)p(x)$, $q_0:=\lim\limits_{x\to x_0}(x-x_0)^2q(x)$.

Finding a Second Linearly Independent Solution

If the two roots r_1, r_2 of the indicial equation (4.132) do not differ by an integer, then a second linearly independent solution to equation (4.130) can be found by taking

$r=r_2$ in the method of Frobenius. However, if $r_1=r_2$ or r_1-r_2 is a positive integer, then discovering a second solution requires Liouville formula.

Laplace Transforms

The use of the Laplace transform helps to simplify the process of solving initial value problems for certain differential and integral equations, especially when a forcing function with jump discontinuities is involved. The Laplace transform $\mathscr{L}\{f\}$ of a function $f(t)$ is defined by

$$\mathscr{L}\{f\}(s)=\int_0^{+\infty} e^{-st} f(t)\,dt$$

for all values of s for which the improper integral exists. If $f(t)$ is piecewise continuous on $[0,+\infty)$ and of exponential order α [that is, $|f(t)|$ grows no faster than a constant times $e^{\alpha t}$ as $t\to+\infty$], then $\mathscr{L}\{f\}(s)$ exists for all $s>\alpha$.

Linearity: $\mathscr{L}\{af+bg\}=a\mathscr{L}\{f\}+b\mathscr{L}\{g\}$.

One reason for the usefulness of the Laplace transform lies in the simple formula for the transform of the derivative f':

$$\mathscr{L}\{f'\}(s)=sF(s)-f(0),\ \text{where } F=\mathscr{L}\{f\}. \tag{4.133}$$

This formula shows that by using the Laplace transform, "differentiation with respect to t" can essentially be replaced by the simple operation of "multiplication by s". The extension of equation (4.133) to higher-order derivatives is

$$\mathscr{L}\{f^{(n)}\}(s)=s^nF(s)-s^{n-1}f(0)-s^{n-2}f'(0)-\dots-f^{(n-1)}(0). \tag{4.134}$$

To solve an initial value problem of the form

$$ay''+by'+cy=f(t);\quad y(0)=\alpha,\ y'(0)=\beta \tag{4.135}$$

via the Laplace transform method, one takes the transform of both sides of the differential equation in (4.135). Using the linearity of $\mathscr{L}$ and formula (4.134) leads to an equation involving the Laplace transform $F(s)$ of the (unknown) solution $y(t)$. The next step is to solve this simpler equation for $F(s)$. Finally, one computes the inverse Laplace transform of $F(s)$ to obtain the desired solution.

Review Problems

1. Determine the intervals for which Theorem 4.1 guarantees the existence of a solution in that interval.

(1) $y^{(4)}-(\ln x)y''+xy'+2y=\cos 3x$.

(2) $(x^2-1)y'''+(\sin x)y''+\sqrt{x+4}\,y'+e^x y=x^2+3$.

2. Determine whether the given functions are linearly dependent or linearly independent on the interval $(0, +\infty)$.

(1) $\{e^{2x}, x^2 e^{2x}, e^{-x}\}$.

(2) $\{e^x \sin 2x, x e^x \sin 2x, e^x, x e^x\}$.

(3) $\{2e^{2x} - e^x, e^{2x} + 1, e^{2x} - 3, e^x + 1\}$.

3. Show that the set of functions $\{\sin x, x \sin x, x^2 \sin x, x^3 \sin x\}$ is linearly independent on $(-\infty, +\infty)$.

4. Find a general solution for the given differential equation.

(1) $y^{(4)} + 2y''' - 4y'' - 2y' + 3y = 0$.

(2) $y''' + 3y'' - 5y' + y = 0$.

(3) $y^{(5)} - y^{(4)} + 2y''' - 2y'' + y' - y = 0$.

(4) $y''' - 2y'' - y' + 2y = e^x + x$.

5. Find a general solution for the homogeneous linear differential equation with constant coefficients whose auxiliary equation is

(1) $(r+5)^2 (r-2)^3 (r^2+1)^2 = 0$.

(2) $r^4 (r-1)^2 (r^2+2r+4)^2 = 0$.

6. Given that $y_p = \sin(x^2)$ is a particular solution to $y^{(4)} + y = (16x^4 - 11)\sin(x^2) - 48x^2 \cos(x^2)$ on $(0, +\infty)$, find a general solution.

7. Find a differential operator that annihilates the given function.

(1) $x^2 - 2x + 5$.

(2) $e^{3x} + x - 1$.

(3) $x \sin 2x$.

(4) $x^2 e^{-2x} \cos 3x$.

(5) $x^2 - 2x + x e^{-x} + \sin 2x - \cos 3x$.

8. Use the annihilator method to determine the form of a particular solution for the given equation.

(1) $y'' + 6y' + 5y = e^{-x} + x^2 - 1$.

(2) $y''' + 2y'' - 19y' - 20y = x e^{-x}$.

(3) $y^{(4)} + 6y'' + 9y = x^2 - \sin 3x$.

(4) $y''' - y'' + 2y = x \sin x$.

9. Find a general solution to the Cauchy-Euler equation

$$x^3 y''' - 2x^2 y'' - 5xy' + 5y = x^{-2}, \quad x > 0,$$

given that $\{x, x^5, x^{-1}\}$ is a fundamental solution set to the corresponding homogeneous equation.

10. Find a general solution to the given Cauchy-Euler equation.

(1) $4x^3y'''+8x^2y''-xy'+y=0, \quad x>0.$

(2) $x^3y'''+2x^2y''+2xy'+4y=0, \quad x>0.$

11. Find at least the first four nonzero terms in a power series expansion about $x=0$ for a general solution to the given equation.

(1) $y''+x^2y'-2y=0.$

(2) $y''+e^{-x}y'-y=0.$

In Problems 12-17, solve the given initial value problem for $y(t)$ *using the method of Laplace transforms.*

12. $y''-7y'+10y=0$; $\quad y(0)=0,\ y'(0)=-3.$

13. $y''+6y'+9y=0$; $\quad y(0)=-3,\ y'(0)=10.$

14. $y''+2y'+2y=t^2+4t$; $\quad y(0)=0,\ y'(0)=-1.$

15. $y''+9y=10e^{2t}$; $\quad y(0)=-1,\ y'(0)=5.$

16. $y''+3y'+4y=u(t-1)$; $\quad y(0)=0,\ y'(0)=1.$

17. $y''-4y'+4y=t^2e^t$; $\quad y(0)=0,\ y'(0)=0.$

Chapter 5

Systems of Differential Equations

To this point, we've only looked as solving single differential equations. However, many "real life" situations are governed by a system of differential equations. In this chapter, we begin to analyze the systems of differential equations. When the equations in the system are linear, matrix algebra provides a compact notation for expressing the system. In fact, the notation itself suggests new and elegant ways of characterizing the solution properties, as well as novel, efficient techniques for explicitly obtaining solutions.

5.1 Introduction

Two large tanks, each holding 24 L of a brine solution, are interconnected by pipes as shown in Figure 5.1. Fresh water flows into tank A at a rate of 6 L/min, and fluid is drained out of tank B at the same rate; also 8 L/min of fluid is pumped from tank A to tank B, and 2 L/min from tank B to tank A. The liquids inside each tank are kept well stirred so that each mixture is homogeneous. If, initially, the brine solution in tank A contains x_0 kg of salt and that in tank B initially contains y_0 kg of salt, determine the mass of salt in each tank at time $t>0$.

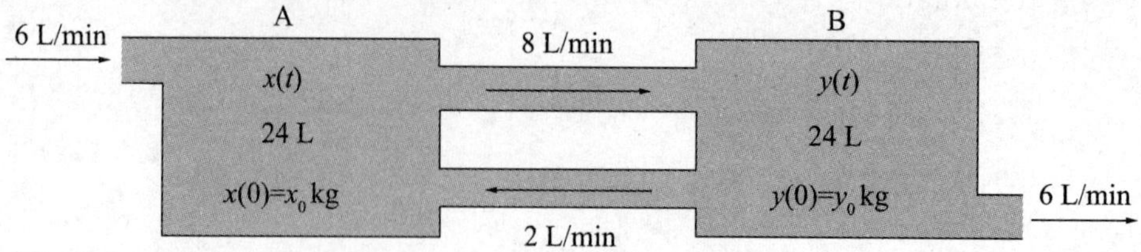

Figure 5.1 Interconnected fluid tanks

Note that the volume of liquid in each tank remains constant at 24 L because of the balance between the inflow and outflow volume rates. Hence, we have two unknown functions of t: the mass of salt $x(t)$ in tank A and the mass of salt $y(t)$ in tank B. By focusing attention on one tank at a time, we can derive two equations relating these unknowns. Since the system is being flushed with fresh water, we expect that the salt content of each tank will diminish to zero as $t \to +\infty$.

To formulate the equations for this system, we equate the rate of change of salt in each tank with the net rate at which salt is transferred to that tank. The salt concentration in tank A is $x(t)/24$ kg/L, so the upper interconnecting pipe carries salt out of tank A at a rate of $8x/24$ kg/min; similarly, the lower interconnecting pipe brings salt into tank A at the rate of $2y/24$ kg/min (the concentration of salt in tank B is $y/24$ kg/L). The fresh water inlet, of course, transfers no salt (it simply maintains the volume in tank A at 24 L). From our premise,

$$\frac{dx}{dt} = \text{input rate} - \text{output rate},$$

so the rate of change of the mass of salt in tank A is

$$\frac{dx}{dt} = \frac{2}{24}y - \frac{8}{24}x = \frac{1}{12}y - \frac{1}{3}x.$$

The rate of change of salt in tank B is determined by the same interconnecting pipes and by the drain pipe, carrying away $6y/24$ kg/min:

$$\frac{dy}{dt} = \frac{8}{24}x - \frac{2}{24}y - \frac{6}{24}y = \frac{1}{3}x - \frac{1}{3}y.$$

The interconnected tanks are thus governed by a system of differential equations:

$$\begin{cases} \dfrac{dx}{dt} = -\dfrac{1}{3}x + \dfrac{1}{12}y, \\ \dfrac{dy}{dt} = \dfrac{1}{3}x - \dfrac{1}{3}y. \end{cases} \tag{5.1}$$

Using the notation for the matrix product, we can write the system (5.1) for the interconnected tanks as

$$\begin{pmatrix} \dfrac{dx}{dt} \\ \dfrac{dy}{dt} \end{pmatrix} = \begin{pmatrix} -\dfrac{1}{3} & \dfrac{1}{12} \\ \dfrac{1}{3} & \dfrac{1}{3} \end{pmatrix} \begin{pmatrix} x \\ y \end{pmatrix}.$$

Thus, we express this system in matrix notation as a single equation.

5.2 Existence and Uniqueness Theorem for Linear Systems

A system of n differential equations in the n unknown functions $x_1(t)$,

$x_2(t), \ldots, x_n(t)$ expressed as

$$\begin{cases} x_1' = f_1(t, x_1, x_2, \ldots, x_n) \\ x_2' = f_2(t, x_1, x_2, \ldots, x_n) \\ \cdots\cdots \\ x_n' = f_n(t, x_1, x_2, \ldots, x_n) \end{cases} \tag{5.2}$$

is said to be in **normal form.** Notice that system (5.1) consists of n first-order equations that collectively look like a vectorized version of the single first-order equation $x'(t)=f(t,x)$.

An initial value problem for system (5.2) entails finding a solution to this system that satisfies the initial conditions

$$x_1(t_0)=x_{01}, \quad x_2(t_0)=x_{02}, \quad \ldots, \quad x_n(t_0)=x_{0n}$$

for prescribed values $x_{01}, x_{02}, \ldots, x_{0n}$.

If a system of differential equations is expressed as

$$\begin{cases} \dfrac{dx_1}{dt} = a_{11}(t)x_1 + a_{12}(t)x_2 + \ldots + a_{1n}(t)x_n + f_1(t), \\ \dfrac{dx_2}{dt} = a_{21}(t)x_1 + a_{22}(t)x_2 + \ldots + a_{2n}(t)x_n + f_2(t), \\ \cdots\cdots \\ \dfrac{dx_n}{dt} = a_{n1}(t)x_1 + a_{n2}(t)x_2 + \ldots + a_{nn}(t)x_n + f_n(t), \end{cases}$$

it is said to be a linear system in **normal form.** The matrix formulation of such a system is then

$$\boldsymbol{x}'(t) = \boldsymbol{A}(t)\boldsymbol{x}(t) + \boldsymbol{f}(t), \tag{5.3}$$

where $\boldsymbol{x}(t) = \begin{pmatrix} x_1(t) \\ x_2(t) \\ \vdots \\ x_n(t) \end{pmatrix}$ is the solution vector, $\boldsymbol{A}(t) = \begin{pmatrix} a_{11}(t) & a_{12}(t) & \ldots & a_{1n}(t) \\ a_{21}(t) & a_{22}(t) & \ldots & a_{2n}(t) \\ \vdots & \vdots & & \vdots \\ a_{n1}(t) & a_{n2}(t) & \ldots & a_{nn}(t) \end{pmatrix}$ is the **coefficient matrix**, $\boldsymbol{f}(t) = \begin{pmatrix} f_1(t) \\ f_2(t) \\ \vdots \\ f_n(t) \end{pmatrix}$.

As with a scalar linear differential equation, a system is called **homogeneous** when $f(t) \equiv 0$; otherwise, it is called **nonhomogeneous.** When the elements of $\boldsymbol{A}$ are all constants, the system is said to have **constant coefficients.**

The importance of the normal form is underscored by the fact that most professional codes for initial value problems presume that the system is written in this form. Furthermore, for a linear system in normal form, the powerful machinery of linear algebra can be readily applied.

For these reasons, it is gratifying to note that a (single) higher-order equation can always be converted to an equivalent system of first-order equations.

To convert an nth-order differential equation

$$y^{(n)}(t)=f(t, y, y', \dots, y^{(n-1)}) \tag{5.4}$$

into a first-order system, we introduce, as additional unknowns, the sequence of derivatives of y:

$$x_1(t)=y(t), \quad x_2(t)=y'(t), \quad \dots, \quad x_n(t)=y^{(n-1)}(t).$$

With this scheme, we obtain n first-order equations quite trivially:

$$\begin{cases} x_1'(t)=x_2(t), \\ x_2'(t)=x_3(t), \\ \cdots\cdots \\ x'_n(t)=x_{n-1}(t). \end{cases} \tag{5.5}$$

If equation (5.4) has initial conditions

$$y(t_0)=\eta_1, \quad y'(t_0)=\eta_2, \quad \dots, \quad y^{(n-1)}(t_0)=\eta_n,$$

then the system (5.5) has initial conditions $x_1(t_0)=\eta_1$, $x_2(t_0)=\eta_2$, $\dots$, $x_n(t_0)=\eta_n$.

Example 5.1 Convert the initial value problem

$$y''(t)+3ty'(t)+y^2(t)=\sin t, \quad y(0)=1, \quad y'(0)=5 \tag{5.6}$$

into an initial value problem for a system in normal form.

Solution We first express the differential equation in (5.6) as

$$y''(t)=-3ty'(t)-y^2(t)+\sin t.$$

Setting $x_1(t)=y(t)$ and $x_2(t)=y'(t)$, we obtain

$$x_1'(t)=x_2(t),$$
$$x_2'(t)=-3tx_2(t)-x_1^2(t)+\sin t.$$

The initial conditions transform to

$$x_1(0)=1, \quad x_2(0)=5.$$

An nth-order linear differential equation

$$y^{(n)}(t)+a_1(t)y^{(n-1)}(t)+\dots+a_{n-1}(t)y'(t)+a_n(t)y(t)=g(t) \tag{5.7}$$

can be rewritten as a first-order system in normal form using the substitution

$$x_1(t)=y(t),\quad x_2(t)=y'(t),\quad \ldots,\quad x_n(t)=y^{(n-1)}(t).$$

Indeed, equation (5.7) is equivalent to

$$\boldsymbol{x}'(t)=\boldsymbol{A}(t)\boldsymbol{x}(t)+\boldsymbol{f}(t),$$

where $\boldsymbol{x}(t)=\mathrm{col}(x_1(t), x_2(t), \ldots, x_n(t))$, $\boldsymbol{f}(t)=\mathrm{col}(0, 0, \ldots, g(t))$, and

$$\boldsymbol{A}(t)=\begin{bmatrix} 0 & 1 & 0 & \ldots & 0 \\ 0 & 0 & 1 & \ldots & 0 \\ \vdots & \vdots & \vdots & & \vdots \\ -a_n & -a_{n-1} & -a_{n-2} & \ldots & -a_1 \end{bmatrix}.$$

The theory for systems in normal form parallels very closely the theory of linear differential equations presented in Chapter 4. In many cases, the proofs for scalar linear differential equations carry over to normal systems with appropriate modifications. Conversely, results for normal systems apply to scalar linear equations since, as we showed, any scalar linear equation can be expressed as a normal system. This is the case with the existence and uniqueness theorems for linear differential equations.

The initial value problem for the normal system (5.3) is the problem of finding a differentiable vector function $\boldsymbol{x}(t)$ that satisfies the system on an interval I and also satisfies the initial condition $\boldsymbol{x}(t_0)=\boldsymbol{x}_0$, where t_0 is a given point of I and $\boldsymbol{x}_0=\mathrm{col}(x_{01}, x_{02}, \ldots, x_{0n})$ is a given vector. ◆

Theorem 5.1 (Existence and Uniqueness Theorem for Linear Systems) Suppose $\boldsymbol{A}(t)$ and $\boldsymbol{f}(t)$ are continuous on an open interval I that contains the point t_0. Then, for any choice of the initial vector $\boldsymbol{x}_0$, there exists a unique solution $x(t)$ on the whole interval I to the initial value problem

$$\boldsymbol{x}'(t)=\boldsymbol{A}(t)\boldsymbol{x}(t)+\boldsymbol{f}(t),\quad \boldsymbol{x}(t_0)=\boldsymbol{x}_0. \tag{5.8}$$

Proof Define the Picard approximations as

$$\begin{cases} \boldsymbol{x}_0(t)\equiv\boldsymbol{x}_0, \\ \boldsymbol{x}_n(t)=\boldsymbol{x}_0+\int_{t_0}^{t}[\boldsymbol{A}(s)x_{n-1}(s)+\boldsymbol{f}(s)]\mathrm{d}s, \\ n=1, 2, \ldots \end{cases} \tag{5.9}$$

We show that the Picard iterations converges uniformly to a unique solution on the interval $[a,b]$.

The approximations are well defined. Since $\boldsymbol{x}_0(t)\equiv\boldsymbol{x}_0$ is continuous on $[a,b]$ and

$\boldsymbol{A}$ and $\boldsymbol{f}$ are continuous on the region $R=\{(t,x)\mid a\leqslant t\leqslant b, x \text{ is arbitrary}\}$, it follows that

$$\boldsymbol{x}_1(t)=\boldsymbol{x}_0+\int_{t_0}^{t}[\boldsymbol{A}(s)x_0+\boldsymbol{f}(s)]\mathrm{d}s$$

is well defined and continuous on $[a,b]$. By induction, we find that $\boldsymbol{x}_n(t)$ is well defined and continuous on $[a,b]$.

The approximations converge uniformly on $[a,b]$. Recall that the sequence of successive approximations $\{x_n(t)\}$ consists of the partial sums of the series

$$\boldsymbol{x}_0+\sum_{n=1}^{\infty}[\boldsymbol{x}_n(t)-\boldsymbol{x}_{n-1}(t)]. \tag{5.10}$$

Thus, if this series converges uniformly on $[a,b]$, then so does the sequence.

Since $\boldsymbol{A}(t)$ and $\boldsymbol{f}(t)$ are continuous on $[a,b]$, they are bounded; that is, there are some constants M and L such that $\|\boldsymbol{A}(t)\boldsymbol{x}_0+\boldsymbol{f}(t)\|\leqslant M$ and $\|\boldsymbol{A}(t)\|t\leqslant L$ on $[a,b]$. To simplify the computations, let's consider x in the interval $[t_0,b]$(the case when $t\in[a,t_0]$ can be handled in a similar fashion). For such t,

$$\|\boldsymbol{x}_1(t)-\boldsymbol{x}_0\|\leqslant\int_{t_0}^{t}\|\boldsymbol{A}(s)y_0+\boldsymbol{f}(s)\|\mathrm{d}s\leqslant M(t-t_0),$$

and

$$\|\boldsymbol{x}_2(t)-\boldsymbol{x}_1(t)\|\leqslant\int_{t_0}^{t}\|\boldsymbol{A}(s)[\boldsymbol{x}_1(s)-\boldsymbol{x}_0(s)]\|\mathrm{d}s\leqslant L\int_{t_0}^{t}\|\boldsymbol{x}_1(s)-\boldsymbol{x}_0(s)\|\mathrm{d}s$$

$$\leqslant LM\int_{t_0}^{t}(s-t_0)\mathrm{d}s=\frac{LM}{2}(t-t_0)^2$$

for $t\in[t_0,b]$. Similarly, for $n=1, 2, \ldots$, we find by induction that

$$\|\boldsymbol{x}_{n+1}(t)-\boldsymbol{x}_n(t)\|\leqslant\int_{t_0}^{t}\|\boldsymbol{A}(s)[\boldsymbol{x}_n(s)-\boldsymbol{x}_{n-1}(s)]\|\mathrm{d}s\leqslant L\int_{x_0}^{x}\|\boldsymbol{x}_n(s)-\boldsymbol{x}_{n-1}(s)\|\mathrm{d}s$$

$$\leqslant\frac{L^nM}{n!}\int_{t_0}^{t}(s-x_0)^n\mathrm{d}s=\frac{L^nM}{(n+1)!}(t-t_0)^{n+1}.$$

From the last inequalities, we see that the terms in (5.10) are bounded in absolute value by the terms in the positive series $\sum_{n=0}^{\infty}\frac{L^nM}{(n+1)!}(b-x_0)^{n+1}$ which is convergence.

Therefore, by the Weierstrass M test, the series in (5.10) converges uniformly on $[t_0,b]$ to a continuous function $\boldsymbol{x}(t)$. Hence, the sequence $\{\boldsymbol{x}_n(t)\}$ converges uniformly on $[t_0,b]$ to $\boldsymbol{x}(t)$.

A similar argument establishes convergence on $[a,t_0]$, so $\{\boldsymbol{x}_n(t)\}$ converges uniformly on $[a,t_0]$ to $\boldsymbol{x}(t)$.

$\boldsymbol{x}(t)$ is a solution. Since $\{\boldsymbol{x}_n(t)\}$ converges uniformly to $\boldsymbol{x}(t)$, we can take the limit inside the integral. Thus, using equation (5.9), we obtain

$$\begin{aligned}\boldsymbol{x}(t)&=\lim_{n\to\infty}\boldsymbol{x}_n(t)=\boldsymbol{x}_0+\lim_{n\to\infty}\int_{t_0}^{t}[\boldsymbol{A}(s)\boldsymbol{x}_{n-1}(s)+\boldsymbol{f}(s)]\mathrm{d}s\\&=\boldsymbol{x}_0+\int_{t_0}^{t}[\boldsymbol{A}(s)\lim_{n\to\infty}\boldsymbol{x}_{n-1}(s)+\boldsymbol{f}(s)]\mathrm{d}s\\&=\boldsymbol{x}_0+\int_{t_0}^{t}[\boldsymbol{A}(s)\boldsymbol{x}(s)+\boldsymbol{f}(s)]\mathrm{d}s.\end{aligned}$$

Since $\boldsymbol{x}(t)$ satisfies the equivalent integral equation, it must satisfy the initial value problem (5.8).

Uniqueness. Let y be a solution to (5.8) on $[a,b]$. Then z must satisfy the integral equation

$$\boldsymbol{y}(t)=\boldsymbol{x}_0+\int_{t_0}^{t}[\boldsymbol{A}(s)\boldsymbol{y}(s)+\boldsymbol{f}(s)]\mathrm{d}t$$

just as $\boldsymbol{x}(t)$ does. Therefore, for t in $[t_0,b]$,

$$\begin{aligned}\|\boldsymbol{y}(t)-\boldsymbol{x}(t)\|&=\left\|\int_{t_0}^{t}\{\boldsymbol{A}(s)[\boldsymbol{y}(s)-\boldsymbol{x}(s)]\}\mathrm{d}s\right\|\\&\leqslant L\int_{t_0}^{t}\|\boldsymbol{y}(s)-\boldsymbol{x}(s)\|\mathrm{d}s\leqslant LM(t-t_0),\end{aligned}\tag{5.11}$$

where $M=\max\limits_{t_0\leqslant t\leqslant b}|\boldsymbol{y}(t)-\boldsymbol{x}(t)|$. Using (5.11), we next show by a similar string of inequalities that

$$\begin{aligned}\|\boldsymbol{y}(t)-\boldsymbol{x}(t)\|&=\left\|\int_{t_0}^{t}\{\boldsymbol{A}(s)[\boldsymbol{y}(s)-\boldsymbol{x}(s)]\}\mathrm{d}s\right\|\leqslant L\int_{t_0}^{t}\|\boldsymbol{y}(s)-\boldsymbol{x}(s)\|\mathrm{d}s\\&\leqslant L^2M\int_{t_0}^{t}(s-t_0)\mathrm{d}s=\frac{L^2M}{2}(t-t_0)^2.\end{aligned}$$

By induction, we find

$$\|\boldsymbol{y}(t)-\boldsymbol{x}(t)\|\leqslant\frac{ML^n(t-t_0)^n}{n!}$$

for t in $[t_0,b]$. Taking the limit as $n\to\infty$, we deduce that $\|\boldsymbol{y}(t)-\boldsymbol{x}(t)\|\equiv 0$. That is, $\boldsymbol{y}$ and $\boldsymbol{x}$ agree on $[t_0,b]$. A similar argument shows that they agree on $[a,x_0]$, so the initial value problem has a unique solution on the interval $[a,b]$.

Corollary 5.1 Suppose $p_1(x)$, $p_2(x)$, ..., $p_n(x)$ and $g(x)$ are each continuous on an interval (a,b) that contains the point x_0. Then, for any choice of the initial values γ_0, γ_1, ..., γ_{n-1}, there exists a unique solution $y(x)$ on the whole interval (a,b) to the initial value problem

$$y^{(n)}(x)+p_1(x)y^{(n-1)}(x)+\ldots+p_n(x)y(x)=g(x), \tag{5.12}$$

$$y(x_0)=\gamma_0,\quad y'(x_0)=\gamma_1,\quad \ldots,\quad y^{(n-1)}(x_0)=\gamma_{n-1}. \tag{5.13}$$

5.3 Properties of Solutions of First-Order Linear Systems

If we rewrite system

$$\boldsymbol{x}'(t)=\boldsymbol{A}(t)\boldsymbol{x}(t)+\boldsymbol{f}(t) \tag{5.14}$$

as

$$\boldsymbol{x}'(t)-\boldsymbol{A}(t)\boldsymbol{x}(t)=\boldsymbol{f}(t)$$

and define the operator $L(\boldsymbol{x})=\boldsymbol{x}'-\boldsymbol{A}\boldsymbol{x}$, then we can express system (5.14) in the operator form $L(\boldsymbol{x})=\boldsymbol{f}$. Here, the operator L maps vector functions into vector functions. Moreover, L is a linear operator in the sense that for any scalars a, b and differentiable vector functions $\boldsymbol{x}$, $\boldsymbol{y}$, we have $L(a\boldsymbol{x}+b\boldsymbol{y})=aL(\boldsymbol{x})+bL(\boldsymbol{y})$. The proof of this linearity follows from the properties of matrix multiplication.

As a consequence of the linearity of L, if they are solutions to the homogeneous system $\boldsymbol{x}'=\boldsymbol{A}(t)\boldsymbol{x}$ or in operator notation $L(\boldsymbol{x})=\boldsymbol{0}$, then any linear combination of these vectors, $c_1\boldsymbol{x}_1+c_2\boldsymbol{x}_2+\ldots+c_n\boldsymbol{x}_n$, is also a solution. Moreover, we will see that if the solutions $\boldsymbol{x}_1, \boldsymbol{x}_2, \ldots, \boldsymbol{x}_n$ are linearly independent, then every solution to $L(\boldsymbol{x})=\boldsymbol{0}$ can be expressed as $c_1\boldsymbol{x}_1+c_2\boldsymbol{x}_2+\ldots+c_n\boldsymbol{x}_n$ for an appropriate choice of the constants $c_1, c_2, \ldots, c_n$.

Definition 5.1 The n vector functions $\boldsymbol{x}_1, \boldsymbol{x}_2, \ldots, \boldsymbol{x}_n$ are said to be **linearly dependent on an interval** I if there exist constants $c_1, c_2, \ldots, c_n$, not all zero, such that $c_1\boldsymbol{x}_1+c_2\boldsymbol{x}_2+\ldots+c_n\boldsymbol{x}_n=\boldsymbol{0}$ for all t in I. If the vectors are not linearly dependent, they are said to be **linearly independent on** I.

Example 5.2 Show that the vector functions

$$\boldsymbol{x}_1(t)=\mathrm{col}(\mathrm{e}^t,0,\mathrm{e}^t),\ \boldsymbol{x}_2(t)=\mathrm{col}(3\mathrm{e}^t,0,3\mathrm{e}^t),\ \text{and}\ \boldsymbol{x}_3(t)=\mathrm{col}(t,1,0)$$

are linearly dependent on $(-\infty,+\infty)$.

Proof Notice that $\boldsymbol{x}_2(t)$ is just 3 times $\boldsymbol{x}_1(t)$ and therefore

$$3\boldsymbol{x}_1(t)-\boldsymbol{x}_2(t)+0\cdot\boldsymbol{x}_3(t)=\boldsymbol{0}\ \text{for all}\ t.$$

Hence, $\boldsymbol{x}_1$, $\boldsymbol{x}_2$, and $\boldsymbol{x}_3$ are linearly dependent on $(-\infty,+\infty)$. ◆

Example 5.3 Show that the vector functions

$$\boldsymbol{x}_1(t)=\mathrm{col}(\mathrm{e}^{2t},0,\mathrm{e}^{2t}),\ \boldsymbol{x}_2(t)=\mathrm{col}(\mathrm{e}^{2t},\mathrm{e}^{2t},-\mathrm{e}^{2t})\ \text{and}\ \boldsymbol{x}_3(t)=\mathrm{col}(\mathrm{e}',2\mathrm{e}',\mathrm{e}')$$

are linearly independent on $(-\infty,+\infty)$.

Proof To prove independence, we assume that constants c_1, c_2, and c_3 for which

$$c_1\boldsymbol{x}_1(t)+c_2\boldsymbol{x}_2(t)+c_3\boldsymbol{x}_3(t)=\boldsymbol{0}, \tag{5.15}$$

holds at every t and show that this forces $c_1=c_2=c_3=0$. In particular, when $t=0$, we obtain $c_1\begin{pmatrix}1\\0\\1\end{pmatrix}+c_2\begin{pmatrix}1\\1\\-1\end{pmatrix}+c_3\begin{pmatrix}1\\2\\1\end{pmatrix}=0$, which is equivalent to the system of linear equations

$$\begin{cases}c_1+c_2+c_3=0,\\ c_2+2c_3=0,\\ c_1-c_2+c_3=0.\end{cases} \tag{5.16}$$

Either by solving system (5.16) or by checking that the determinant of its coefficients is nonzero, we can verify that equation (5.15) has only the trivial solution $c_1=c_2=c_3=0$. Therefore, the vector functions $\boldsymbol{x}_1$, $\boldsymbol{x}_2$, and $\boldsymbol{x}_3$ are linearly independent on $(-\infty,+\infty)$. ◆

As Example 5.3 illustrates, if $\boldsymbol{x}_1$, $\boldsymbol{x}_2$, …, $\boldsymbol{x}_n$ are n vector functions each having n components, we can establish their linear independence on an interval I if we can find one point t_0 in I where the determinant $\det[\boldsymbol{x}_1(t_0), \boldsymbol{x}_2(t_0), \ldots, \boldsymbol{x}_n(t_0))]$ is not zero. ◆

Because of the analogy with scalar equations, we call this determinant the **Wronskian.**

Definition 5.2 The **Wronskian** of n vector functions $\boldsymbol{x}_1=\mathrm{col}(x_{11}, x_{12}, \ldots, x_{1n})$, $\boldsymbol{x}_2=\mathrm{col}(x_{21}, x_{22}, \ldots, x_{2n})$, …, $\boldsymbol{x}_n=\mathrm{col}(x_{n1}, x_{n2}, \ldots, x_{nn})$ is defined to be the real-valued function

$$W[\boldsymbol{x}_1, \boldsymbol{x}_2, \ldots, \boldsymbol{x}_n](t)=\begin{bmatrix}x_{11}(t) & x_{12}(t) & \ldots & x_{1n}(t)\\ x_{21}(t) & x_{22}(t) & \ldots & x_{2n}(t)\\ \vdots & \vdots & & \vdots\\ x_{n1}(t) & x_{n2}(t) & \ldots & x_{nn}(t)\end{bmatrix}.$$

Theorem 5.2 If $\boldsymbol{x}_1(t)$, $\boldsymbol{x}_2(t)$, …, $\boldsymbol{x}_n(t)$ are linearly dependent on I, then the Wronskian $W[\boldsymbol{x}_1, \boldsymbol{x}_2, \ldots, \boldsymbol{x}_n](t)\equiv\boldsymbol{0}$ on I.

Proof Assume $\boldsymbol{x}_1(t)$, $\boldsymbol{x}_2(t)$, …, $\boldsymbol{x}_n(t)$ are linearly dependent on I. Then there exist constants c_1, c_2, …, c_n not all zero, such that

$$c_1\boldsymbol{x}_1(t)+c_2\boldsymbol{x}_2(t)+\ldots+c_n\boldsymbol{x}_n(t)\equiv\boldsymbol{0} \text{ for all } t\in I,$$

that is

$$\begin{cases} c_1 x_{11} + c_2 x_{12} + \dots + c_n x_{1n} \equiv 0, \\ c_1 x_{21} + c_2 x_{22} + \dots + c_n x_{2n} \equiv 0, \\ \dots\dots \\ c_1 x_{n1} + c_2 x_{n2} + \dots + c_n x_{nn} \equiv 0. \end{cases}$$

By assumption, these equations have a nontrivial solution at each $t \in I$. Therefore,

$$W[\boldsymbol{x}_1, \boldsymbol{x}_2, \dots, \boldsymbol{x}_n](t) \equiv 0 \text{ on } I.$$

Theorem 5.3 If $\boldsymbol{x}_1, \boldsymbol{x}_2, \dots, \boldsymbol{x}_n$ are linearly independent solutions on I to the homogeneous system $\boldsymbol{x}' = \boldsymbol{A}(t)\boldsymbol{x}$, where $\boldsymbol{A}$ is an $n \times n$ matrix of continuous functions, then the Wronskian $W(t) := \det(\boldsymbol{x}_1, \boldsymbol{x}_2, \dots, \boldsymbol{x}_n)$ is never zero on I.

Proof For suppose to the contrary that $W(t_0) = 0$ at some point t_0. Then the vanishing of the determinant implies that the column vectors $\boldsymbol{x}_1(t_0), \boldsymbol{x}_2(t_0), \dots, \boldsymbol{x}_n(t_0)$ are linearly dependent. Thus, there exist scalars $c_1, c_2, \dots, c_n$ not all zero, such that

$$c_1 \boldsymbol{x}_1(t_0) + c_2 \boldsymbol{x}_2(t_0) + \dots + c_n \boldsymbol{x}_n(t_0) = \boldsymbol{0}.$$

However, $c_1 \boldsymbol{x}_1(t) + c_2 \boldsymbol{x}_2(t) + \dots + c_n \boldsymbol{x}_n(t)$ and the vector function $\boldsymbol{z}(t) \equiv \boldsymbol{0}$ are both solutions to $\boldsymbol{x}' = \boldsymbol{A}(t)\boldsymbol{x}$ on I, and they agree at the point t_0. So these solutions must be identical on I according to the existence-uniqueness theorem. That is, $c_1 \boldsymbol{x}_1(t) + c_2 \boldsymbol{x}_2(t) + \dots + c_n \boldsymbol{x}_n(t) = \boldsymbol{0}$ for all t in I. But this contradicts the given information that $\boldsymbol{x}_1, \boldsymbol{x}_2, \dots, \boldsymbol{x}_n$ are linearly independent on I. We have shown that $W(t_0) \neq 0$ and since t_0 is an arbitrary point, it follows that $W(t) \neq 0$ for all $t \in I$.

The preceding argument has two important implications that parallel the scalar case. First, the Wronskian of solutions to $\boldsymbol{x}' = \boldsymbol{A}(t)\boldsymbol{x}$ is either identically zero or never zero on I. Second, a set of n solutions on I is linearly independent on I if and only if their Wronskian is never zero on I. With these facts in hand, we can imitate the proof given for the scalar case to obtain the following representation theorem for the solutions to $\boldsymbol{x}' = \boldsymbol{A}(t)\boldsymbol{x}$.

Representation of Solutions (Homogeneous Case)

Theorem 5.4 Let $\boldsymbol{x}_1, \boldsymbol{x}_2, \dots, \boldsymbol{x}_n$ be n linearly independent solutions to the homogeneous system

$$\boldsymbol{x}'(t) = \boldsymbol{A}(t)\boldsymbol{x}(t) \tag{5.17}$$

on the interval I, where $\boldsymbol{A}(t)$ is an matrix function continuous on I. Then every solution to system (5.17) on I can be expressed in the form

$$\boldsymbol{x}(t)=c_1\boldsymbol{x}_1(t)+c_2\boldsymbol{x}_2(t)+\ldots+c_n\boldsymbol{x}_n(t), \tag{5.18}$$

where $c_1, c_2, \ldots, c_n$ are constants.

A set of solutions $\{\boldsymbol{x}_1, \boldsymbol{x}_2, \ldots, \boldsymbol{x}_n\}$ that are linearly independent on I or, equivalently, whose Wronskian does not vanish on I, is called a **fundamental solution set** for system (5.17) on I. The linear combination in equation (5.18), written with arbitrary constants, is referred to as a **general solution** to (5.17).

If we take the vectors in a fundamental solution set and let them form the columns of a matrix $\boldsymbol{X}(t)$, that is,

$$\boldsymbol{X}(t)=[\boldsymbol{x}_1(t),\boldsymbol{x}_2(t),\ldots,\boldsymbol{x}_n(t)]=\begin{bmatrix} x_{11}(t) & x_{12}(t) & \ldots & x_{1n}(t) \\ x_{21}(t) & x_{22}(t) & \ldots & x_{2n}(t) \\ \vdots & \vdots & & \vdots \\ x_{n1}(t) & x_{n2}(t) & \ldots & x_{nn}(t) \end{bmatrix},$$

then the matrix $\boldsymbol{X}(t)$ is called a **fundamental matrix** for system (5.17). We can use it to express the general solution (5.18) as $\boldsymbol{x}(t)=\boldsymbol{X}(t)\boldsymbol{c}$, where $\boldsymbol{c}=\mathrm{col}(c_1, c_2, \ldots, c_n)$ is an arbitrary constant vector. Since $\det \boldsymbol{X}(t)=W[\boldsymbol{x}_1(t), \boldsymbol{x}_2(t), \ldots, \boldsymbol{x}_n(t)]$ is never zero on I, it follows that $\boldsymbol{X}(t)$ is invertible for every t in I.

Another consequence of the linearity of the operator L defined by $L(\boldsymbol{x}):=\boldsymbol{x}'-\boldsymbol{A}\boldsymbol{x}$ is the **superposition principle** for linear systems. It states that if $\boldsymbol{x}_1$ and $\boldsymbol{x}_2$ are solutions, respectively, to the nonhomogeneous systems $L(\boldsymbol{x})=\boldsymbol{g}_1$ and $L(\boldsymbol{x})=\boldsymbol{g}_2$, Then $c_1\boldsymbol{x}_1+c_2\boldsymbol{x}_2$ is a solution to $L(\boldsymbol{x})=c_1\boldsymbol{g}_1+c_2\boldsymbol{g}_2$.

Using the superposition principle and the representation theorem for homogeneous systems, we can prove the following theorem.

Theorem 5.5 (Representation of Solutions for Nonhomogeneous linear Systems) Let $\boldsymbol{x}_p$ be a particular solution to the nonhomogeneous system

$$\boldsymbol{x}'(t)=\boldsymbol{A}(t)\boldsymbol{x}(t)+\boldsymbol{f}(t) \tag{5.19}$$

on the interval I, and let $\{\boldsymbol{x}_1, \boldsymbol{x}_2, \ldots, \boldsymbol{x}_n\}$ be a fundamental solution set on I for the corresponding homogeneous system $\boldsymbol{x}'(t)=\boldsymbol{A}(t)\boldsymbol{x}(t)$. Then every solution to system (5.19) on I can be expressed in the form

$$\boldsymbol{x}(t)=c_1\boldsymbol{x}_1(t)+c_2\boldsymbol{x}_2(t)+\ldots+c_n\boldsymbol{x}_n(t)+\boldsymbol{x}_p(t) \tag{5.20}$$

where $c_1, c_2, \ldots, c_n$ are constants.

The linear combination of $\boldsymbol{x}_p, \boldsymbol{x}_1, \boldsymbol{x}_2, \ldots, \boldsymbol{x}_n$ in equation (5.20) written with arbitrary constants $c_1, c_2, \ldots, c_n$ is called a **general solution** of system (5.19). This general solution can also be expressed as $\boldsymbol{x}=\boldsymbol{X}\boldsymbol{c}+\boldsymbol{x}_p$, where $\boldsymbol{X}$ is a fundamental matrix for the homogeneous system and $\boldsymbol{c}$ is an arbitrary constant vector.

We now summarize the results of this section as they apply to the problem of finding a general solution to a system of n linear first-order differential equations in normal form.

Approach to Solving Normal Systems

1. To determine a general solution to the $n\times n$ homogeneous system $\boldsymbol{x}'(t)=\boldsymbol{A}(t)\boldsymbol{x}(t)$:

(1) Find a fundamental solution set that consists of n linearly independent solutions $\{\boldsymbol{x}_1, \boldsymbol{x}_2, \ldots, \boldsymbol{x}_n\}$ to the homogeneous system.

(2) Form the linear combination $\boldsymbol{x}(t)=\boldsymbol{X}\boldsymbol{c}=c_1\boldsymbol{x}_1(t)+c_2\boldsymbol{x}_2(t)+\ldots+c_n\boldsymbol{x}_n(t)$, where $\boldsymbol{c}=\mathrm{col}(c_1, c_2, \ldots, c_n)$ is any constant vector and $\boldsymbol{X}=[\boldsymbol{x}_1, \boldsymbol{x}_2, \ldots, \boldsymbol{x}_n]$ is the fundamental matrix, to obtain a general solution.

2. To determine a general solution to the nonhomogeneous system $\boldsymbol{x}'(t)=\boldsymbol{A}(t)\boldsymbol{x}(t)+\boldsymbol{f}(t)$:

(1) Find a particular solution $\boldsymbol{x}_p$ to the nonhomogeneous system.

(2) Form the sum of the particular solution and the general solution $\boldsymbol{X}\boldsymbol{c}=c_1\boldsymbol{x}_1(t)+c_2\boldsymbol{x}_2(t)+\ldots+c_n\boldsymbol{x}_n(t)$ to the corresponding homogeneous system in part 1, to obtain a general solution to the given system.

We devote the rest of this chapter to methods for finding fundamental solution sets for homogeneous systems and particular solutions for nonhomogeneous systems.

Exercises 5.3

In Problems 1-4, write the given system in the matrix form $\mathbf{x}'=\mathbf{A}\mathbf{x}+\boldsymbol{f}$.

1. $x'(t)=3x(t)-y(t)+t^2$, $\quad y'(t)=-x(t)+2y(t)+e^t$.

2. $r'(t)=2r(t)+\sin t$, $\quad \theta'(t)=r(t)-\theta(t)+1$.

3. $\dfrac{dx}{dt}=t^2x-y-z+t$, $\quad \dfrac{dy}{dt}=e^tz+5$, $\quad \dfrac{dz}{dt}=tx-y+3z-e^t$.

4. $\dfrac{dx}{dt}=x+y+z$, $\quad \dfrac{dy}{dt}=2x-y+3z$, $\quad \dfrac{dz}{dt}=x+5z$.

In Problems 5-8, rewrite the given scalar equation as a first-order system in normal form. Express the system in the matrix form $\mathbf{x}'=\mathbf{A}\mathbf{x}+\boldsymbol{f}$.

5. $y''(t)-3y'(t)-10y(t)=\sin t$.

6. $x''(t)+x(t)=t^2$.

7. $\frac{d^4 w}{dt^4}+w=t^2$.

8. $\frac{d^3 y}{dt^3}-\frac{dy}{dt}+y=\cos t$.

In Problems 9-12, write the given system as a set of scalar equations.

9. $\boldsymbol{x}'=\begin{bmatrix} 5 & 0 \\ -2 & 4 \end{bmatrix}\boldsymbol{x}+e^{-2t}\begin{bmatrix} 2 \\ -3 \end{bmatrix}$.

10. $\boldsymbol{x}'=\begin{bmatrix} 2 & 1 \\ -1 & 3 \end{bmatrix}\boldsymbol{x}+e^{t}\begin{bmatrix} t \\ 1 \end{bmatrix}$.

11. $\boldsymbol{x}'=\begin{bmatrix} 1 & 0 & 1 \\ -1 & 2 & 5 \\ 0 & 5 & 1 \end{bmatrix}\boldsymbol{x}+e^{t}\begin{bmatrix} 1 \\ 0 \\ 0 \end{bmatrix}+t\begin{bmatrix} 0 \\ 1 \\ 0 \end{bmatrix}$.

12. $\boldsymbol{x}'=\begin{bmatrix} 0 & 1 & 0 \\ 0 & 0 & 1 \\ -1 & 1 & 2 \end{bmatrix}\boldsymbol{x}+t\begin{bmatrix} 1 \\ -1 \\ 2 \end{bmatrix}+\begin{bmatrix} 3 \\ 1 \\ 0 \end{bmatrix}$.

In Problems 13-19, determine whether the given vector functions are linearly dependent (LD) or linearly independent (LI) on the interval $(-\infty,+\infty)$.

13. $\begin{bmatrix} t \\ 3 \end{bmatrix}, \quad \begin{bmatrix} 4 \\ 1 \end{bmatrix}$.

14. $\begin{bmatrix} te^{-t} \\ e^{-t} \end{bmatrix}, \quad \begin{bmatrix} e^{-t} \\ e^{-t} \end{bmatrix}$.

15. $e^{t}\begin{bmatrix} 1 \\ 5 \end{bmatrix}, \quad e^{t}\begin{bmatrix} -3 \\ -15 \end{bmatrix}$.

16. $\begin{bmatrix} \sin t \\ \cos t \end{bmatrix}, \quad \begin{bmatrix} \sin 2t \\ \cos 2t \end{bmatrix}$.

17. $e^{2t}\begin{bmatrix} 1 \\ 0 \\ 5 \end{bmatrix}, \quad e^{2t}\begin{bmatrix} 1 \\ 1 \\ -1 \end{bmatrix}, \quad e^{3t}\begin{bmatrix} 0 \\ 1 \\ 0 \end{bmatrix}$.

18. $\begin{bmatrix} \sin t \\ \cos t \end{bmatrix}, \quad \begin{bmatrix} \sin t \\ \sin t \end{bmatrix}, \quad \begin{bmatrix} \cos t \\ \cos t \end{bmatrix}$.

19. $\begin{bmatrix} 1 \\ 0 \\ 1 \end{bmatrix}, \quad \begin{bmatrix} t \\ 0 \\ t \end{bmatrix}, \quad \begin{bmatrix} t^2 \\ 0 \\ t^2 \end{bmatrix}$.

20. Let

$$x_1=\begin{bmatrix}\cos t\\0\\0\end{bmatrix},\quad x_2=\begin{bmatrix}\sin t\\\cos t\\\cos t\end{bmatrix},\quad x_3=\begin{bmatrix}\cos t\\\sin t\\\cos t\end{bmatrix}.$$

(1) Compute the Wronskian.

(2) Are these vector functions linearly independent on $(-\infty,+\infty)$?

(3) Is there a homogeneous linear system for which these functions are solutions?

In Problems 21-24, the given vector functions are solutions to the system $\mathbf{x}'(t)=\mathbf{A}\mathbf{x}(t)$. *Determine whether they form a fundamental solution set. If they do, find a fundamental matrix for the system and give a general solution.*

21. $x_1=e^{2t}\begin{bmatrix}1\\-2\end{bmatrix},\quad x_2=e^{2t}\begin{bmatrix}-2\\4\end{bmatrix}.$

22. $x_1=e^{-t}\begin{bmatrix}3\\2\end{bmatrix},\quad x_2=e^{4t}\begin{bmatrix}1\\-1\end{bmatrix}.$

23. $x_1=\begin{bmatrix}e^{-t}\\2e^{-t}\\e^{-t}\end{bmatrix},\quad x_2=\begin{bmatrix}e^{t}\\0\\e^{t}\end{bmatrix},\quad x_3=\begin{bmatrix}e^{3t}\\-e^{3t}\\2e^{3t}\end{bmatrix}.$

24. $x_1=\begin{bmatrix}e^{t}\\e^{t}\\e^{t}\end{bmatrix},\quad x_2=\begin{bmatrix}\sin t\\\cos t\\-\sin t\end{bmatrix},\quad x_3=\begin{bmatrix}-\cos t\\\sin t\\\cos t\end{bmatrix}$

25. Verify that the vector functions

$$x_1=\begin{bmatrix}e^{t}\\e^{t}\end{bmatrix}\text{ and } x_2=\begin{bmatrix}e^{-t}\\3e^{-t}\end{bmatrix}$$

are solutions to the homogeneous system

$$x'=\mathbf{A}x=\begin{bmatrix}2&-1\\3&-2\end{bmatrix}x$$

on $(-\infty,+\infty)$, and that

$$x_p=\frac{3}{2}\begin{bmatrix}te^{t}\\te^{t}\end{bmatrix}-\frac{1}{4}\begin{bmatrix}e^{t}\\3e^{t}\end{bmatrix}+\begin{bmatrix}t\\2t\end{bmatrix}-\begin{bmatrix}0\\1\end{bmatrix}$$

is a particular solution to the nonhomogeneous system $x'=\mathbf{A}x+f(t)$, where $f(t)=\text{col}(e^{t},t)$. Find a general solution to $x'=\mathbf{A}x+f(t)$.

26. Verify that the vector functions

$$x_1=\begin{bmatrix} e^{3t} \\ 0 \\ e^{3t} \end{bmatrix}, \quad x_2=\begin{bmatrix} -e^{3t} \\ e^{3t} \\ 0 \end{bmatrix}, \quad x_3=\begin{bmatrix} -e^{-3t} \\ -e^{-3t} \\ e^{-3t} \end{bmatrix}$$

are solutions to the homogeneous system

$$x'=Ax=\begin{bmatrix} 1 & -2 & 2 \\ -2 & 1 & 2 \\ 2 & 2 & 1 \end{bmatrix} x$$

on $(-\infty, +\infty)$, and that

$$x_p=\begin{bmatrix} 5t+1 \\ 2t \\ 4t+2 \end{bmatrix}$$

is a particular solution to $x'=Ax+f(t)$, where $f(t)=\mathrm{col}(-9t,\ 0,\ -18t)$. Find a general solution to $x'=Ax+f(t)$.

27. Prove that the operator defined by $L[x]:=x'-Ax$, where A is an $n\times n$ matrix function and x is an $n\times 1$ differentiable vector function, is a linear operator.

28. Let $X(t)$ be a fundamental matrix for the system $x'=Ax$. Show that $x(t)=X(t)X^{-1}(t_0)x_0$ is the solution to the initial value problem $x'=Ax$, $x(t_0)=x_0$.

In Problems 29 and 30, verify that $X(t)$ *is a fundamental matrix for the given system and compute* $X^{-1}(t)$. *Use the result of Problem 28 to find the solution to the given initial value problem.*

29. $x'=\begin{bmatrix} 0 & 6 & 0 \\ 1 & 0 & 1 \\ 1 & 1 & 0 \end{bmatrix} x$, $x(0)=\begin{bmatrix} -1 \\ 0 \\ 1 \end{bmatrix}$; $X(t)=\begin{bmatrix} 6e^{-t} & -3e^{-2t} & 2e^{3t} \\ -e^{-t} & e^{-2t} & e^{3t} \\ -5e^{-t} & e^{-2t} & e^{3t} \end{bmatrix}$.

30. $x'=\begin{bmatrix} 2 & 3 \\ 3 & 2 \end{bmatrix} x$, $x(0)=\begin{bmatrix} 3 \\ -1 \end{bmatrix}$; $X(t)=\begin{bmatrix} e^{-t} & e^{5t} \\ -e^{-t} & e^{5t} \end{bmatrix}$.

31. Show that

$$\begin{vmatrix} t^2 & t|t| \\ 2t & 2|t| \end{vmatrix} \equiv 0$$

on $(-\infty, +\infty)$, but that the two vector functions

$$\begin{bmatrix} t^2 \\ 2t \end{bmatrix}, \quad \begin{bmatrix} t|t| \\ 2|t| \end{bmatrix}$$

are linearly independent on $(-\infty, +\infty)$.

32. **Abel's Formula.** If $\boldsymbol{x}_1, \boldsymbol{x}_2, \dots, \boldsymbol{x}_n$ are any n solutions to the $n\times n$ system $\boldsymbol{x}'(t)=\boldsymbol{A}(t)\boldsymbol{x}(t)$, then Abel's formula gives a representation for the Wronskian $W(t):=W[\boldsymbol{x}_1, \boldsymbol{x}_2, \dots, \boldsymbol{x}_n](t)$. Namely,

$$W(t)=W(t_0)\exp\left\{\int_{t_0}^{t}[a_{11}(s)+a_{22}(s)+\dots+a_{nn}(s)]\mathrm{d}s\right\},$$

where $a_{11}(s), a_{22}(s), \dots, a_{nn}(s)$ are the main diagonal elements of $\boldsymbol{A}(s)$. Prove this formula in the special case when $n=3$.

33. Using Abel's formula (Problem 32), prove that the Wronskian of n solutions to $\boldsymbol{x}'=\boldsymbol{A}\boldsymbol{x}$ on the interval I is either identically zero on I or never zero on I.

34. Prove that a fundamental solution set for the homogeneous system $\boldsymbol{x}'(t)=\boldsymbol{A}(t)\boldsymbol{x}(t)$ always exists on an interval I, provided $\boldsymbol{A}(t)$ is continuous on I. [Hint: Use the existence and uniqueness theorem (Theorem 5.1) and make judicious choices for $\boldsymbol{x}_0$.]

35. Prove Theorem 5.4 on the representation of solutions of the homogeneous system.

36. Prove Theorem 5.5 on the representation of solutions of the nonhomogeneous system.

37. To illustrate the connection between a higher-order equation and the equivalent first-order system, consider the equation

$$y'''(t)-6y''(t)+11y'(t)-6y(t)=0. \tag{5.21}$$

(1) Show that $\{e^t, e^{2t}, e^{3t}\}$ is a fundamental solution set for equation (5.21).

(2) Using the definition of Section 5.2, compute the Wronskian of $\{e^t, e^{2t}, e^{3t}\}$.

(3) Setting $x_1=y$, $x_2=y'$, $x_3=y''$, show that equation (5.21) is equivalent to the first-order system

$$\boldsymbol{x}'=\boldsymbol{A}\boldsymbol{x}, \tag{5.22}$$

where

$$\boldsymbol{A}=\begin{bmatrix}0 & 1 & 0\\ 0 & 0 & 1\\ 6 & -11 & 6\end{bmatrix}.$$

(4) The substitution used in part (3) suggests that

$$S=\left\{\begin{bmatrix}e^t\\ e^t\\ e^t\end{bmatrix}, \begin{bmatrix}e^{2t}\\ 2e^{2t}\\ 4e^{2t}\end{bmatrix}, \begin{bmatrix}e^{3t}\\ 3e^{3t}\\ 9e^{3t}\end{bmatrix}\right\}$$

is a fundamental solution set for system (5.22). Verify that this is the case.

(5) Compute the Wronskian of S. How does it compare with the Wronskian computed in part (2)?

38. Define $\boldsymbol{x}_1(t)$, $\boldsymbol{x}_2(t)$, and $\boldsymbol{x}_3(t)$, for $-\infty<t<+\infty$, by

$$\boldsymbol{x}_1(t)=\begin{bmatrix}\sin t\\ \sin t\\ 0\end{bmatrix},\quad \boldsymbol{x}_2(t)=\begin{bmatrix}\sin t\\ 0\\ \sin t\end{bmatrix},\quad \boldsymbol{x}_3(t)=\begin{bmatrix}0\\ \sin t\\ \sin t\end{bmatrix}.$$

(1) Show that for the three scalar functions in each individual row there are nontrivial linear combinations that sum to zero for all t.

(2) Show that, nonetheless, the three vector functions are linearly independent. (No single nontrivial combination works for each row, for all t.)

(3) Calculate the Wronskian $W[\boldsymbol{x}_1, \boldsymbol{x}_2, \boldsymbol{x}_3](t)$.

(4) Is there a linear third-order homogeneous differential equation system having $\boldsymbol{x}_1(t)$, $\boldsymbol{x}_2(t)$, and $\boldsymbol{x}_3(t)$ as solutions?

5.4 Homogeneous Linear Systems with Constant Coefficients

In this section, we discuss a procedure for obtaining a general solution for the homogeneous system

$$\boldsymbol{x}'(t)=\boldsymbol{A}\boldsymbol{x}(t) \tag{5.23}$$

where $\boldsymbol{A}$ is a (real) $n\times n$ constant matrix. The general solution we seek will be defined for all t because the elements of $\boldsymbol{A}$ are just constant functions, which are continuous on $(-\infty,+\infty)$. In Section 5.3, we showed that a general solution to system (5.23) can be constructed from a fundamental solution set consisting of n linearly independent solutions to (5.23). Thus, our goal is to find n such vector solutions.

In Chapter 4, we were successful in solving homogeneous linear equations with constant coefficients by guessing that the equation had a solution of the form e^{rt}. Because any scalar linear equation can be expressed as a system, it is reasonable to expect system (5.23) to have solutions of the form

$$\boldsymbol{x}(t)=e^{rt}\boldsymbol{u},$$

where r is a constant and $\boldsymbol{u}$ is a constant vector, both of which must be determined. Substituting $e^{rt}\boldsymbol{u}$ for $\boldsymbol{x}(t)$ in system (5.23) gives

$$re^{rt}\boldsymbol{u}=\boldsymbol{A}e^{rt}\boldsymbol{u}=e^{rt}\boldsymbol{A}\boldsymbol{u}.$$

Canceling the factor e^{rt} and rearranging terms, we find that

$$(\boldsymbol{A}-r\boldsymbol{I})\boldsymbol{u}=\boldsymbol{0}, \tag{5.24}$$

where $r\boldsymbol{I}$ denotes the diagonal matrix with r's along its main diagonal.

The preceding calculation shows that $\boldsymbol{x}(t)=e^{rt}\boldsymbol{u}$ is a solution to system (5.23) if and only if r and $\boldsymbol{u}$ satisfy equation (5.24). Since the trivial case, $\boldsymbol{u}=\boldsymbol{0}$, is of no help in finding linearly independent solutions to (5.23), we require that $\boldsymbol{u}\neq\boldsymbol{0}$. Such vectors are given a special name, as follows.

From the theory of linear algebra, a linear homogeneous system of n algebraic equation in n unknowns has a nontrivial solution if and only if the determinant of its coefficients is zero. Hence, a necessary and sufficient condition for equation (5.24) to have a nontrivial solution is that

$$|\boldsymbol{A}-r\boldsymbol{I}|=0. \tag{5.25}$$

Expanding the determinant of $\boldsymbol{A}-r\boldsymbol{I}$ in terms of its cofactors, we find that it is an n-th degree polynomial in r; that is,

$$|\boldsymbol{A}-r\boldsymbol{I}|=p(r). \tag{5.26}$$

Therefore, finding the eigenvalues of a matrix $\boldsymbol{A}$ is equivalent to finding the zeros of the polynomial $p(r)$. Equation (5.25) is called the **characteristic equation** of $\boldsymbol{A}$, and $p(r)$ in equation (5.26) is the **characteristic polynomial** of $\boldsymbol{A}$. The characteristic equation plays a role for systems similar to the role played by the auxiliary equation for scalar equations.

Let's return to the problem of finding a general solution to a homogeneous system of differential equations. We have already shown that $e^{rt}\boldsymbol{u}$ is a solution to system (5.23) if r is an eigenvalue and $\boldsymbol{u}$ a corresponding eigenvector. The question is: Can we obtain n linearly independent solutions to the homogeneous system by finding all the eigenvalues and eigenvectors of $\boldsymbol{A}$? The answer is yes, if $\boldsymbol{A}$ has n linearly independent eigenvectors.

n Linearly Independent Eigenvectors

Theorem 5.6 Suppose the $n\times n$ constant matrix $\boldsymbol{A}$ has n linearly independent eigenvectors $\boldsymbol{u}_1, \boldsymbol{u}_2, \ldots, \boldsymbol{u}_n$. Let r_i be the eigenvalue corresponding to $\boldsymbol{u}_i$. Then

$$\{e^{r_1 t}\boldsymbol{u}_1, e^{r_2 t}\boldsymbol{u}_2, \ldots, e^{r_n t}\boldsymbol{u}_n\} \tag{5.27}$$

is a fundamental solution set (and $\boldsymbol{X}(t)=[e^{r_1 t}\boldsymbol{u}_1, e^{r_2 t}\boldsymbol{u}_2, \ldots, e^{r_n t}\boldsymbol{u}_n]$ is a fundamental matrix) on $(-\infty, +\infty)$ for the homogeneous system $\boldsymbol{x}'=\boldsymbol{A}\boldsymbol{x}$. Consequently, a general solution of $\boldsymbol{x}'=\boldsymbol{A}\boldsymbol{x}$ is

$$x(t)=c_1e^{r_1t}u_1+c_2e^{r_2t}u_2+\ldots+c_ne^{r_nt}u_n \tag{5.28}$$

where $c_1, c_2, \ldots, c_n$ are arbitrary constants.

Proof As we have seen, the vector functions listed in (5.27) are solutions to the homogeneous system. Moreover, their Wronskian is

$$W(t)=\det[e^{r_1t}u_1, e^{r_2t}u_2, \ldots, e^{r_nt}u_n]=e^{(r_1+r_2+\cdots+r_n)t}\det[u_1, u_2, \ldots, u_n].$$

Since the eigenvectors are assumed to be linearly independent, it follows that

$$\det[u_1, u_2, \ldots, u_n]$$

is not zero. Hence, the Wronskian $W(t)$ shows that (5.27) is a fundamental solution set, and consequently a general solution is given by equation (5.28).

n Distinct Eigenvalues

Corollary 5.2 If the $n\times n$ constant matrix A has n distinct eigenvalues $r_1, r_2, \ldots, r_n$ and u_i is an eigenvector associated with r_i, then

$$\{e^{r_1t}u_1, e^{r_2t}u_2, \ldots, e^{r_nt}u_n\}$$

is a fundamental solution set for the homogeneous system $x'=Ax$.

Applications of Theorem 5.6 and Corollary 5.2 are given in next examples.

Example 5.4 Find a general solution of

$$x'(t)=Ax(t), \text{ where } A=\begin{bmatrix}2 & -3\\ 1 & -2\end{bmatrix}. \tag{5.29}$$

Solution The characteristic equation for A is

$$|A-rI|=\begin{vmatrix}2-r & -3\\ 1 & -2-r\end{vmatrix}=(2-r)(-2-r)+3=r^2-1=0.$$

Hence, the eigenvalues of A are $r_1=1$, $r_2=-1$. To find the eigenvectors corresponding to $r_1=1$, we must solve $(A-r_1I)u=0$. Substituting for A and r_1 gives

$$\begin{bmatrix}1 & -3\\ 1 & -3\end{bmatrix}\begin{bmatrix}u_1\\ u_2\end{bmatrix}=\begin{bmatrix}0\\ 0\end{bmatrix}.$$

Notice that this matrix equation is equivalent to the single scalar equation $u_1-3u_2=0$. Therefore, the solutions to (5.29) are obtained by assigning an arbitrary value for u_2 (say $u_2=1$) and setting $u_1=3u_2=3$. Consequently, the eigenvectors associated with $r_1=1$ can be expressed as

$$u_1=\begin{bmatrix}3\\ 1\end{bmatrix}.$$

For $r_2=-1$, the equation $(A-r_2I)u=0$ becomes

$$\begin{bmatrix} 3 & -3 \\ 1 & -1 \end{bmatrix}\begin{bmatrix} u_1 \\ u_2 \end{bmatrix}=\begin{bmatrix} 0 \\ 0 \end{bmatrix}.$$

Solving, we obtain $u_1=s$ and $u_2=s$, with s arbitrary. Therefore, the eigenvectors associated with the eigenvalue $r_2=-1$ are

$$\boldsymbol{u}_2=\begin{bmatrix} 1 \\ 1 \end{bmatrix}.$$

Because u_1 and u_2 are linearly independent, it follows from Theorem 5. 6 that a general solution to (5. 29) is

$$\boldsymbol{x}(t)=c_1 e^t\begin{bmatrix} 3 \\ 1 \end{bmatrix}+c_2 e^{-t}\begin{bmatrix} 1 \\ 1 \end{bmatrix}. \tag{5.30}$$

If we sum the vectors on the right-hand side of equation (5. 30) and then write out the expressions for the components of $\boldsymbol{x}(t)=\text{col}(x_1(t), x_2(t))$, we get

$$x_1(t)=3c_1 e^t+c_2 e^{-t},$$

$$x_2(t)=c_1 e^t+c_2 e^{-t}. \blacklozenge$$

Example 5. 5 Solve the initial value problem

$$\boldsymbol{x}'(t)=\begin{bmatrix} 1 & 2 & -1 \\ 1 & 0 & 1 \\ 4 & -4 & 5 \end{bmatrix}\boldsymbol{x}(t), \quad \boldsymbol{x}(0)=\begin{bmatrix} -1 \\ 0 \\ 0 \end{bmatrix}. \tag{5.31}$$

Solution The characteristic equation for $\boldsymbol{A}$ is

$$|\boldsymbol{A}-r\boldsymbol{I}|=\begin{vmatrix} 1-r & 2 & -1 \\ 1 & -r & 1 \\ 4 & -4 & 5-r \end{vmatrix}=0,$$

which simplifies to $(r-1)(r-2)(r-3)=0$. Hence, the eigenvalues of $\boldsymbol{A}$ are $r_1=1$, $r_2=2$, and $r_3=3$. To find the eigenvectors corresponding to $r_1=1$, we set $r=1$ in $(\boldsymbol{A}-r\boldsymbol{I})\boldsymbol{u}=\boldsymbol{0}$. This gives

$$\begin{bmatrix} 0 & 2 & -1 \\ 1 & -1 & 1 \\ 4 & -4 & 4 \end{bmatrix}\begin{bmatrix} u_1 \\ u_2 \\ u_3 \end{bmatrix}=\begin{bmatrix} 0 \\ 0 \\ 0 \end{bmatrix}. \tag{5.32}$$

Using elementary row operations (Gaussian elimination), we see that equation (5. 32) is equivalent to the two equations:

$$u_1-u_2+u_3=0,$$

$$2u_2-u_3=0.$$

Thus, we can obtain the solutions to equation (5.30) by assigning an arbitrary value to u_2 (say, $u_2=s$), solving $2u_2-u_3=0$ for u_3 to get $u_3=2s$, and then solving $u_1-u_2+u_3=0$ for u_1 to get $u_1=-s$. Hence, the eigenvectors associated with $r_1=1$ are

$$\boldsymbol{u}_1=s\begin{bmatrix}-1\\1\\2\end{bmatrix}. \tag{5.33}$$

For $r_2=2$, we solve

$$\begin{bmatrix}-1 & 2 & -1\\1 & -2 & 1\\4 & -4 & 3\end{bmatrix}\begin{bmatrix}u_1\\u_2\\u_3\end{bmatrix}=\begin{bmatrix}0\\0\\0\end{bmatrix}$$

in a similar fashion to obtain the eigenvectors

$$\boldsymbol{u}_2=s\begin{bmatrix}-2\\1\\4\end{bmatrix}. \tag{5.34}$$

Finally, for $r_3=3$, we solve

$$\begin{bmatrix}-2 & 2 & -1\\1 & -3 & 1\\4 & -4 & 2\end{bmatrix}\begin{bmatrix}u_1\\u_2\\u_3\end{bmatrix}=\begin{bmatrix}0\\0\\0\end{bmatrix}$$

and get the eigenvectors

$$\boldsymbol{u}_3=s\begin{bmatrix}-1\\1\\4\end{bmatrix}. \tag{5.35}$$

The matrix $\boldsymbol{A}$ has the three distinct eigenvalues $r_1=1$, $r_2=2$, $r_3=3$. If we set $s=1$ in equations (5.33), (5.34), and (5.35), we obtain the corresponding eigenvectors

$$\boldsymbol{u}_1=\begin{bmatrix}-1\\1\\2\end{bmatrix},\quad \boldsymbol{u}_2=\begin{bmatrix}-2\\1\\4\end{bmatrix},\quad \boldsymbol{u}_3=\begin{bmatrix}-1\\1\\4\end{bmatrix},$$

which are linear independent. Hence, a general solution to (5.31) is

$$\boldsymbol{x}(t)=c_1\mathrm{e}^{t}\begin{bmatrix}-1\\1\\2\end{bmatrix}+c_2\mathrm{e}^{2t}\begin{bmatrix}-2\\1\\4\end{bmatrix}+c_3\mathrm{e}^{3t}\begin{bmatrix}-1\\1\\4\end{bmatrix}$$

$$=\begin{bmatrix} -e^{t} & -2e^{2t} & -e^{3t} \\ e^{t} & e^{2t} & e^{3t} \\ 2e^{t} & 4e^{2t} & 4e^{3t} \end{bmatrix}\begin{bmatrix} c_1 \\ c_2 \\ c_3 \end{bmatrix}. \tag{5.36}$$

To satisfy the initial condition in (5.31), we solve

$$\boldsymbol{x}(0)=\begin{bmatrix} -1 & -2 & -1 \\ 1 & 1 & 1 \\ 2 & 4 & 4 \end{bmatrix}\begin{bmatrix} c_1 \\ c_2 \\ c_3 \end{bmatrix}=\begin{bmatrix} -1 \\ 0 \\ 0 \end{bmatrix}$$

and find that $c_1=0$, $c_2=1$, and $c_3=-1$. Inserting these values into equation (5.36) gives the desired solution. ◆

If $\boldsymbol{A}$ is an $n\times n$ real symmetric matrix, it is known that there always exist n linearly independent eigenvectors. Thus, Theorem 5.6 applies and a general solution to $\boldsymbol{x}'=\boldsymbol{A}\boldsymbol{x}$ is given by equation (5.28).

Example 5.6 Find a general solution of

$$\boldsymbol{x}'(t)=\boldsymbol{A}\boldsymbol{x}(t), \text{ where } \boldsymbol{A}=\begin{bmatrix} -1 & -2 & 2 \\ -2 & 1 & 2 \\ 2 & 2 & 1 \end{bmatrix}. \tag{5.37}$$

Solution Because $\boldsymbol{A}$ is symmetric, so we are assured that $\boldsymbol{A}$ has three linearly independent eigenvectors. To find them, we first compute the characteristic equation for $\boldsymbol{A}$:

$$|\boldsymbol{A}-r\boldsymbol{I}|=\begin{vmatrix} 1-r & -2 & 2 \\ -2 & 1-r & 2 \\ 2 & 2 & 1-r \end{vmatrix}=-(r-3)^2(r+3)=0.$$

Thus, the eigenvalues of $\boldsymbol{A}$ are $r_1=r_2=3$, and $r_3=-3$.

Notice that the eigenvalue $r=3$ has multiplicity 2 when considered as a root of the characteristic equation. Therefore, we must find two linearly independent eigenvectors associated with $r=3$. Substituting $r=3$ in $(\boldsymbol{A}-r\boldsymbol{I})\boldsymbol{u}=\boldsymbol{0}$ gives

$$\begin{bmatrix} -2 & -2 & 2 \\ -2 & -2 & 2 \\ 2 & 2 & -2 \end{bmatrix}\begin{bmatrix} u_1 \\ u_2 \\ u_3 \end{bmatrix}=\begin{bmatrix} 0 \\ 0 \\ 0 \end{bmatrix}.$$

This system is equivalent to the single equation $-u_1-u_2+u_3=0$, so we can obtain its solutions by assigning an arbitrary value to u_2, say $u_2=v$, and an arbitrary value to u_3, say $u_3=s$.

Solving for u_1, we find $u_1 = u_3 - u_2 = s - v$. Therefore, the eigenvectors associated with $r_1 = r_2 = 3$ can be expressed as

$$\boldsymbol{u} = \begin{bmatrix} s-v \\ v \\ s \end{bmatrix} = s\begin{bmatrix} 1 \\ 0 \\ 1 \end{bmatrix} + v\begin{bmatrix} -1 \\ 1 \\ 0 \end{bmatrix}.$$

By first taking $s=1$, $v=0$ and then taking $s=0$, $v=1$, we get the two linearly independent eigenvectors:

$$\boldsymbol{u}_1 = \begin{bmatrix} 1 \\ 0 \\ 1 \end{bmatrix}, \quad \boldsymbol{u}_2 = \begin{bmatrix} -1 \\ 1 \\ 0 \end{bmatrix}.$$

For $r_3 = 3$, we solve

$$(\boldsymbol{A}+3\boldsymbol{I})\boldsymbol{u} = \begin{bmatrix} 4 & -2 & 2 \\ -2 & 4 & 2 \\ 2 & 2 & 4 \end{bmatrix}\begin{bmatrix} u_1 \\ u_2 \\ u_3 \end{bmatrix} = \begin{bmatrix} 0 \\ 0 \\ 0 \end{bmatrix}$$

to obtain the eigenvectors $\operatorname{col}(-s, -s, s)$. Taking $s=1$ gives

$$\boldsymbol{u}_3 = \begin{bmatrix} -1 \\ -1 \\ 1 \end{bmatrix}.$$

Since the eigenvectors $\boldsymbol{u}_1$, $\boldsymbol{u}_2$, and $\boldsymbol{u}_3$ are linearly independent, a general solution to (5.37) is

$$\boldsymbol{x}(t) = c_1 e^{3t}\begin{bmatrix} 1 \\ 0 \\ 1 \end{bmatrix} + c_2 e^{3t}\begin{bmatrix} -1 \\ 1 \\ 0 \end{bmatrix} + c_3 e^{-3t}\begin{bmatrix} -1 \\ -1 \\ 1 \end{bmatrix}. \blacklozenge$$

If a matrix $\boldsymbol{A}$ is not symmetric, it is possible for $\boldsymbol{A}$ to have a repeated eigenvalue but not to have two linearly independent corresponding eigenvectors. In particular, the matrix

$$\boldsymbol{A} = \begin{bmatrix} 1 & -1 \\ 4 & -3 \end{bmatrix} \tag{5.38}$$

has the repeated eigenvalue $r_1 = r_2 = -1$, but Problem 35 in Exercises 5.4 shows that all the eigenvectors associated with $r = -1$ are of the form $\boldsymbol{u} = s \operatorname{col}(1, 2)$. Consequently, no two eigenvectors are linearly independent.

Exercises 5.4

In Problems 1-8, find the eigenvalues and eigenvectors of the given matrix.

1. $\begin{bmatrix} -4 & 2 \\ 2 & -1 \end{bmatrix}$.

2. $\begin{bmatrix} 6 & -3 \\ 2 & 1 \end{bmatrix}$.

3. $\begin{bmatrix} 1 & -1 \\ 2 & 4 \end{bmatrix}$.

4. $\begin{bmatrix} 1 & 5 \\ 1 & -3 \end{bmatrix}$.

5. $\begin{bmatrix} 1 & 0 & 0 \\ 0 & 0 & 2 \\ 0 & 2 & 0 \end{bmatrix}$.

6. $\begin{bmatrix} 0 & 1 & 1 \\ 1 & 0 & 1 \\ 1 & 1 & 0 \end{bmatrix}$.

7. $\begin{bmatrix} 1 & 0 & 0 \\ 2 & 3 & 1 \\ 0 & 2 & 4 \end{bmatrix}$.

8. $\begin{bmatrix} -3 & 1 & 0 \\ 0 & -3 & 1 \\ 4 & -8 & 2 \end{bmatrix}$.

In Problems 9 and 10, some of the eigenvalues of the given matrix are complex. Find all the eigenvalues and eigenvectors.

9. $\begin{bmatrix} 0 & -1 \\ 1 & 0 \end{bmatrix}$.

10. $\begin{bmatrix} 1 & 2 & -1 \\ 0 & 1 & 1 \\ 0 & -1 & 1 \end{bmatrix}$.

In Problems 11-16, find a general solution of the system $\mathbf{x}'(t)=\mathbf{A}\mathbf{x}(t)$ *for the given matrix* $\mathbf{A}$.

11. $\mathbf{A}=\begin{bmatrix} -1 & \frac{3}{4} \\ -5 & 3 \end{bmatrix}$.

12. $\mathbf{A}=\begin{bmatrix}1 & 3\\ 12 & 1\end{bmatrix}$.

13. $\mathbf{A}=\begin{bmatrix}1 & 2 & 2\\ 2 & 0 & 3\\ 2 & 3 & 0\end{bmatrix}$.

14. $\mathbf{A}=\begin{bmatrix}-1 & 1 & 0\\ 1 & 2 & 1\\ 0 & 3 & -1\end{bmatrix}$.

15. $\mathbf{A}=\begin{bmatrix}1 & 2 & 3\\ 0 & 1 & 0\\ 2 & 1 & 2\end{bmatrix}$.

16. $\mathbf{A}=\begin{bmatrix}-7 & 0 & 6\\ 0 & 5 & 0\\ 6 & 0 & 2\end{bmatrix}$.

17. Consider the system $\boldsymbol{x}'(t)=\mathbf{A}\boldsymbol{x}(t)$, $t\geqslant 0$, with

$$\mathbf{A}=\begin{bmatrix}1 & \sqrt{3}\\ \sqrt{3} & -1\end{bmatrix}.$$

(1) Show that the matrix $\boldsymbol{A}$ has eigenvalues $r_1=2$ and $r_2=-2$ with corresponding eigenvectors $\boldsymbol{u}_1=\mathrm{col}(\sqrt{3},1)$ and $\boldsymbol{u}_2=\mathrm{col}(1,-\sqrt{3})$.

(2) Sketch the trajectory of the solution having initial vector $\boldsymbol{x}(0)=\boldsymbol{u}_1$.

(3) Sketch the trajectory of the solution having initial vector $\boldsymbol{x}(0)=\boldsymbol{u}_2$.

(4) Sketch the trajectory of the solution having initial vector $\boldsymbol{x}(0)=\boldsymbol{u}_2-\boldsymbol{u}_1$.

18. Consider the system $\boldsymbol{x}'(t)=\mathbf{A}\boldsymbol{x}(t)$, $t\geqslant 0$, with

$$\mathbf{A}=\begin{bmatrix}-2 & 1\\ 1 & -2\end{bmatrix}.$$

(1) Show that the matrix $\boldsymbol{A}$ has eigenvalues $r_1=-1$ and $r_2=-3$ with corresponding eigenvectors $\boldsymbol{u}_1=\mathrm{col}(1,1)$ and $\boldsymbol{u}_2=\mathrm{col}(1,-1)$.

(2) Sketch the trajectory of the solution having initial vector $\boldsymbol{x}(0)=\boldsymbol{u}_1$.

(3) Sketch the trajectory of the solution having initial vector $\boldsymbol{x}(0)=-\boldsymbol{u}_2$.

(4) Sketch the trajectory of the solution having initial vector $\boldsymbol{x}(0)=\boldsymbol{u}_1-\boldsymbol{u}_2$.

In Problems 19-24, find a fundamental matrix for the system $\mathbf{x}'(t)=\mathbf{A}\mathbf{x}(t)$ for the given matrix $\mathbf{A}$.

19. $\mathbf{A}=\begin{bmatrix}-1 & 1\\ 8 & 1\end{bmatrix}$.

20. $\mathbf{A}=\begin{bmatrix}5 & 4\\ -1 & 0\end{bmatrix}$.

21. $\mathbf{A}=\begin{bmatrix}0 & 1 & 0\\ 0 & 0 & 1\\ 8 & -14 & 7\end{bmatrix}$.

22. $\mathbf{A}=\begin{bmatrix}3 & 1 & -1\\ 1 & 3 & -1\\ 3 & 3 & -1\end{bmatrix}$.

23. $\mathbf{A}=\begin{bmatrix}2 & 1 & 1 & -1\\ 0 & -1 & 0 & 1\\ 0 & 0 & 3 & 1\\ 0 & 0 & 0 & 7\end{bmatrix}$.

24. $\mathbf{A}=\begin{bmatrix}4 & -1 & 0 & 0\\ 0 & 0 & 0 & 0\\ 0 & 0 & 2 & -3\\ 0 & 0 & 1 & -2\end{bmatrix}$.

25. Using matrix algebra techniques, find a general solution of the system

$$\begin{cases}x'=x+2y-z,\\ y'=x+z,\\ z'=4x-4y+5z.\end{cases}$$

26. Using matrix algebra techniques, find a general solution of the system

$$\begin{cases}x'=3x-4y,\\ y'=4x-7y.\end{cases}$$

In Problems 27-30, use a linear algebra software package such as MATLAB, *Maple, or Mathematica to compute the required eigenvalues and eigenvectors and then give a fundamental matrix for the system* $\mathbf{x}'(t)=\mathbf{A}\mathbf{x}(t)$ *for the given matrix* $\mathbf{A}$.

27. $\mathbf{A}=\begin{bmatrix}0 & 1.1 & 0\\ 0 & 0 & 1.3\\ 0.9 & 1.1 & -6.9\end{bmatrix}$.

28. $\boldsymbol{A}=\begin{bmatrix} 2 & 1 & 1 \\ -1 & 1 & 0 \\ 3 & 3 & 3 \end{bmatrix}$.

29. $\boldsymbol{A}=\begin{bmatrix} 0 & 1 & 0 & 0 \\ 0 & 0 & 1 & 0 \\ 0 & 0 & 0 & 1 \\ 2 & -6 & 3 & 3 \end{bmatrix}$.

30. $\boldsymbol{A}=\begin{bmatrix} 0 & 1 & 0 & 0 \\ 1 & -1 & 0 & 0 \\ 0 & 0 & 0 & 1 \\ 0 & 0 & -2 & 4 \end{bmatrix}$.

In Problems 31-34, solve the given initial value problem.

31. $\boldsymbol{x}'(t)=\begin{bmatrix} 1 & 3 \\ 3 & 1 \end{bmatrix}\boldsymbol{x}(t), \quad \boldsymbol{x}(0)=\begin{bmatrix} 3 \\ 1 \end{bmatrix}$.

32. $\boldsymbol{x}'(t)=\begin{bmatrix} 6 & -3 \\ 2 & 1 \end{bmatrix}\boldsymbol{x}(t), \quad \boldsymbol{x}(0)=\begin{bmatrix} -10 \\ -6 \end{bmatrix}$.

33. $\boldsymbol{x}'(t)=\begin{bmatrix} 1 & -2 & 2 \\ -2 & 1 & -2 \\ 2 & -2 & 1 \end{bmatrix}\boldsymbol{x}(t), \quad \boldsymbol{x}(0)=\begin{bmatrix} -2 \\ -3 \\ 2 \end{bmatrix}$.

34. $\boldsymbol{x}'(t)=\begin{bmatrix} 0 & 1 & 1 \\ 1 & 0 & 1 \\ 1 & 1 & 0 \end{bmatrix}\boldsymbol{x}(t), \quad \boldsymbol{x}(0)=\begin{bmatrix} -1 \\ 4 \\ 0 \end{bmatrix}$.

35. (1) Show that the matrix

$$\boldsymbol{A}=\begin{bmatrix} 1 & -1 \\ 4 & -3 \end{bmatrix}$$

has the repeated eigenvalue $r=-1$ and that all the eigenvectors are of the form $\boldsymbol{u}=s\,\mathrm{col}(1,2)$.

(2) Use the result of part (1) to obtain a nontrivial solution $\boldsymbol{x}_1(t)$ to the system $\boldsymbol{x}'=\boldsymbol{A}\boldsymbol{x}$.

(3) To obtain a second linearly independent solution to $\boldsymbol{x}'=\boldsymbol{A}\boldsymbol{x}$, try $\boldsymbol{x}_2(t)=te^{-t}\boldsymbol{u}_1+e^{-t}\boldsymbol{u}_2$. [Hint: Substitute $\boldsymbol{x}_2$ into the system $\boldsymbol{x}'=\boldsymbol{A}\boldsymbol{x}$ and derive the relations

$$(\boldsymbol{A}+\boldsymbol{I})\boldsymbol{u}_1=0, \quad (\boldsymbol{A}+\boldsymbol{I})\boldsymbol{u}_2=\boldsymbol{u}_1.$$

Since $\boldsymbol{u}_1$ must be an eigenvector, set $\boldsymbol{u}_1=\text{col}(1,2)$ and solve for $\boldsymbol{u}_2$.]

(4) What is $(\boldsymbol{A}+\boldsymbol{I})^2\boldsymbol{u}_2$? ($\boldsymbol{u}_2$ will be identified as a generalized eigenvector.)

36. Use the method discussed in Problem 35 to find a general solution to the system

$$\boldsymbol{x}'(t)=\begin{bmatrix}5 & -3\\ 3 & -1\end{bmatrix}\boldsymbol{x}(t).$$

37. (1) Show that the matrix

$$\boldsymbol{A}=\begin{bmatrix}2 & 1 & 6\\ 0 & 2 & 5\\ 0 & 0 & 2\end{bmatrix}$$

has the repeated eigenvalue $r=2$ with multiplicity 3 and that all the eigenvectors of $\boldsymbol{A}$ are of the form $\boldsymbol{u}=s\,\text{col}(1, 0, 0)$.

(2) Use the result of part (1) to obtain a solution to the system $\boldsymbol{x}'=\boldsymbol{A}\boldsymbol{x}$ of the form $\boldsymbol{x}_1(t)=e^{2t}\boldsymbol{u}_1$.

(3) To obtain a second linearly independent solution to $\boldsymbol{x}'=\boldsymbol{A}\boldsymbol{x}$, try $\boldsymbol{x}_2(t)=te^{2t}\boldsymbol{u}_1+e^{2t}\boldsymbol{u}_2$. [Hint: Show that $\boldsymbol{u}_1$ and $\boldsymbol{u}_2$ must satisfy

$$(\boldsymbol{A}-2\boldsymbol{I})\boldsymbol{u}_1=\boldsymbol{0},\ (\boldsymbol{A}-2\boldsymbol{I})\boldsymbol{u}_2=\boldsymbol{u}_1.]$$

(4) To obtain a third linearly independent solution to $\boldsymbol{x}'=\boldsymbol{A}\boldsymbol{x}$, try

$$\boldsymbol{x}_3(t)=\frac{t^2}{2}e^{2t}\boldsymbol{u}_1+te^{2t}\boldsymbol{u}_2+e^{2t}\boldsymbol{u}_3.$$

[Hint: Show that $\boldsymbol{u}_1$, $\boldsymbol{u}_2$, and $\boldsymbol{u}_3$ must satisfy

$$(\boldsymbol{A}-2\boldsymbol{I})\boldsymbol{u}_1=0,(\boldsymbol{A}-2\boldsymbol{I})\boldsymbol{u}_2=\boldsymbol{u}_1,(\boldsymbol{A}-2\boldsymbol{I})\boldsymbol{u}_3=\boldsymbol{u}_2.]$$

(5) Show that $(\boldsymbol{A}-2\boldsymbol{I})^2\boldsymbol{u}_2=(\boldsymbol{A}-2\boldsymbol{I})^3\boldsymbol{u}_3=\boldsymbol{0}$.

38. Use the method discussed in Problem 37 to find a general solution to the system

$$\boldsymbol{x}'(t)=\begin{bmatrix}3 & -2 & 1\\ 2 & -1 & 1\\ -4 & 4 & 1\end{bmatrix}\boldsymbol{x}(t).$$

39. (1) Show that the matrix

$$\boldsymbol{A}=\begin{bmatrix}2 & 1 & 1\\ 1 & 2 & 1\\ -2 & -2 & -1\end{bmatrix}$$

has the repeated eigenvalue $r=1$ of multiplicity 3 and that all the eigenvectors of $\boldsymbol{A}$ are of the form $\boldsymbol{u}=s\,\text{col}(-1,1,0)+v\,\text{col}(-1,0,1)$.

(2) Use the result of part (1) to obtain two linearly independent solutions to the system $\boldsymbol{x}'=\boldsymbol{A}\boldsymbol{x}$ of the form

$$\boldsymbol{x}_1(t)=e^t\boldsymbol{u}_1 \text{ and } \boldsymbol{x}_2(t)=e^t\boldsymbol{u}_2.$$

(3) To obtain a third linearly independent solution to $\boldsymbol{x}'=\boldsymbol{A}\boldsymbol{x}$, try $\boldsymbol{x}_3(t)=te^t\boldsymbol{u}_3+e^t\boldsymbol{u}_4$. [Hint: Show that $\boldsymbol{u}_3$ and $\boldsymbol{u}_4$ must satisfy

$$(\boldsymbol{A}-\boldsymbol{I})\boldsymbol{u}_3=\boldsymbol{0}, \quad (\boldsymbol{A}-\boldsymbol{I})\boldsymbol{u}_4=\boldsymbol{u}_3.$$

Choose $\boldsymbol{u}_3$, an eigenvector of $\boldsymbol{A}$, so that you can solve for $\boldsymbol{u}_4$.]

(4) What is $(\boldsymbol{A}-\boldsymbol{I})^2\boldsymbol{u}_4$?

40. Use the method discussed in Problem 39 to find a general solution to the system

$$\boldsymbol{x}'(t)=\begin{bmatrix}1 & 3 & -2\\ 0 & 7 & -4\\ 0 & 9 & -5\end{bmatrix}\boldsymbol{x}(t).$$

41. Use the substitution $x_1=y$, $x_2=y'$ to convert the linear equation $ay''+by'+cy=0$, where a, b, and c are constants, into a normal system. Show that the characteristic equation for this system is the same as the auxiliary equation for the original equation.

42. (1) Show that the Cauchy-Euler equation $at^2y''+bty'+cy=0$ can be written as a Cauchy-Euler system

$$t\boldsymbol{x}'=\boldsymbol{A}\boldsymbol{x} \tag{5.39}$$

with a constant coefficient matrix $\boldsymbol{A}$, by setting $x_1=y/t$ and $x_2=y'$.

(2) Show that for $t>0$, any system of the form (5.39) with $\boldsymbol{A}$ an $n\times n$ constant matrix has nontrivial solutions of the form $\boldsymbol{x}(t)=t^r\boldsymbol{u}$ if and only if r is an eigenvalue of $\boldsymbol{A}$ and $\boldsymbol{u}$ is a corresponding eigenvector.

In Problems 43 and 44, use the result of Problem 42 to find a general solution to the given system.

43. $t\boldsymbol{x}'(t)=\begin{bmatrix}1 & 3\\ -1 & 5\end{bmatrix}\boldsymbol{x}(t), \quad t>0.$

44. $t\boldsymbol{x}'(t)=\begin{bmatrix}-4 & 2\\ 2 & -1\end{bmatrix}\boldsymbol{x}(t), \quad t>0.$

45. **Mixing between Interconnected Tanks.** Two tanks, each holding 50 L of liquid, are interconnected by pipes with liquid flowing from tank A into tank B at a rate of 4

L/min and from tank B into tank A at 1 L/min (see Figure 5.2). The liquid inside each tank is kept well stirred. Pure water flows into tank A at a rate of 3 L/min, and the solution flows out of tank B at 3 L/min. If, initially, tank A contains 2.5 kg of salt and tank B contains no salt (only water), determine the mass of salt in each tank at time $t \geqslant 0$. Graph on the same axes the two quantities $x_1(t)$ and $x_2(t)$, where $x_1(t)$ is the mass of salt in tank A and $x_2(t)$ is the mass in tank B.

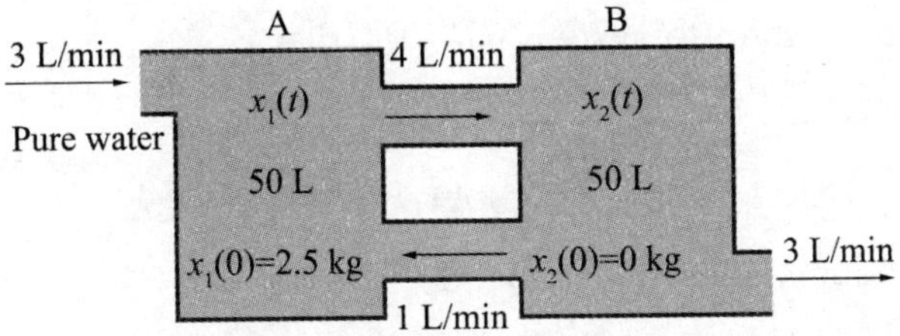

Figure 5.2 Mixing problem for interconnected tanks

5.5 Complex Eigenvalues

In the previous section, we showed that the homogeneous system

$$\boldsymbol{x}'(t) = \boldsymbol{A}\boldsymbol{x}(t), \tag{5.40}$$

where $\boldsymbol{A}$ is an $n \times n$ constant matrix, has a solution of the form $\boldsymbol{x}(t) = \mathrm{e}^{rt}\boldsymbol{u}$ if and only if r is an eigenvalue of $\boldsymbol{A}$ and $\boldsymbol{u}$ is a corresponding eigenvector. In this section, we will show how to obtain two real vector solutions to system (5.40) when $\boldsymbol{A}$ is real and has a pair of complex conjugate eigenvalues $\alpha + \mathrm{i}\beta$ and $\alpha - \mathrm{i}\beta$.

Suppose $r_1 = \alpha + \mathrm{i}\beta$ (α and β are real numbers) is an eigenvalue of $\boldsymbol{A}$ with corresponding eigenvector $\boldsymbol{z} = \boldsymbol{a} + \mathrm{i}\boldsymbol{b}$, where $\boldsymbol{a}$ and $\boldsymbol{b}$ are real constant vectors. We first observe that the complex conjugate of $\boldsymbol{z}$, namely $\overline{\boldsymbol{z}} = \boldsymbol{a} - \mathrm{i}\boldsymbol{b}$, is an eigenvector associated with the eigenvalue $r_2 = \alpha - \mathrm{i}\beta$. To see this, note that taking the complex conjugate of $(\boldsymbol{A} - r_1\boldsymbol{I})\boldsymbol{z} = \boldsymbol{0}$ yields $(\boldsymbol{A} - \overline{r}_1\boldsymbol{I})\overline{\boldsymbol{z}} = \boldsymbol{0}$ because the conjugate of the product is the product of the conjugates and $\boldsymbol{A}$ and $\boldsymbol{I}$ have real entries ($\overline{\boldsymbol{A}} = \boldsymbol{A}$, $\overline{\boldsymbol{I}} = \boldsymbol{I}$). Since $r_2 - \overline{r}_1$, we see that $\overline{\boldsymbol{z}}$ is an eigenvector associated with r_2. Therefore, two linearly independent complex vector solutions to system (5.40) are

$$\boldsymbol{w}_1(t) = \mathrm{e}^{r_1 t}\boldsymbol{z} = \mathrm{e}^{(\alpha + \mathrm{i}\beta)t}(\boldsymbol{a} + \mathrm{i}\boldsymbol{b}), \tag{5.41}$$

$$\boldsymbol{w}_2(t) = \mathrm{e}^{r_2 t}\overline{\boldsymbol{z}} = \mathrm{e}^{(\alpha - \mathrm{i}\beta)t}(\boldsymbol{a} - \mathrm{i}\boldsymbol{b}). \tag{5.42}$$

With the aid of Euler's formula, we rewrite $\boldsymbol{w}_1(t)$ as

$$\boldsymbol{w}_1(t)=e^{\alpha t}(\cos\beta t+i\sin\beta t)(\boldsymbol{a}+i\boldsymbol{b})$$
$$=e^{\alpha t}\{(\cos\beta t\boldsymbol{a}-\sin\beta t\boldsymbol{b})+i(\sin\beta t\boldsymbol{a}+\cos\beta t\boldsymbol{b})\}.$$

We have thereby expressed $\boldsymbol{w}_1(t)$ in the form $\boldsymbol{w}_1(t)=\boldsymbol{x}_1+i\boldsymbol{x}_2(\mathrm{t})$, where $\boldsymbol{x}_1(t)$ and $\boldsymbol{x}_2(t)$ are the two real vector functions

$$\boldsymbol{x}_1(t):=e^{\alpha t}\cos\beta t\boldsymbol{a}-e^{\alpha t}\sin\beta t\boldsymbol{b},\tag{5.43}$$
$$\boldsymbol{x}_2(t):=e^{\alpha t}\sin\beta t\boldsymbol{a}+e^{\alpha t}\cos\beta t\boldsymbol{b}.\tag{5.44}$$

Since $\boldsymbol{w}_1(t)$ is a solution to system (5.40), then

$$\boldsymbol{w}_1'(t)=\boldsymbol{A}\boldsymbol{w}_1(t),$$
$$\boldsymbol{x}_1'+i\boldsymbol{x}_2'=\boldsymbol{A}\boldsymbol{x}_1+i\boldsymbol{A}\boldsymbol{x}_2.$$

Equating the real and imaginary parts yields

$$\boldsymbol{x}_1'(t)=\boldsymbol{A}\boldsymbol{x}_1(t) \text{ and } \boldsymbol{x}_2'(t)=\boldsymbol{A}\boldsymbol{x}_2(t).$$

Hence, $\boldsymbol{x}_1(t)$ and $\boldsymbol{x}_2(t)$ are real vector solutions to system (5.40) associated with the complex conjugate eigenvalues $\alpha\pm i\beta$. Because $\boldsymbol{a}$ and $\boldsymbol{b}$ are not both the zero vector, it can be shown that $\boldsymbol{x}_1(t)$ and $\boldsymbol{x}_2(t)$ are linearly independent vector functions on $(-\infty,+\infty)$.

Let's summarize our findings.

Complex Eigenvalues

If the real matrix $\boldsymbol{A}$ has complex conjugate eigenvalues $\alpha\pm i\beta$ with corresponding eigenvectors $\boldsymbol{a}\pm i\boldsymbol{b}$, then two linearly independent real vector solutions to $\boldsymbol{x}'(t)=\boldsymbol{A}\boldsymbol{x}(t)$ are

$$e^{\alpha t}\cos\beta t\boldsymbol{a}-e^{\alpha t}\sin\beta t\boldsymbol{b},\tag{5.45}$$
$$e^{\alpha t}\sin\beta t\boldsymbol{a}+e^{\alpha t}\cos\beta t\boldsymbol{b}.\tag{5.46}$$

Example 5.7 Find a general solution of

$$\boldsymbol{x}'(t)=\begin{bmatrix}-1 & 2\\ -1 & -3\end{bmatrix}\boldsymbol{x}(t).\tag{5.47}$$

Solution The characteristic equation for $\boldsymbol{A}$ is

$$|\boldsymbol{A}-r\boldsymbol{I}|=\begin{vmatrix}-1-r & 2\\ -1 & -3-r\end{vmatrix}=r^2+4r+5=0.$$

Hence, $\boldsymbol{A}$ has eigenvalues $r=-2\pm i$.

To find a general solution, we need only find an eigenvector associated with the eigenvalue $r=-2+i$. Substituting $r=-2+i$ into $(\boldsymbol{A}-r\boldsymbol{I})\boldsymbol{z}=\boldsymbol{0}$ gives

$$\begin{bmatrix} 1-\mathrm{i} & 2 \\ -1 & -1-\mathrm{i} \end{bmatrix}\begin{bmatrix} z_1 \\ z_2 \end{bmatrix}=\begin{bmatrix} 0 \\ 0 \end{bmatrix}.$$

The solutions can be expressed as $z_1=2s$ and $z_2=(-1+\mathrm{i})s$, with s arbitrary. Hence, the eigenvectors associated with $r=-2+\mathrm{i}$ are $\boldsymbol{z}=s\ \mathrm{col}(2,-1+\mathrm{i})$. Taking $s=1$ gives the eigenvector

$$\boldsymbol{z}=\begin{bmatrix} 2 \\ -1+\mathrm{i} \end{bmatrix}=\begin{bmatrix} 2 \\ -1 \end{bmatrix}+\mathrm{i}\begin{bmatrix} 0 \\ 1 \end{bmatrix}.$$

We have found that $\alpha=-2$, $\beta=1$, and $\boldsymbol{z}=\boldsymbol{a}+\mathrm{i}\boldsymbol{b}$ with $\boldsymbol{a}=\mathrm{col}(2,-1)$, and $\boldsymbol{b}=\mathrm{col}(0,1)$, so a general solution to (5.47) is

$$\boldsymbol{x}(t)=c_1\left\{\mathrm{e}^{-2t}\cos t\begin{bmatrix} 2 \\ -1 \end{bmatrix}-\mathrm{e}^{-2t}\sin t\begin{bmatrix} 0 \\ 1 \end{bmatrix}\right\}+c_2\left\{\mathrm{e}^{-2t}\sin t\begin{bmatrix} 2 \\ -1 \end{bmatrix}+\mathrm{e}^{-2t}\cos t\begin{bmatrix} 0 \\ 1 \end{bmatrix}\right\}$$

$$=c_1\begin{bmatrix} 2\mathrm{e}^{-2t}\cos t \\ -\mathrm{e}^{-2t}(\cos t+\sin t) \end{bmatrix}+c_2\begin{bmatrix} 2\mathrm{e}^{-2t}\sin t \\ \mathrm{e}^{-2t}(\cos t-\sin t) \end{bmatrix}. \blacklozenge \tag{5.48}$$

Complex eigenvalues occur in modeling coupled mass-spring systems. For example, the motion of the mass-spring system illustrated in Figure 5.3 is governed by the second-order system

$$\begin{cases} m_1 x_1''=-k_1 x_1+k_2(x_2-x_1), \\ m_2 x_2''=-k_2(x_2-x_1)-k_3 x_2, \end{cases} \tag{5.49}$$

where x_1 and x_2 represent the displacements of the masses m_1 and m_2 to the right of their equilibrium positions and k_1, k_2, k_3 are the spring constants of the three springs (see the discussion in Section 5.6). If we introduce the new variables $y_1:=x_1$, $y_2:=x_1'$, $y_3:=x_2$, $y_4:=x_2'$, then we can rewrite the system in the normal form

$$\boldsymbol{y}'(t)=\boldsymbol{A}\boldsymbol{y}(t)=\begin{bmatrix} 0 & 1 & 0 & 0 \\ -(k_1+k_2)/m_1 & 0 & k_2/m_1 & 0 \\ 0 & 0 & 0 & 1 \\ k_2/m_2 & 0 & -(k_2+k_3)/m_2 & 0 \end{bmatrix}\boldsymbol{y}(t). \tag{5.50}$$

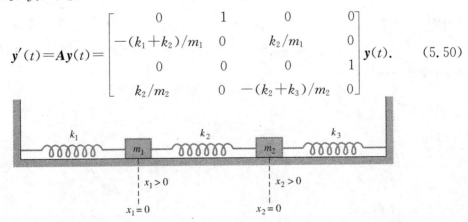

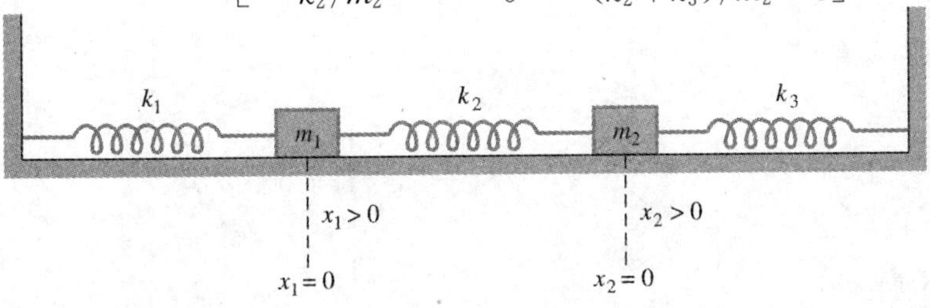

Figure 5.3 Coupled mass-spring system with fixed ends

For such a system, it turns out that $\boldsymbol{A}$ has only imaginary eigenvalues and they occur in complex conjugate pairs: $\pm i\beta_1$, $\pm i\beta_2$. Hence, any solution will consist of sums of sine and cosine functions. The frequencies of these functions

$$f_1 := \frac{\beta_1}{2\pi} \text{ and } f_2 := \frac{\beta_2}{2\pi}$$

are called the **normal** or **natural frequencies** of the system (β_1 and β_2 are the **angular frequencies** of the system).

In some engineering applications, the only information that is required about a particular device is a knowledge of its normal frequencies; one must ensure that they are far from the frequencies that occur naturally in the device's operating environment (so that no resonances will be excited).

Example 5.8 Determine the normal frequencies for the coupled mass-spring system governed by system (5.50) when $m_1 = m_2 = 1$ kg, $k_1 = 1$ N/m, $k_2 = 2$ N/m, and $k_3 = 3$ N/m.

Solution To find the eigenvalues of $\boldsymbol{A}$, we must solve the characteristic equation

$$|\boldsymbol{A} - r\boldsymbol{I}| = \begin{vmatrix} -r & 1 & 0 & 0 \\ -3 & -r & 2 & 0 \\ 0 & 0 & -r & 1 \\ 2 & 0 & -5 & -r \end{vmatrix} = r^4 + 8r^2 + 11 = 0.$$

From the quadratic formula, we find $r^2 = -4 \pm \sqrt{5}$, so the four eigenvalues of $\boldsymbol{A}$ are $\pm i\sqrt{4-\sqrt{5}}$ and $\pm i\sqrt{4+\sqrt{5}}$. Hence, the two normal frequencies for this system are $\frac{\sqrt{4-\sqrt{5}}}{2\pi} \approx 0.211$ and $\frac{\sqrt{4+\sqrt{5}}}{2\pi} \approx 0.397$ cycles per second. ◆

Exercises 5.5

In Problems 1-4, find a general solution of the system $\boldsymbol{x}'(t) = \boldsymbol{A}\boldsymbol{x}(t)$ for the given matrix $\boldsymbol{A}$.

1. $\boldsymbol{A} = \begin{bmatrix} 2 & -4 \\ 2 & -2 \end{bmatrix}$.

2. $\boldsymbol{A} = \begin{bmatrix} -2 & -5 \\ 1 & 2 \end{bmatrix}$.

3. $\boldsymbol{A} = \begin{bmatrix} 1 & 2 & -1 \\ 0 & 1 & 1 \\ 0 & -1 & 1 \end{bmatrix}$.

4. $\mathbf{A}=\begin{bmatrix} 5 & -5 & -5 \\ -1 & 4 & 2 \\ 3 & -5 & -3 \end{bmatrix}$.

In Problems 5-8, find a fundamental matrix for the system $\mathbf{x}'(t)=\mathbf{A}\mathbf{x}(t)$ *for the given matrix* $\mathbf{A}$.

5. $\mathbf{A}=\begin{bmatrix} -1 & -2 \\ 8 & -1 \end{bmatrix}$.

6. $\mathbf{A}=\begin{bmatrix} -2 & -2 \\ 4 & 2 \end{bmatrix}$.

7. $\mathbf{A}=\begin{bmatrix} 0 & 0 & 1 \\ 0 & 0 & -1 \\ 0 & 1 & 0 \end{bmatrix}$.

8. $\mathbf{A}=\begin{bmatrix} 0 & 1 & 0 & 0 \\ 1 & 0 & 0 & 0 \\ 0 & 0 & 0 & 1 \\ 0 & 0 & -13 & 4 \end{bmatrix}$.

In Problems 9-12, use a linear algebra software package to compute the required eigenvalues and eigenvectors for the given matrix $\mathbf{A}$ *and then give a fundamental matrix for the system* $\mathbf{x}'(t)=\mathbf{A}\mathbf{x}(t)$.

9. $\mathbf{A}=\begin{bmatrix} 0 & 1 & 1 \\ -1 & 0 & 1 \\ -1 & -1 & 0 \end{bmatrix}$.

10. $\mathbf{A}=\begin{bmatrix} 0 & 1 & 0 & 0 \\ 0 & 0 & 1 & 0 \\ 0 & 0 & 0 & 1 \\ 13 & -4 & -12 & 4 \end{bmatrix}$.

11. $\mathbf{A}=\begin{bmatrix} 0 & 1 & 0 & 0 \\ 0 & 0 & 1 & 0 \\ 0 & 0 & 0 & 1 \\ -2 & 2 & -3 & 2 \end{bmatrix}$.

12. $\boldsymbol{A}=\begin{bmatrix}1&0&0&0&0\\0&0&1&0&0\\0&1&0&0&0\\0&0&0&0&1\\0&0&0&-29&-4\end{bmatrix}$.

In Problems 13 and 14, find the solution to the given system that satisfies the given initial condition.

13. $\boldsymbol{x}'[t]=\begin{bmatrix}-3&-1\\2&-1\end{bmatrix}\boldsymbol{x}(t)$.

(1) $\boldsymbol{x}(0)=\begin{bmatrix}-1\\0\end{bmatrix}$. (2) $\boldsymbol{x}(\pi)=\begin{bmatrix}1\\-1\end{bmatrix}$.

(3) $\boldsymbol{x}(-2\pi)=\begin{bmatrix}2\\1\end{bmatrix}$. (4) $\boldsymbol{x}(\pi/2)=\begin{bmatrix}0\\1\end{bmatrix}$.

14. $\boldsymbol{x}'(t)=\begin{bmatrix}1&0&-1\\0&2&0\\1&0&1\end{bmatrix}\boldsymbol{x}(t)$.

(1) $\boldsymbol{x}(0)=\begin{bmatrix}-2\\2\\-1\end{bmatrix}$. (2) $\boldsymbol{x}(-\pi)=\begin{bmatrix}0\\1\\1\end{bmatrix}$.

15. Show that $\boldsymbol{x}_1(t)$ and $\boldsymbol{x}_2(t)$ given by equations (5.43) and (5.44) are linearly independent on $(-\infty,+\infty)$, provided $\beta\neq0$ and $\boldsymbol{a}$ and $\boldsymbol{b}$ are not both the zero vector.

16. Show that $\boldsymbol{x}_1(t)$ and $\boldsymbol{x}_2(t)$ given by equations (5.43) and (5.44) can be obtained as linear combinations of $\boldsymbol{w}_1(t)$ and $\boldsymbol{w}_2(t)$ given by equations (5.41) and (5.42). [Hint: Show that

$$\boldsymbol{x}_1(t)=\frac{\boldsymbol{w}_1(t)+\boldsymbol{w}_2(t)}{2},\quad \boldsymbol{x}_2(t)=\frac{\boldsymbol{w}_1(t)-\boldsymbol{w}_2(t)}{2\mathrm{i}}.\ ]$$

In Problems 17 and 18, use the results of Problem 42 in Exercises 5.4 to find a general solution to the given Cauchy-Euler system for $t>0$.

17. $t\boldsymbol{x}'(t)=\begin{bmatrix}-1&-1&0\\2&-1&1\\0&1&-1\end{bmatrix}\boldsymbol{x}(t)$.

18. $t\boldsymbol{x}'(t)=\begin{bmatrix}-1&-1\\9&-1\end{bmatrix}\boldsymbol{x}(t)$.

19. For the coupled mass-spring system governed by system (5.49), assume $m_1 = m_2 = 1$ kg, $k_1 = k_2 = 2$ N/m, and $k_3 = 3$ N/m. Determine the normal frequencies for this coupled mass-spring system.

20. For the coupled mass-spring system governed by system (5.49), assume $m_1 = m_2 = 1$ kg, $k_1 = k_2 = k_3 = 1$ N/m, and assume initially that $x_1(0) = 0$ m, $x_1'(0) = 0$ m/s, $x_2(0) = 2$ m, and $x_2'(0) = 0$ m/s. Using matrix algebra techniques, solve this initial value problem.

21. ***RLC* Network.** The currents in the *RLC* network given by the schematic diagram in Figure 5.4 are governed by the following equations:

$$\begin{cases} 4I_2'(t) + 52q_1(t) = 10, \\ 13I_3(t) + 52q_1(t) = 10, \\ I_1(t) = I_2(t) + I_3(t), \end{cases}$$

where $q_1(t)$ is the charge on the capacitor, $I_1(t) = q_1'(t)$, and initially $q_1(0) = 0$ coulombs and $I_1(0) = 0$ amps. Solve for the currents I_1, I_2, and I_3. [Hint: Differentiate the first two equations, eliminate I_1, and form a normal system with $x_1 = I_2$, $x_2 = I_2'$, and $x_3 = I_3$.]

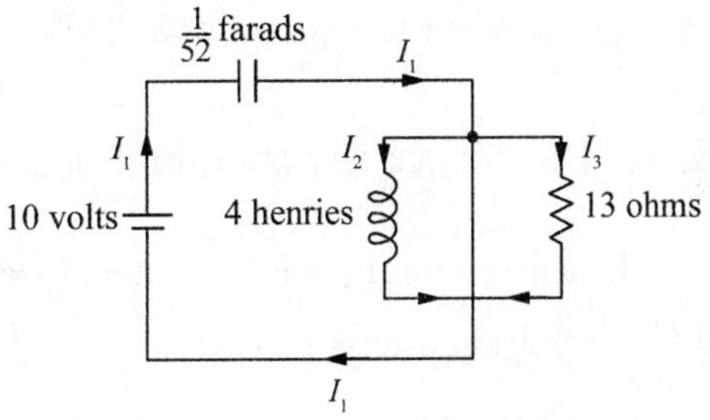

Figure 5.4 *RLC* network for Problem 21

22. ***RLC* Network.** The currents in the *RLC* network given by the schematic diagram in Figure 5.5 are governed by the following equations:

$$\begin{cases} 50I_1'(t) + 80I_2(t) = 160, \\ 50I_1'(t) + 800q_3(t) = 160, \\ I_1(t) = I_2(t) + I_3(t), \end{cases}$$

where $q_3(t)$ is the charge on the capacitor, $I_3(t) = q_3'(t)$, and initially $q_3(0) = 0.5$ coulombs and $I_3(0) = 0$ amp. Solve for the currents I_1, I_2, and I_3.

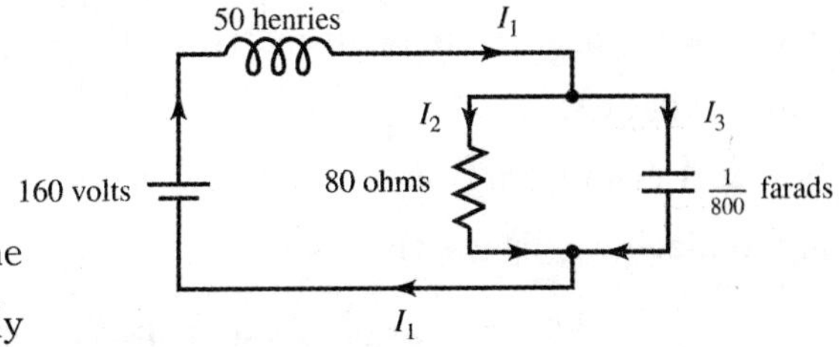

Figure 5.5 *RLC* network for Problem 22

[Hint: Differentiate the first two equations, use the third equation to eliminate I_3, and form a normal system with $x_1 = I_1$, $x_2 = I_1'$, and $x_3 = I_2$.]

23. **Stability.** We discussed the notion of stability and asymptotic stability for a linear system of the form $\mathbf{x}'(t) = \mathbf{A}\mathbf{x}(t)$. Assume that $\mathbf{A}$ has all distinct eigenvalues (real or complex).

(1)Show that the system is stable if and only if all the eigenvalues of $\boldsymbol{A}$ have nonpositive real part.

(2)Show that the system is asymptotically stable if and only if all the eigenvalues of A have negative real part.

24. (1)For Example 5.7, verify that

$$\boldsymbol{x}(t)=c_1\begin{bmatrix}-e^{-2t}\cos t+e^{-2t}\sin t\\ e^{-2t}\cos t\end{bmatrix}+c_2\begin{bmatrix}-e^{-2t}\sin t-e^{-2t}\cos t\\ e^{-2t}\sin t\end{bmatrix}$$

is another general solution to equation (5.47).

(2)How can the general solution of part (1) be directly obtained from the general solution derived in equation (5.48)?

5.6 The Matrix Exponential Function

In this chapter, we have developed various ways to extend techniques for scalar differential equations to systems. In this section, we take a substantial step further by showing that with the right notation, the formulas for solving normal systems with constant coefficents are identical to the formulas for solving first-order equations with constant coefficents. For example, we know that a general solution to the equation $x'(t)=ax(t)$, where a is a constant, $x(t)=ce^{at}$. Analogously, we show that a general solution to the normal system

$$\boldsymbol{x}'(t)=\boldsymbol{A}\boldsymbol{x}(t), \tag{5.51}$$

where $\boldsymbol{A}$ is an $n\times n$ constant matrix, $\boldsymbol{x}(t)=e^{\boldsymbol{A}t}\boldsymbol{c}$. Our first task is to define the matrix exponential $e^{\boldsymbol{A}t}$.

If $\boldsymbol{A}$ is an $n\times n$ constant matrix, we define $e^{\boldsymbol{A}t}$ by taking the series expansion for e^{at} and replacing a by $\boldsymbol{A}$; that is,

$$e^{\boldsymbol{A}t}:=\boldsymbol{I}+\boldsymbol{A}t+\boldsymbol{A}^2\frac{t^2}{2!}+\dots+\boldsymbol{A}^n\frac{t^n}{n!}+\dots \tag{5.52}$$

(Note that we also replace 1 by $\boldsymbol{I}$.) By the right-hand side of equation (5.52), we mean the $n\times n$ matrix whose elements are power series with coefficients given by the corresponding entries in the matrices $\boldsymbol{I}$, $\boldsymbol{A}$, $\boldsymbol{A}^2/2!$, ...

If $\boldsymbol{A}$ is a diagonal matrix, then the computation of $e^{\boldsymbol{A}t}$ is straightforward. For example, if

$$\boldsymbol{A}=\begin{bmatrix}-1 & 0\\ 0 & 2\end{bmatrix},$$

then

$$A^2=AA=\begin{bmatrix}1 & 0\\0 & 4\end{bmatrix},\quad A^3=\begin{bmatrix}-1 & 0\\0 & 8\end{bmatrix},\quad \ldots,\quad A^n=\begin{bmatrix}(-1)^n & 0\\0 & 2^n\end{bmatrix},$$

and so

$$e^{At}=\sum_{n=0}^{\infty}A^n\frac{t^n}{n!}=\begin{bmatrix}\sum_{n=0}^{\infty}(-1)^n\frac{t^n}{n!} & 0\\0 & \sum_{n=0}^{\infty}2^n\frac{t^n}{n!}\end{bmatrix}=\begin{bmatrix}e^{-t} & 0\\0 & e^{2t}\end{bmatrix}.$$

More generally, if A is an $n\times n$ diagonal matrix with r_1, r_2, ..., r_n down its main diagonal, then e^{At} is the diagonal matrix with e^{r_1t}, e^{r_2t}, ..., e^{r_nt} down its main diagonal (see Problem 26). If A is not a diagonal matrix, the computation of e^{At} is more involved. We deal with this important problem later in this section.

It can be shown that the series (5. 52) converges for all t and has many of the same properties as the scalar exponential e^{at}.

Properties of the matrix Exponential Function

Theorem 5. 7 Let A and B be $n\times n$ constant matrices and r, s, and t be real (or complex) numbers. Then,

(i) $e^{A0}=e^{0}=I$.

(ii) $e^{A(t+s)}=e^{At}e^{As}$.

(iii) $(e^{At})^{-1}=e^{-At}$.

(iv) $e^{(A+B)t}=e^{At}e^{Bt}$, provided that $AB=BA$.

(v) $e^{rIt}=e^{rt}I$.

Property (iii) has profound implications. First, it asserts that for any matrix A, the matrix e^{At} has an inverse for all t. Moreover, this inverse is obtained by simply replacing t by $-t$. [Note that (iii) follows from (i) and (ii) with $s=-t$.] In applying property (iv) (the law of exponents), one must exercise care because of the stipulation that the matrices A and B commute.

Another important property of the matrix exponential arises from the fact that we can differentiate the series in (5. 52) term by term. This gives

$$\frac{d}{dt}(e^{At}=\frac{d}{dt}\left(I+At+A^2\frac{t^2}{2}+\ldots+A^n\frac{t^n}{n!}+\ldots\right)$$

$$=A+A^2t+A^3\frac{t^2}{2}+\ldots+A^n\frac{t^{n-1}}{(n-1)!}+\ldots$$

$$=A\left[I+At+A^2\frac{t^2}{2}+\ldots+A^{n-1}\frac{t^{n-1}}{(n-1)!}+\ldots\right].$$

Hence,

$$\frac{\mathrm{d}}{\mathrm{d}t}(\mathrm{e}^{At}) = A\mathrm{e}^{At},$$

and so e^{At} is a solution to the matrix differential equation $X' = AX$. Since e^{At} is invertible [property (iii)], it follows that the columns of e^{At} are linearly independent solutions to system (5.51). Combining these facts we have the following.

e^{At} is a Fundamental Matrix

Theorem 5.8 If A is an $n \times n$ constant matrix, then the columns of the matrix exponential e^{At} form a fundamental solution set for the system $x'(t) = Ax(t)$. Therefore, e^{At} is a fundamental matrix for the system, and a general solution is $x(t) = \mathrm{e}^{At}c$.

If a fundamental matrix $X(t)$ for the system $x' = Ax$ has somehow been determined, it is easy to compute e^{At}, as the next theorem describes.

Relationship between Fundamental Matrices

Theorem 5.9 Let $X(t)$ and $Y(t)$ be two fundamental matrices for the same system $x' = Ax$. Then there exists a constant matrix C such that $Y(t) = X(t)C$ for all t. In particular,

$$\mathrm{e}^{At} = X(t)X(0)^{-1}. \tag{5.53}$$

Proof Since $X(t)$ is a fundamental matrix, every column of $Y(t)$ can be expressed as $X(t)c$ for a suitable constant vector c, so column-by-column we have

$$\begin{bmatrix} \\ Y(t) \\ \\ \end{bmatrix} = \begin{bmatrix} \\ X(t) \\ \\ \end{bmatrix} \begin{bmatrix} \vdots & \vdots & \cdots & \vdots \\ c_1 & c_2 & \cdots & c_n \\ \vdots & \vdots & \cdots & \vdots \end{bmatrix} = X(t)C.$$

If we choose $Y(t) = \mathrm{e}^{At} = x(t)C$, then (5.53) follows by setting $t=0$. ◆

If the $n \times n$ matrix A has n linearly independent eigenvectors u_i, then Theorem 5.8 provides us with $X(t)$ and formula (5.53) gives us

$$\mathrm{e}^{At} = [\mathrm{e}^{r_1 t}u_1, \mathrm{e}^{r_2 t}u_2, \ldots, \mathrm{e}^{r_n t}u_n][u_1, u_2, \ldots, u_n]^{-1}. \tag{5.54}$$

Are there any other ways that we can compute e^{At}? As we observed, if A is a diagonal matrix, then we simply exponentiate the diagonal elements of At to obtain e^{At}. Also, if A is a nilpotent matrix, that is, $A^k = 0$ for some positive integer k, then the series for e^{At} has only a finite number of terms—it "truncates"—since $A^k = A^{k+1} = \ldots = 0$. In such cases, e^{At} reduces to

$$e^{At}=I+At+\ldots+A^{k-1}\frac{t^{k-1}}{(k-1)!}+0+0+\ldots=I+At+\ldots+A^{k-1}\frac{t^{k-1}}{(k-1)!}.$$

Thus e^{At} can be calculated in finite terms if A is diagonal or nilpotent. Can we take this any further? Yes; a consequence of the Cayley-Hamilton theorem is that when the characteristic polynomial for A has the form $p(r)=(r_1-r)^n$, then $(r_1 I-A)^n=0=(-1)^n(A-r_1 I)^n$. So if A has only one (multiple) eigenvalue r_1, then $A-r_1 I$ is nilpotent, and we exploit that by writing $A=r_1 I+A-r_1 I$:

$$e^{At}=e^{r_1 It}e^{(A-r_1 I)t}=e^{r_1 t}\left[I+(A-r_1 I)t+\ldots+(A-r_1 I)^{n-1}\frac{t^{n-1}}{(n-1)!}\right]$$

Example 5.9 Find the fundamental matrix e^{At} for the system

$$x'=Ax,\text{ where }A=\begin{bmatrix}2 & 1 & 1\\ 1 & 2 & 1\\ -2 & -2 & -1\end{bmatrix}.$$

Solution We begin by computing the characteristic polynomial for A:

$$p(r)=|A-rI|=\begin{vmatrix}2-r & 1 & 1\\ 1 & 2-r & 1\\ -2 & -2 & -1-r\end{vmatrix}=-r^3+3r^2-3r+1=-(r-1)^3.$$

Thus, $r=1$ is an eigenvalue of A with multiplicity 3. By the Cayley-Hamilton theorem, $(A-I)^3=0$, and so

$$e^{At}=e^t e^{(A-I)t}=e^t\left[I+(A-I)t+(A-I)^2\frac{t^2}{2}\right].\tag{5.55}$$

Computing, we find

$$A-I=\begin{bmatrix}1 & 1 & 1\\ 1 & 1 & 1\\ -2 & -2 & -2\end{bmatrix}\text{ and }(A-I)^2=\begin{bmatrix}0 & 0 & 0\\ 0 & 0 & 0\\ 0 & 0 & 0\end{bmatrix}.$$

Substituting into (5.55) yields

$$e^{At}=e^t\begin{bmatrix}1 & 0 & 0\\ 0 & 1 & 0\\ 0 & 0 & 1\end{bmatrix}+te^t\begin{bmatrix}1 & 1 & 1\\ 1 & 1 & 1\\ -2 & -2 & -2\end{bmatrix}=\begin{bmatrix}e^t+te^t & te^t & te^t\\ te^t & e^t+te^t & te^t\\ -2te^t & -2te^t & e^t-2te^t\end{bmatrix}.\tag{5.56}$$

We are not through with nilpotency yet. What if we have a nonzero vector u, an exponent m, and a scalar r satisfying $(A-rI)^m u=0$, so that $A-rI$ is "nilpotent when restricted to u"? Such vectors are given a (predictable) name. ◆

Generalized Eigenvectors

Definition 5.3 Let $\boldsymbol{A}$ be a square matrix. A nonzero vector $\boldsymbol{u}$ satisfying

$$(\boldsymbol{A}-r\boldsymbol{I})^m\boldsymbol{u}=\boldsymbol{0} \tag{5.57}$$

for some scalar r and some positive integer m is called a **generalized eigenvector** associated with r. [Note that r must be an eigenvalue of $\boldsymbol{A}$, since the final nonzero vector in the list $\boldsymbol{u}$, $(\boldsymbol{A}-r\boldsymbol{I})\boldsymbol{u}$, $(\boldsymbol{A}-r\boldsymbol{I})^2\boldsymbol{u}$, ..., $(\boldsymbol{A}-r\boldsymbol{I})^{m-1}\boldsymbol{u}$ is a "regular" eigenvector.]

A valuable feature of generalized eigenvectors $\boldsymbol{u}$ is that we can compute $e^{\boldsymbol{A}t}\boldsymbol{u}$ in finite terms without knowing $e^{\boldsymbol{A}t}$, because

$$\begin{aligned} e^{\boldsymbol{A}t}\boldsymbol{u} &= e^{r\boldsymbol{I}t}e^{(\boldsymbol{A}-r\boldsymbol{I})t}\boldsymbol{u} \\ &= e^{rt}\left[\boldsymbol{I}\boldsymbol{u}+t(\boldsymbol{A}-r\boldsymbol{I})\boldsymbol{u}+\ldots+\frac{t^{m-1}}{(m-1)!}(\boldsymbol{A}-r\boldsymbol{I})^{m-1}\boldsymbol{u}+\frac{t^m}{m!}(\boldsymbol{A}-r\boldsymbol{I})^m\boldsymbol{u}+\ldots\right] \\ &= e^{rt}\left[\boldsymbol{u}+t(\boldsymbol{A}-r\boldsymbol{I})\boldsymbol{u}+\ldots+\frac{t^{m-1}}{(m-1)!}(\boldsymbol{A}-r\boldsymbol{I})^{m-1}\boldsymbol{u}+\boldsymbol{0}+\ldots\right]. \end{aligned} \tag{5.58}$$

Moreover, $e^{\boldsymbol{A}t}\boldsymbol{u}$ is a solution to the system $\boldsymbol{x}'=\boldsymbol{A}\boldsymbol{x}$ (recall Theorem 5.8). Hence, if we can find n generalized eigenvectors $\boldsymbol{u}_i$ for the $n\times n$ matrix $\boldsymbol{A}$ that are linearly independent, the corresponding solutions $\boldsymbol{x}_i(t)=e^{\boldsymbol{A}t}\boldsymbol{u}_i$ will form a fundamental solution set and can be assembled into a fundamental matrix $\boldsymbol{X}(t)$. (Since the $\boldsymbol{x}_i$'s are solutions that reduce to the linearly independent $\boldsymbol{u}_i$'s at $t=0$, they are always linearly independent.) Finally, we get the matrix exponential by applying (5.53) from Theorem 5.9:

$$e^{\boldsymbol{A}t}=\boldsymbol{X}(t)\boldsymbol{X}(0)^{-1}=[e^{\boldsymbol{A}t}\boldsymbol{u}_1, e^{\boldsymbol{A}t}\boldsymbol{u}_2, \ldots, e^{\boldsymbol{A}t}\boldsymbol{u}_n][\boldsymbol{u}_1, \boldsymbol{u}_2, \ldots, \boldsymbol{u}_n]^{-1}, \tag{5.59}$$

computed in a finite number of steps.

Of course, any (regular) eigenvector is a generalized eigenvector (corresponding to $m=1$), and if $\boldsymbol{A}$ has a full set of n linearly independent eigenvectors, then (5.59) is simply the representation of (5.54). But what if $\boldsymbol{A}$ is defective, that is, possesses fewer than n linearly independent eigenvectors? Luckily, the primary decomposition theorem in advanced linear algebra guarantees that when the characteristic polynomial of $\boldsymbol{A}$ is

$$p(r)=(r_1-r)^{m_1}(r_2-r)^{m_2}\ldots(r_k-r)^{m_k}, \tag{5.60}$$

where the r_i's are the distinct eigenvalues of $\boldsymbol{A}$ and m_i is the multiplicity of the eigenvalue r_i, then for each i there exist m_i linearly independent generalized eigenvectors satisfying

$$(\boldsymbol{A}-r_i\boldsymbol{I})^{m_i}\boldsymbol{u}=\boldsymbol{0} \tag{5.61}$$

Furthermore, the conglomeration of these $n = m_1 + m_2 + \ldots + m_k$ generalized eigenvectors is linearly independent.

We're home! The following scheme will always yield a fundamental solution set, for any square matrix **A**.

Solving $x' = Ax$

To obtain a fundamental solution set for $\boldsymbol{x}' = \boldsymbol{A}\boldsymbol{x}$ for any constant square matrix **A**:

(i) Compute the characteristic polynomial $p(r) = |\boldsymbol{A} - r\boldsymbol{I}|$.

(ii) Find the zeros of $p(r)$ and express it as $p(r) = (r_1 - r)^{m_1}(r_2 - r)^{m_2}\ldots(r_k - r)^{m_k}$, where $r_1, r_2, \ldots, r_k$ are the distinct zeros (i. e., eigenvalues) and $m_1, m_2, \ldots, m_k$ are their multiplicities.

(iii) For each eigenvalue r_i, find m_i linearly independent generalized eigenvectors by applying the Gauss-Jordan algorithm to the system $(\boldsymbol{A} - r_i\boldsymbol{I})^{m_i}\boldsymbol{u} = \boldsymbol{0}$.

(iv) Form $n = m_1 + m_2 + \ldots + m_k$ linearly independent solutions to $\boldsymbol{x}' = \boldsymbol{A}\boldsymbol{x}$ by computing

$$\boldsymbol{x}(t) = e^{\boldsymbol{A}t}\boldsymbol{u} = e^{rt}\left[\boldsymbol{u} + t(\boldsymbol{A} - r\boldsymbol{I})\boldsymbol{u} + \frac{t^2}{2!}(\boldsymbol{A} - r\boldsymbol{I})^2\boldsymbol{u} + \ldots\right]$$

for each generalized eigenvector $\boldsymbol{u}$ found in part (iii) and corresponding eigenvalue r.

If r has multiplicity m, this series terminates after m or fewer terms.

We can then, if desired, assemble the fundamental matrix $\boldsymbol{X}(t)$ from the n solutions and obtain the matrix exponential $e^{\boldsymbol{A}t}$ using (5.59).

Example 5.10 Find the fundamental matrix $e^{\boldsymbol{A}t}$ for the system

$$\boldsymbol{x}' = \boldsymbol{A}\boldsymbol{x}, \text{ where } \boldsymbol{A} = \begin{bmatrix} 1 & 0 & 0 \\ 1 & 3 & 0 \\ 0 & 1 & 1 \end{bmatrix}. \tag{5.62}$$

Solution We begin by finding the characteristic polynomial for **A**:

$$p(r) = |\boldsymbol{A} - r\boldsymbol{I}| = \begin{vmatrix} 1-r & 0 & 0 \\ 1 & 3-r & 0 \\ 0 & 1 & 1-r \end{vmatrix} = -(r-1)^2(r-3).$$

Hence, the eigenvalues of **A** are $r=1$ with multiplicity 2 and $r=3$ with multiplicity 1.

Since $r=1$ has multiplicity 2, we must determine two linearly independent associated generalized eigenvectors satisfying $(\boldsymbol{A} - \boldsymbol{I})^2\boldsymbol{u} = \boldsymbol{0}$. From

$$(\boldsymbol{A} - \boldsymbol{I})^2\boldsymbol{u} = \begin{bmatrix} 0 & 0 & 0 \\ 2 & 4 & 0 \\ 1 & 2 & 0 \end{bmatrix}\begin{bmatrix} u_1 \\ u_2 \\ u_3 \end{bmatrix} = \begin{bmatrix} 0 \\ 0 \\ 0 \end{bmatrix},$$

we find $u_2=s$, $u_1=2u_2=-2s$, and $u_3=v$, where s and v are arbitrary.

Taking $s=0$ and $v=1$, we obtain the generalized eigenvector $\boldsymbol{u}_1=\operatorname{col}(0,0,1)$. The corresponding solution to (5.62) is

$$\boldsymbol{x}_1(t)=\mathrm{e}^t\{\boldsymbol{u}_1+t(\boldsymbol{A}-\boldsymbol{I})\boldsymbol{u}_1\}=\mathrm{e}^t\begin{bmatrix}0\\0\\1\end{bmatrix}+t\mathrm{e}^t\begin{bmatrix}0&0&0\\1&2&0\\0&1&0\end{bmatrix}\begin{bmatrix}0\\0\\1\end{bmatrix}=\begin{bmatrix}0\\0\\\mathrm{e}^t\end{bmatrix} \tag{5.63}$$

($\boldsymbol{u}_1$ is, in fact, a regular eigenvector).

Next we take $s=1$ and $v=0$ to derive the second linearly independent generalized eigenvector $\boldsymbol{u}_2=\operatorname{col}(-2,1,0)$ and (linearly independent) solution

$$\boldsymbol{x}_2(t)=\mathrm{e}^{\boldsymbol{A}t}\boldsymbol{u}_2=\mathrm{e}^t\{\boldsymbol{u}_2+t(\boldsymbol{A}-\boldsymbol{I})\boldsymbol{u}_2\} \tag{5.64}$$

$$=\mathrm{e}^t\begin{bmatrix}-2\\1\\0\end{bmatrix}+t\mathrm{e}^t\begin{bmatrix}0&0&0\\1&2&0\\0&1&0\end{bmatrix}\begin{bmatrix}-2\\1\\0\end{bmatrix}$$

$$=\mathrm{e}^t\begin{bmatrix}-2\\1\\0\end{bmatrix}+t\mathrm{e}^t\begin{bmatrix}0\\0\\1\end{bmatrix}=\begin{bmatrix}-2\mathrm{e}^t\\\mathrm{e}^t\\t\mathrm{e}^t\end{bmatrix}.$$

For the eigenvalue $r=3$, we solve $(\boldsymbol{A}-3\boldsymbol{I})\boldsymbol{u}=\boldsymbol{0}$, that is,

$$\begin{bmatrix}-2&0&0\\1&0&0\\0&1&-2\end{bmatrix}\begin{bmatrix}u_1\\u_2\\u_3\end{bmatrix}=\begin{bmatrix}0\\0\\0\end{bmatrix},$$

to obtain the eigenvector $\boldsymbol{u}_3=\operatorname{col}(0,2,1)$. Hence, a third linearly independent solution to (5.62) is

$$\boldsymbol{x}_3(t)=\mathrm{e}^{3t}\boldsymbol{u}_3=\mathrm{e}^{3t}\begin{bmatrix}0\\2\\1\end{bmatrix}=\begin{bmatrix}0\\2\mathrm{e}^{3t}\\\mathrm{e}^{3t}\end{bmatrix}. \tag{5.65}$$

The matrix $\boldsymbol{X}(t)$ whose columns are the vectors $\boldsymbol{x}_1(t)$, $\boldsymbol{x}_2(t)$, and $\boldsymbol{x}_3(t)$,

$$\boldsymbol{X}(t)=\begin{bmatrix}0&-2\mathrm{e}^t&0\\0&\mathrm{e}^t&2\mathrm{e}^{3t}\\\mathrm{e}^t&t\mathrm{e}^t&\mathrm{e}^{3t}\end{bmatrix},$$

is a fundamental matrix for (5.62). Setting $t=0$ and then computing $\boldsymbol{X}^{-1}(0)$, we find

$$\boldsymbol{X}(0)=\begin{bmatrix}0 & -2 & 0\\ 0 & 1 & 2\\ 1 & 0 & 1\end{bmatrix} \text{ and } \boldsymbol{X}^{-1}(0)=\begin{bmatrix}-\frac{1}{4} & -\frac{1}{2} & 1\\ -\frac{1}{2} & 0 & 0\\ \frac{1}{4} & \frac{1}{2} & 0\end{bmatrix}.$$

It now follows from formula (5.53) that

$$e^{At}=\boldsymbol{X}(t)\boldsymbol{X}^{-1}(0)=\begin{bmatrix}0 & -2e^{t} & 0\\ 0 & e^{t} & 2e^{3t}\\ e^{t} & te^{t} & e^{3t}\end{bmatrix}\begin{bmatrix}-\frac{1}{4} & -\frac{1}{2} & 1\\ -\frac{1}{2} & 0 & 0\\ \frac{1}{4} & \frac{1}{2} & 0\end{bmatrix}$$

$$=\begin{bmatrix}e^{t} & 0 & 0\\ -\frac{1}{2}e^{t}+\frac{1}{2}e^{3t} & e^{3t} & 0\\ -\frac{1}{4}e^{t}-\frac{1}{2}te^{t}+\frac{1}{4}e^{3t} & -\frac{1}{2}e^{t}+\frac{1}{2}e^{3t} & e^{t}\end{bmatrix}. \quad \blacklozenge$$

Exercises 5.6

In Problems 1-6, (a) show that the given matrix $\mathbf{A}$ *satisfies* $(\mathbf{A}-r\boldsymbol{I})^{k}=\mathbf{0}$ for some number r and some positive integer k, and (b) use this fact to determine the matrix e^{At}. [Hint: *Compute the characteristic polynomial and use the Cayley-Hamilton theorem.*]

1. $\mathbf{A}=\begin{bmatrix}3 & -2\\ 0 & 3\end{bmatrix}$.

2. $\mathbf{A}=\begin{bmatrix}1 & -1\\ 1 & 3\end{bmatrix}$.

3. $\mathbf{A}=\begin{bmatrix}2 & 1 & -1\\ -3 & -1 & 1\\ 9 & 3 & -4\end{bmatrix}$.

4. $\mathbf{A}=\begin{bmatrix}2 & 1 & 3\\ 0 & 2 & -1\\ 0 & 0 & 2\end{bmatrix}$.

5. $\mathbf{A}=\begin{bmatrix}-2 & 0 & 0\\ 4 & -2 & 0\\ 1 & 0 & -2\end{bmatrix}$.

6. $\mathbf{A}=\begin{bmatrix}0 & 1 & 0\\ 0 & 0 & 1\\ -1 & -3 & -3\end{bmatrix}$.

In Problems 7-10, determine $e^{\mathbf{A}t}$ *by first finding a fundamental matrix* $\mathbf{X}(t)$ *for* $\mathbf{x}'=\mathbf{A}\mathbf{x}$ *and then using formula* (5.63).

7. $\mathbf{A}=\begin{bmatrix}0 & 1\\ -1 & 0\end{bmatrix}$.

8. $\mathbf{A}=\begin{bmatrix}1 & 1\\ 4 & 1\end{bmatrix}$.

9. $\mathbf{A}=\begin{bmatrix}0 & 1 & 0\\ 0 & 0 & 1\\ 1 & -1 & 1\end{bmatrix}$.

10. $\mathbf{A}=\begin{bmatrix}0 & 2 & 2\\ 2 & 0 & 2\\ 2 & 2 & 0\end{bmatrix}$.

In Problems 11 and 12, determine $e^{\mathbf{A}t}$ *by using generalized eigenvectors to find a fundamental matrix and then using formula* (5.63).

11. $\mathbf{A}=\begin{bmatrix}5 & -4 & 0\\ 1 & 0 & 2\\ 0 & 2 & 5\end{bmatrix}$.

12. $\mathbf{A}=\begin{bmatrix}1 & 1 & 1\\ 2 & 1 & -1\\ 0 & -1 & 1\end{bmatrix}$.

In Problems 13-16, use a linear algebra software package for help in determining $e^{\mathbf{A}t}$.

13. $\mathbf{A}=\begin{bmatrix}0 & 1 & 0 & 0 & 0\\ 0 & 0 & 1 & 0 & 0\\ 1 & -3 & 3 & 0 & 0\\ 0 & 0 & 0 & 0 & 1\\ 0 & 0 & 0 & -1 & 0\end{bmatrix}$.

14. $\mathbf{A}=\begin{bmatrix} 1 & 0 & 0 & 0 & 0 \\ 0 & 0 & 1 & 0 & 0 \\ 0 & -1 & -2 & 0 & 0 \\ 0 & 0 & 0 & 0 & 1 \\ 0 & 0 & 0 & -1 & 0 \end{bmatrix}$.

15. $\mathbf{A}=\begin{bmatrix} 0 & 1 & 0 & 0 & 0 \\ 0 & 0 & 1 & 0 & 0 \\ -1 & -3 & -3 & 0 & 0 \\ 0 & 0 & 0 & 0 & 1 \\ 0 & 0 & 0 & -4 & -4 \end{bmatrix}$.

16. $\mathbf{A}=\begin{bmatrix} -1 & 0 & 0 & 0 & 0 \\ 0 & 0 & 1 & 0 & 0 \\ 0 & -1 & -2 & 0 & 0 \\ 0 & 0 & 0 & 0 & 1 \\ 0 & 0 & 0 & -4 & -4 \end{bmatrix}$.

In Problems 17-20, use the generalized eigenvectors of $\mathbf{A}$ *to find a general solution to the system* $\mathbf{x}'(t)=\mathbf{A}\mathbf{x}(t)$, *where* $\mathbf{A}$ *is given.*

17. $\mathbf{A}=\begin{bmatrix} 0 & 1 & 0 \\ 0 & 0 & 1 \\ -2 & -5 & -4 \end{bmatrix}$.

18. $\mathbf{A}=\begin{bmatrix} 0 & 0 & 1 \\ 0 & 1 & 2 \\ 0 & 0 & 1 \end{bmatrix}$.

19. $\mathbf{A}=\begin{bmatrix} 1 & 0 & 1 & 2 \\ 1 & 1 & 2 & 1 \\ 0 & 0 & 2 & 0 \\ 0 & 0 & 1 & 1 \end{bmatrix}$.

20. $\mathbf{A}=\begin{bmatrix} -1 & -8 & 1 \\ -1 & -3 & 2 \\ -4 & -16 & 7 \end{bmatrix}$.

21. Use the results of Problem 5 to find the solution to the initial value problem

$$x'(t)=\begin{bmatrix}-2 & 0 & 0\\ 4 & -2 & 0\\ 1 & 0 & -2\end{bmatrix}x(t),\quad x(0)=\begin{bmatrix}1\\1\\-1\end{bmatrix}.$$

22. Use your answer to Problem 12 to find the solution to the initial value problem

$$x'(t)=\begin{bmatrix}1 & 1 & 1\\ 2 & 1 & -1\\ 0 & -1 & 1\end{bmatrix}x(t),\quad x(0)=\begin{bmatrix}-1\\0\\3\end{bmatrix}.$$

23. Use the results of Problem 3 and the variation of parameters formula to find the solution to the initial value problem

$$x'(t)=\begin{bmatrix}2 & 1 & -1\\ -3 & -1 & 1\\ 9 & 3 & -4\end{bmatrix}x(t)+\begin{bmatrix}0\\t\\0\end{bmatrix},\quad x(0)=\begin{bmatrix}0\\3\\0\end{bmatrix}.$$

24. Use your answer to Problem 9 and the variation of parameters formula to find the solution to the initial value problem

$$x'(t)=\begin{bmatrix}0 & 1 & 0\\ 0 & 0 & 1\\ 1 & -1 & 1\end{bmatrix}x(t)+\begin{bmatrix}0\\0\\t\end{bmatrix},\quad x(0)=\begin{bmatrix}1\\-1\\0\end{bmatrix}.$$

25. Let

$$A=\begin{bmatrix}5 & 2 & -4\\ 0 & 3 & 0\\ 4 & -5 & -5\end{bmatrix}.$$

(1) Find a general solution to $x'=Ax$.

(2) Determine which initial conditions $x(0)=x_0$ yield a solution $x(t)=\mathrm{col}(x_1(t),x_2(t),x_3(t))$ that remains bounded for all $t\geqslant 0$; that is, satisfies

$$\|x(t)\|:=\sqrt{x_1^2(t)+x_2^2(t)+x_3^2(t)}\leqslant M$$

for some constant M and all $t\geqslant 0$.

26. For the matrix A in Problem 25, solve the initial value problem

$$x'(t)=Ax(t)+\sin(2t)\begin{bmatrix}2\\0\\4\end{bmatrix},\quad x(0)=\begin{bmatrix}0\\1\\-1\end{bmatrix}.$$

5.7 Nonhomogeneous Linear Systems

The techniques discussed in Chapter 4 for finding a particular solution to the higher order nonhomogeneous linear equation have natural extensions to nonhomogeneous linear systems.

Undetermined Coefficients

The method of undetermined coefficients can be used to find a particular solution to the nonhomogeneous linear system

$$\boldsymbol{x}'(t)=\boldsymbol{A}\boldsymbol{x}(t)+\boldsymbol{f}(t)$$

when $\boldsymbol{A}$ is an $n\times n$ constant matrix and the entries of $\boldsymbol{f}(t)$ are polynomials, exponential functions, sines and cosines, or finite sums and products of these functions.

Example 5.11 Find a general solution of

$$\boldsymbol{x}'(t)=\boldsymbol{A}\boldsymbol{x}(t)+t\boldsymbol{g}, \tag{5.66}$$

where $\boldsymbol{A}=\begin{bmatrix}1 & -2 & 2\\ -2 & 1 & 2\\ 2 & 2 & 1\end{bmatrix}$ and $\boldsymbol{g}=\begin{bmatrix}-9\\ 0\\ -18\end{bmatrix}$.

Solution In Example 5.6 in Section 5.4, we found that a general solution to the corresponding homogeneous system $\boldsymbol{x}'=\boldsymbol{A}\boldsymbol{x}$ is

$$\boldsymbol{x}_h(t)=c_1\mathrm{e}^{3t}\begin{bmatrix}1\\ 0\\ 1\end{bmatrix}+c_2\mathrm{e}^{3t}\begin{bmatrix}-1\\ 1\\ 0\end{bmatrix}+c_3\mathrm{e}^{-3t}\begin{bmatrix}-1\\ -1\\ 1\end{bmatrix}. \tag{5.67}$$

Since the entries in $\boldsymbol{f}(t):=t\boldsymbol{g}$ are just linear functions of t, we are inclined to seek a particular solution of the form

$$\boldsymbol{x}_p(t)=t\boldsymbol{a}+\boldsymbol{b}=t\begin{bmatrix}a_1\\ a_2\\ a_3\end{bmatrix}+\begin{bmatrix}b_1\\ b_2\\ b_3\end{bmatrix},$$

where the constant vectors $\boldsymbol{a}$ and $\boldsymbol{b}$ are to be determined. Substituting this expression for $\boldsymbol{x}_p(t)$ into system (5.66) yields

$$\boldsymbol{a}=\boldsymbol{A}(t\boldsymbol{a}+\boldsymbol{b})+t\boldsymbol{g},$$

which can be written as

$$t(\boldsymbol{A}\boldsymbol{a}+\boldsymbol{g})+(\boldsymbol{A}\boldsymbol{b}-\boldsymbol{a})=\boldsymbol{0}.$$

Setting the "coefficients" of this vector polynomial equal to zero yields the following two systems:

$$\boldsymbol{A}\boldsymbol{a}=-\boldsymbol{g}, \tag{5.68}$$

$$\boldsymbol{Ab}=\boldsymbol{a}. \tag{5.69}$$

By Gaussian elimination or by using a linear algebra software package, we can solve system (5. 68) for $\boldsymbol{a}$ and we find $\boldsymbol{a}=\text{col}(5, 2, 4)$. Next we substitute for $\boldsymbol{a}$ in system (5. 69) and solve for $\boldsymbol{b}$ to obtain $\boldsymbol{b}=\text{col}(1, 0, 2)$. Hence, a particular solution for system (5. 66) is

$$\boldsymbol{x}_p(t)=t\boldsymbol{a}+\boldsymbol{b}=t\begin{bmatrix}5\\2\\4\end{bmatrix}+\begin{bmatrix}1\\0\\2\end{bmatrix}=\begin{bmatrix}5t+1\\2t\\4t+2\end{bmatrix}. \tag{5.70}$$

A general solution for system (5. 66) is $\boldsymbol{x}(t)=\boldsymbol{x}_h(t)+\boldsymbol{x}_p(t)$, where $\boldsymbol{x}_h(t)$ is given in (5. 67) and $\boldsymbol{x}_p(t)$ in (5. 70). ◆

In the preceding example, the nonhomogeneous term $\boldsymbol{f}(t)$ was a vector polynomial. If, instead, $\boldsymbol{f}(t)$ has the form

$$\boldsymbol{f}(t)=\text{col}(1, t, \sin t),$$

then, using the superposition principle, we would seek a particular solution of the form

$$\boldsymbol{x}_p(t)=t\boldsymbol{a}+\boldsymbol{b}+(\sin t)\boldsymbol{c}+(\cos t)\boldsymbol{d}.$$

Similarly, if

$$\boldsymbol{f}(t)=\text{col}(t, e^t, t^2),$$

we would take

$$\boldsymbol{x}_p(t)=t^2\boldsymbol{a}+t\boldsymbol{b}+\boldsymbol{c}+e^t\boldsymbol{d}.$$

Of course, we must modify our guess, should one of the terms be a solution to the corresponding homogeneous system. If this is the case, the annihilator method [equations (4. 81) and (4. 82) of Section 4. 3] would appear to suggest that for a nonhomogeneity $\boldsymbol{f}(t)$ of the form $e^{rt}t^m\boldsymbol{g}$, where r is an eigenvalue of $\boldsymbol{A}$, m is a nonnegative integer, and $\boldsymbol{g}$ is a constant vector, a particular solution of $\boldsymbol{x}'=\boldsymbol{Ax}+\boldsymbol{f}$ can be found in the form

$$\boldsymbol{x}_p(t)=e^{rt}\{t^{m+s}\boldsymbol{a}_{m+s}+t^{m+s-1}\boldsymbol{a}_{m+s-1}+\ldots+t\boldsymbol{a}_1+\boldsymbol{a}_0\},$$

for a suitable choice of s. We omit the details.

Variation of Parameters

In Section 4. 4, we discussed the method of variation of parameters for a general constant coefficient second-order linear equation. Simply put, the idea is that if a general solution to the homogeneous equation has the form $x_h(t)=c_1x_1(t)+c_2x_2(t)$, where $x_1(t)$ and $x_2(t)$ are linearly independent solutions to the homogeneous equation, then a particular solution to the nonhomogeneous equation would have the form $x_p(t)=v_1(t)x_1(t)+v_2(t)x_2(t)$, where $v_1(t)$ and $v_2(t)$ are certain functions of t. A similar idea can be used for systems.

Let $\boldsymbol{X}(t)$ be a fundamental matrix for the homogeneous system

$$\boldsymbol{x}'(t)=\boldsymbol{A}(t)\boldsymbol{x}(t), \tag{5.71}$$

where now the entries of $\boldsymbol{A}$ may be any continuous functions of t. Because a general solution to system (5.71) is given by $\boldsymbol{X}(t)\boldsymbol{c}$, where $\boldsymbol{c}$ is an $n\times1$ constant vector, we seek a particular solution to the nonhomogeneous system

$$\boldsymbol{x}'(t)=\boldsymbol{A}(t)\boldsymbol{x}(t)+\boldsymbol{f}(t) \tag{5.72}$$

of the form

$$\boldsymbol{x}_p(t)=\boldsymbol{X}(t)\boldsymbol{v}(t), \tag{5.73}$$

where $\boldsymbol{v}(t)=\text{col}(v_1(t), v_2(t), \ldots, v_n(t))$ is a vector function of t to be determined.

To derive a formula for $\boldsymbol{v}(t)$, we first differentiate equation (5.73) using the matrix version of the product rule to obtain

$$\boldsymbol{x}_p'(t)=\boldsymbol{X}(t)\boldsymbol{v}'(t)+\boldsymbol{X}'(t)\boldsymbol{v}(t).$$

Substituting the expressions for $\boldsymbol{x}_p(t)$ and $\boldsymbol{x}_p'(t)$ into system (5.72) yields

$$\boldsymbol{X}(t)\boldsymbol{v}'(t)+\boldsymbol{X}'(t)\boldsymbol{v}(t)=\boldsymbol{A}(t)\boldsymbol{X}(t)\boldsymbol{v}(t)+\boldsymbol{f}(t). \tag{5.74}$$

Since $\boldsymbol{X}(t)$ satisfies the matrix equation $\boldsymbol{X}'(t)=\boldsymbol{A}(t)\boldsymbol{X}(t)$, equation (5.74) becomes

$$\boldsymbol{X}\boldsymbol{v}'+\boldsymbol{A}\boldsymbol{X}\boldsymbol{v}=\boldsymbol{A}\boldsymbol{X}\boldsymbol{v}+\boldsymbol{f},$$

$$\boldsymbol{X}\boldsymbol{v}'=\boldsymbol{f}.$$

Multiplying both sides of the last equation by $\boldsymbol{X}^{-1}(t)$ [which exists since the columns of $\boldsymbol{X}(t)$ are linearly independent] gives

$$\boldsymbol{v}'(t)=\boldsymbol{X}^{-1}(t)\boldsymbol{f}(t).$$

Integrating, we obtain

$$\boldsymbol{v}(t)=\int \boldsymbol{X}^{-1}(t)\boldsymbol{f}(t)\mathrm{d}t.$$

Hence, a particular solution to system (5.72) is

$$\boldsymbol{x}_p(t)=\boldsymbol{X}(t)\boldsymbol{v}(t)=\boldsymbol{X}(t)\int \boldsymbol{X}^{-1}(t)\boldsymbol{f}(t)\mathrm{d}t. \tag{5.75}$$

Combining formula (5.75) with the solution $\boldsymbol{X}(t)\boldsymbol{c}$ to the homogeneous system yields the following general solution to system (5.72):

$$\boldsymbol{x}(t)=\boldsymbol{X}(t)\boldsymbol{c}+\boldsymbol{X}(t)\int \boldsymbol{X}^{-1}(t)\boldsymbol{f}(t)\mathrm{d}t. \tag{5.76}$$

The elegance of the derivation of the variation of parameters formula (5.75) for systems becomes evident when one compares it with the more lengthy derivations for the scalar case in Section 4.4.

Given an initial value problem of the form

$$\boldsymbol{x}'(t)=\boldsymbol{A}(t)\boldsymbol{x}(t)+\boldsymbol{f}(t),\quad \boldsymbol{x}(t_0)=\boldsymbol{x}_0,\tag{5.77}$$

we can use the initial condition $\boldsymbol{x}(t_0)=\boldsymbol{x}_0$ to solve for $\boldsymbol{c}$ in formula (5.76). Expressing $\boldsymbol{x}(t)$ using a definite integral, we have

$$\boldsymbol{x}(t)=\boldsymbol{X}(t)\boldsymbol{c}+\boldsymbol{X}(t)\int_{t_0}^{t}\boldsymbol{X}^{-1}(s)\boldsymbol{f}(s)\mathrm{d}s.$$

Using the initial condition $\boldsymbol{x}(t_0)=\boldsymbol{x}_0$, we find

$$\boldsymbol{x}_0=\boldsymbol{x}(t_0)=\boldsymbol{X}(t_0)\boldsymbol{c}+\boldsymbol{X}(t_0)\int_{t_0}^{t_0}\boldsymbol{X}^{-1}(s)\boldsymbol{f}(s)\mathrm{d}s=\boldsymbol{X}(t_0)\boldsymbol{c}.$$

Solving for $\boldsymbol{c}$, we have $\boldsymbol{c}=\boldsymbol{X}^{-1}(t_0)\boldsymbol{x}_0$. Thus, the solution to (5.77) is given by the formula

$$\boldsymbol{x}(t)=\boldsymbol{X}(t)\boldsymbol{X}^{-1}(t_0)\boldsymbol{x}_0+\boldsymbol{X}(t)\int_{t_0}^{t}\boldsymbol{X}^{-1}(s)\boldsymbol{f}(s)\mathrm{d}s.\tag{5.78}$$

To apply the variation of parameters formulas, we first must determine a fundamental matrix $\boldsymbol{X}(t)$ for the homogeneous system. In the case when the coefficient matrix $\boldsymbol{A}$ is constant, we have discussed methods for finding $\boldsymbol{X}(t)$. However, if the entries of $\boldsymbol{A}$ depend on t, the determination of $\boldsymbol{X}(t)$ may be extremely difficult (entailing, perhaps, a matrix power series).

Example 5.12 Find the solution to the initial value problem

$$\boldsymbol{x}'(t)=\begin{bmatrix}2 & -3\\ 1 & -2\end{bmatrix}\boldsymbol{x}(t)+\begin{bmatrix}\mathrm{e}^{2t}\\ 1\end{bmatrix},\quad \boldsymbol{x}(0)=\begin{bmatrix}-1\\ 0\end{bmatrix}.\tag{5.79}$$

Solution In Example 5.4 in Section 5.4, we found two linearly independent solutions to the corresponding homogeneous system; namely,

$$\boldsymbol{x}_1(t)=\begin{bmatrix}3\mathrm{e}^{t}\\ \mathrm{e}^{t}\end{bmatrix}\text{ and }\boldsymbol{x}_2(t)=\begin{bmatrix}\mathrm{e}^{-t}\\ \mathrm{e}^{-t}\end{bmatrix}.$$

Hence, a fundamental matrix for the homogeneous system is

$$\boldsymbol{X}(t)=\begin{bmatrix}3\mathrm{e}^{t} & \mathrm{e}^{-t}\\ \mathrm{e}^{t} & \mathrm{e}^{-t}\end{bmatrix}.$$

Although the solution to (5.79) can be found via the method of undetermined coefficients, we shall find it directly from formula (5.78). For this purpose, we need $\boldsymbol{X}^{-1}(t)$. One way to obtain $\boldsymbol{X}^{-1}(t)$ is to form the augmented matrix

$$\left[\begin{array}{cc:cc} 3e^{t} & e^{-t} & 1 & 0 \\ e^{t} & e^{-t} & 0 & 1 \end{array}\right]$$

and row-reduce this matrix to the matrix $[\boldsymbol{I} \vdots \boldsymbol{X}^{-1}(t)]$. This gives

$$\boldsymbol{X}^{-1}(t)=\begin{bmatrix} \frac{1}{2}e^{-t} & -\frac{1}{2}e^{-t} \\ -\frac{1}{2}e^{t} & \frac{3}{2}e^{t} \end{bmatrix}.$$

Substituting into formula (5.78), we obtain the solution

$$\boldsymbol{x}(t)=\begin{bmatrix} 3e^{t} & e^{-t} \\ e^{t} & e^{-t} \end{bmatrix}\begin{bmatrix} \frac{1}{2} & -\frac{1}{2} \\ -\frac{1}{2} & \frac{3}{2} \end{bmatrix}\begin{bmatrix} -1 \\ 0 \end{bmatrix}+\begin{bmatrix} 3e^{t} & e^{-t} \\ e^{t} & e^{-t} \end{bmatrix}\int_0^t \begin{bmatrix} \frac{1}{2}e^{-s} & -\frac{1}{2}e^{-s} \\ -\frac{1}{2}e^{s} & \frac{3}{2}e^{s} \end{bmatrix}\begin{bmatrix} e^{2s} \\ 1 \end{bmatrix}ds$$

$$=\begin{bmatrix} -\frac{3}{2}e^{t}+\frac{1}{2}e^{-t} \\ -\frac{1}{2}e^{t}+\frac{1}{2}e^{-t} \end{bmatrix}+\begin{bmatrix} 3e^{t} & e^{-t} \\ e^{t} & e^{-t} \end{bmatrix}\int_0^t \begin{bmatrix} \frac{1}{2}e^{s}-\frac{1}{2}e^{-s} \\ -\frac{1}{2}e^{3s}+\frac{3}{2}e^{s} \end{bmatrix}ds$$

$$=\begin{bmatrix} -\frac{3}{2}e^{t}+\frac{1}{2}e^{-t} \\ -\frac{1}{2}e^{t}+\frac{1}{2}e^{-t} \end{bmatrix}+\begin{bmatrix} 3e^{t} & e^{-t} \\ e^{t} & e^{-t} \end{bmatrix}\begin{bmatrix} \frac{1}{2}e^{t}+\frac{1}{2}e^{-t}-1 \\ \frac{3}{2}e^{t}-\frac{1}{6}e^{3t}-\frac{4}{3} \end{bmatrix}$$

$$=\begin{bmatrix} -\frac{9}{2}e^{t}-\frac{5}{6}e^{-t}+\frac{4}{3}e^{2t}+3 \\ -\frac{3}{2}e^{t}-\frac{5}{6}e^{-t}+\frac{1}{3}e^{2t}+2 \end{bmatrix}. \blacklozenge$$

Exercises 5.7

In Problems 1-6, use the method of undetermined coefficients to find a general solution to the system $\mathbf{x}'(t)=\mathbf{A}\mathbf{x}(t)+\boldsymbol{f}(t)$, *where* $\mathbf{A}$ *and* $\boldsymbol{f}(t)$ *are given.*

1. $\mathbf{A}=\begin{bmatrix} 6 & 1 \\ 4 & 3 \end{bmatrix}$, $\boldsymbol{f}(t)=\begin{bmatrix} -11 \\ -5 \end{bmatrix}$.

2. $\mathbf{A}=\begin{bmatrix} 1 & 1 \\ 4 & 1 \end{bmatrix}$, $\boldsymbol{f}(t)=\begin{bmatrix} -t-1 \\ -4t-2 \end{bmatrix}$.

3. $\mathbf{A}=\begin{bmatrix} 1 & -2 & 2 \\ -2 & 1 & 2 \\ 2 & 2 & 1 \end{bmatrix}$, $\boldsymbol{f}(t)=\begin{bmatrix} 2e^{t} \\ 4e^{t} \\ -2e^{t} \end{bmatrix}$.

4. $\mathbf{A}=\begin{bmatrix} 2 & 2 \\ 2 & 2 \end{bmatrix}$, $\boldsymbol{f}(t)=\begin{bmatrix} -4\cos t \\ -\sin t \end{bmatrix}$.

5. $\mathbf{A}=\begin{bmatrix}0 & -1 & 0\\ -1 & 0 & 0\\ 0 & 0 & 1\end{bmatrix}$, $\boldsymbol{f}(t)=\begin{bmatrix}e^{2t}\\ \sin t\\ t\end{bmatrix}$.

6. $\mathbf{A}=\begin{bmatrix}1 & 1\\ 0 & 2\end{bmatrix}$, $\boldsymbol{f}(t)=e^{-2t}\begin{bmatrix}t\\ 3\end{bmatrix}$.

In Problems 7-10, use the method of undetermined coefficients to determine only the form of a particular solution for the system $\boldsymbol{x}'(t)=\mathbf{A}\boldsymbol{x}(t)+\boldsymbol{f}(t)$, *where* $\mathbf{A}$ *and* $\boldsymbol{f}(t)$ *are given.*

7. $\mathbf{A}=\begin{bmatrix}0 & 1\\ 2 & 0\end{bmatrix}$, $\boldsymbol{f}(t)=\begin{bmatrix}\sin 3t\\ t\end{bmatrix}$.

8. $\mathbf{A}=\begin{bmatrix}-1 & 0\\ 2 & 2\end{bmatrix}$, $\boldsymbol{f}(t)\begin{bmatrix}t^2\\ t+1\end{bmatrix}$.

9. $\mathbf{A}=\begin{bmatrix}0 & -1 & 0\\ -1 & 0 & 1\\ 0 & 0 & 1\end{bmatrix}$, $\boldsymbol{f}(t)=\begin{bmatrix}e^{2t}\\ \sin t\\ t\end{bmatrix}$.

10. $\mathbf{A}=\begin{bmatrix}2 & -1\\ 1 & 5\end{bmatrix}$, $\boldsymbol{f}(t)=\begin{bmatrix}te^{-t}\\ 3e^{-t}\end{bmatrix}$

In Problems 11-16, use the variation of parameters formula (5.76) to find a general solution of the system $\boldsymbol{x}'(t)=\mathbf{A}\boldsymbol{x}(t)+\boldsymbol{f}(t)$, *where* $\mathbf{A}$ *and* $\boldsymbol{f}(t)$ *are given.*

11. $\mathbf{A}=\begin{bmatrix}0 & 1\\ -1 & 0\end{bmatrix}$, $\boldsymbol{f}(t)=\begin{bmatrix}1\\ 0\end{bmatrix}$.

12. $\mathbf{A}=\begin{bmatrix}1 & 2\\ 3 & 2\end{bmatrix}$, $\boldsymbol{f}(t)=\begin{bmatrix}1\\ -1\end{bmatrix}$.

13. $\mathbf{A}=\begin{bmatrix}2 & 1\\ -3 & -2\end{bmatrix}$, $\boldsymbol{f}(t)=\begin{bmatrix}2e^{t}\\ 4e^{t}\end{bmatrix}$.

14. $\mathbf{A}=\begin{bmatrix}0 & -1\\ 1 & 0\end{bmatrix}$, $\boldsymbol{f}(t)=\begin{bmatrix}t^2\\ 1\end{bmatrix}$.

15. $\mathbf{A}=\begin{bmatrix}-4 & 2\\ 2 & -1\end{bmatrix}$, $\boldsymbol{f}(t)=\begin{bmatrix}t^{-1}\\ 4+2t^{-1}\end{bmatrix}$.

16. $\mathbf{A}=\begin{bmatrix}0 & 1\\ -1 & 0\end{bmatrix}$, $\boldsymbol{f}(t)=\begin{bmatrix}8\sin t\\ 0\end{bmatrix}$.

In Problems 17-20, use the variation of parameters formula (5.76) and possibly a linear algebra software package to find a general solution of the system $\boldsymbol{x}'(t)=\boldsymbol{A}\boldsymbol{x}(t)+\boldsymbol{f}(t)$, *where* $\boldsymbol{A}$ *and* $\boldsymbol{f}(t)$ *are given.*

17. $\boldsymbol{A}=\begin{bmatrix}0 & 1 & 1\\ 1 & 0 & 1\\ 1 & 1 & 0\end{bmatrix}$, $\boldsymbol{f}(t)=\begin{bmatrix}3e^t\\ -e^t\\ -e^t\end{bmatrix}$.

18. $\boldsymbol{A}=\begin{bmatrix}1 & -1 & 1\\ 0 & 0 & 1\\ 0 & -1 & 2\end{bmatrix}$, $\boldsymbol{f}(t)=\begin{bmatrix}0\\ e^t\\ e^t\end{bmatrix}$.

19. $\boldsymbol{A}=\begin{bmatrix}0 & 1 & 0 & 0\\ -1 & 0 & 0 & 0\\ 0 & 0 & 0 & 1\\ 0 & 0 & 1 & 0\end{bmatrix}$, $\boldsymbol{f}(t)=\begin{bmatrix}t\\ 0\\ e^{-t}\\ t\end{bmatrix}$.

20. $\boldsymbol{A}=\begin{bmatrix}0 & 1 & 0 & 0\\ 0 & 0 & 1 & 0\\ 0 & 0 & 0 & 1\\ 8 & -4 & -2 & -1\end{bmatrix}$, $\boldsymbol{f}(t)=\begin{bmatrix}e^t\\ 0\\ 1\\ 0\end{bmatrix}$.

In Problems 21 and 22, find the solution to the given system that satisfies the given initial condition.

21. $\boldsymbol{x}'(t)=\begin{bmatrix}0 & 2\\ -1 & 3\end{bmatrix}\boldsymbol{x}(t)+\begin{bmatrix}e^t\\ -e^t\end{bmatrix}$.

(1) $\boldsymbol{x}(0)=\begin{bmatrix}5\\ 4\end{bmatrix}$. (2) $\boldsymbol{x}(1)=\begin{bmatrix}0\\ 1\end{bmatrix}$.

(3) $\boldsymbol{x}(5)=\begin{bmatrix}1\\ 0\end{bmatrix}$. (4) $\boldsymbol{x}(-1)=\begin{bmatrix}-4\\ 5\end{bmatrix}$.

22. $\boldsymbol{x}'(t)=\begin{bmatrix}0 & 2\\ 4 & -2\end{bmatrix}\boldsymbol{x}(t)+\begin{bmatrix}4t\\ -4t-2\end{bmatrix}$.

(1) $\boldsymbol{x}(0)=\begin{bmatrix}4\\ -5\end{bmatrix}$. (2) $\boldsymbol{x}(2)=\begin{bmatrix}1\\ 1\end{bmatrix}$.

23. Using matrix algebra techniques and the method of undetermined coefficients, find a general solution for

$$x''(t)+y'(t)-x(t)+y(t)=-1,$$
$$x'(t)+y'(t)-x(t)=t^2.$$

24. Using matrix algebra techniques and the method of undetermined coefficients, solve the initial value problem

$$x'(t)-2y(t)=4t, \quad x(0)=4;$$
$$y'(t)+2y(t)-4x(t)=-4t-2, \quad y(0)=-5.$$

25. To find a general solution to the system

$$\mathbf{x}'(t)=\begin{bmatrix} 0 & 1 \\ -2 & 3 \end{bmatrix}\mathbf{x}(t)+\mathbf{f}(t), \text{ where } \mathbf{f}(t)=\begin{bmatrix} e^t \\ 0 \end{bmatrix},$$

proceed as follows:

(1) Find a fundamental solution set for the corresponding homogeneous system.

(2) The obvious choice for a particular solution would be a vector function of the form $\mathbf{x}_p(t)=e^t\mathbf{a}$; however, the homogeneous system has a solution of this form. The next choice would be $\mathbf{x}_p(t)=te^t\mathbf{a}$. Show that this choice t does not work.

(3) For systems, multiplying by t is not always sufficient. The proper guess is

$$\mathbf{x}_p(t)=te^t\mathbf{a}+e^t\mathbf{b}.$$

Use this guess to find a particular solution of the given system.

(4) Use the results of parts (1) and (3) to find a general solution of the given system.

26. For the system of Problem 25, we found that a proper guess for a particular solution is $\mathbf{x}_p(t)=te^t\mathbf{a}+e^t\mathbf{b}$. In some cases, $\mathbf{a}$ or $\mathbf{b}$ may be zero.

(1) Find a particular solution for the system of Problem 25 if $\mathbf{f}(t)=\text{col}(3e^t, 6e^t)$.

(2) Find a particular solution for the system of Problem 25 if $\mathbf{f}(t)=\text{col}(e^t, e^t)$.

27. Find a general solution for the system

$$\mathbf{x}'(t)=\begin{bmatrix} 0 & 1 & 1 \\ 1 & 0 & 1 \\ 1 & 1 & 0 \end{bmatrix}\mathbf{x}(t)+\begin{bmatrix} -1 \\ -1-e^{-t} \\ -2e^{-t} \end{bmatrix}.$$

[Hint: Try $\mathbf{x}_p(t)=e^{-t}\mathbf{a}+te^{-t}\mathbf{b}+\mathbf{c}$.]

28. Find a particular solution for the system

$$\mathbf{x}'(t)=\begin{bmatrix} 1 & -1 \\ -1 & 1 \end{bmatrix}\mathbf{x}(t)+\begin{bmatrix} -3 \\ 1 \end{bmatrix}.$$

[Hint: Try $\mathbf{x}_p(t)=t\mathbf{a}+\mathbf{b}$.]

In Problems 29 and 30, find a general solution to the given Cauchy-Euler system for $t>0$. Remember to express the system in the form $\mathbf{x}'(t)=\mathbf{A}(t)\mathbf{x}(t)+\mathbf{f}(t)$ before using the variation of parameters formula.

29. $t\boldsymbol{x}'(t)=\begin{bmatrix}2 & -1\\ 3 & -2\end{bmatrix}\boldsymbol{x}(t)+\begin{bmatrix}t^{-1}\\ 1\end{bmatrix}.$

30. $t\boldsymbol{x}'(t)=\begin{bmatrix}4 & -3\\ 8 & -6\end{bmatrix}\boldsymbol{x}(t)+\begin{bmatrix}t\\ 2t\end{bmatrix}.$

31. Use the variation of parameters formula (5. 75) to derive a formula for a particular solution y_p to the scalar equation $y''+p(t)y'+q(t)y=g(t)$ in terms of two linearly independent solutions $y_1(t)$, $y_2(t)$ of the corresponding homogeneous equation. Show that your answer agrees with the formulas derived in Section 4. 4. [Hint: First write the scalar equation in system form.]

32. Let $\boldsymbol{U}(t)$ be the invertible 2×2 matrix

$$\boldsymbol{U}(t):=\begin{bmatrix}a(t) & b(t)\\ c(t) & d(t)\end{bmatrix}.$$

Show that

$$\boldsymbol{U}^{-1}(t)=\frac{1}{[a(t)d(t)-b(t)c(t)]}\begin{bmatrix}d(t) & -b(t)\\ -c(t) & a(t)\end{bmatrix}.$$

33. ***RL* Network.** The currents in the *RL* network given by the schematic diagram in Figure 5. 6 are governed by the following equations:

$$2I_1'(t)+90I_2(t)=9,$$
$$I_3'(t)+30I_4(t)-90I_2(t)=0,$$
$$60I_5(t)-30I_4(t)=0,$$
$$I_1(t)=I_2(t)+I_3(t),$$
$$I_3(t)=I_4(t)+I_5(t).$$

Assume the currents are initially zero. Solve for the five currents $I_1, I_2, \ldots, I_5$. [Hint: Eliminate all unknowns except I_2 and I_5, and form a normal system with $x_1=I_2$ and $x_2=I_5$.]

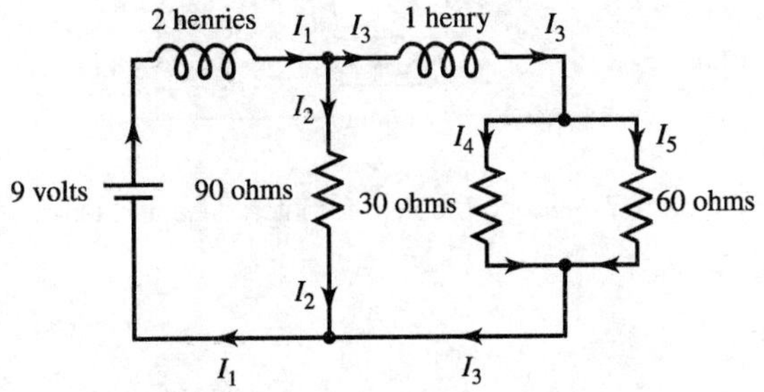

Figure 5.6 *RL* network for Problem 33

34. **Conventional Combat Model.** A simplistic model of a pair of conventional forces in combat yields the following system:

$$\mathbf{x}'=\begin{bmatrix}-a & -b\\ -c & -d\end{bmatrix}\mathbf{x}+\begin{bmatrix}p\\ q\end{bmatrix},$$

where $\mathbf{x}=\operatorname{col}(x_1, x_2)$. The variables $x_1(t)$ and $x_2(t)$ represent the strengths of opposing forces at time t. The terms $-ax_1$ and $-dx_2$ represent the operational loss rates, and the terms $-bx_2$ and $-cx_1$ represent the combat loss rates for the troops x_1 and x_2, respectively. The constants p and q represent the respective rates of reinforcement. Let $a=1$, $b=4$, $c=3$, $d=2$, and $p=q=5$. By solving the appropriate initial value problem, determine which forces will win if

(1) $x_1(0)=20$, $x_2(0)=20$.

(2) $x_1(0)=21$, $x_2(0)=20$.

(3) $x_1(0)=20$, $x_2(0)=21$.

35. **Mixing Problem.** Two tanks A and B, each holding 50 L of liquid, are interconnected by pipes. The liquid flows from tank A into tank B at a rate of 4 L/min and from B into A at a rate of 1 L/min (see Figure 5.7). The liquid inside each tank is kept well stirred. A brine solution that has a concentration of 0.2 kg/L of salt flows into tank A at a rate of 4 L/min. A brine solution that has a concentration of 0.1 kg/L of salt flows into tank B at a rate of 1 L/min. The solutions flow out of the system from both tanks—from tank A at 1 L/min and from tank B at 4 L/min. If, initially, tank A contains pure water and tank B contains 0.5 kg of salt, determine the mass of salt in each tank at time $t\geqslant 0$. After several minutes have elapsed, which tank has the higher concentration of salt? What is its limiting concentration?

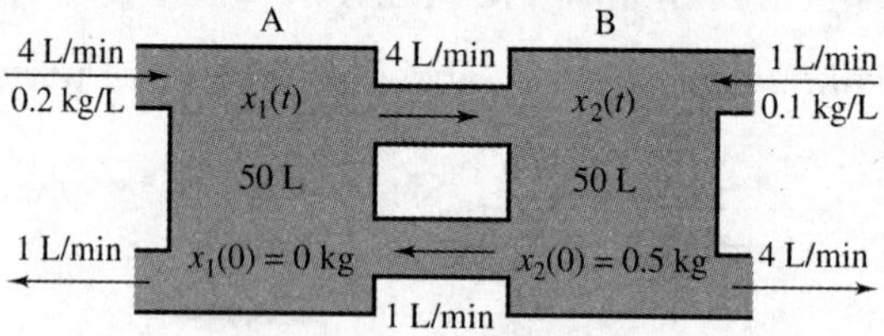

Figure 5.7 Mixing problem for interconnected tanks

Chapter Summary

In this chapter, we discussed the theory of linear systems in normal form and presented methods for solving such systems. The theory and methods are natural extensions of the development for second-order and higher-order linear equations. The important properties and techniques are as follows.

Homogeneous Normal Systems

$$\mathbf{x}'(t)=\mathbf{A}(t)\mathbf{x}(t).$$

The $n\times n$ matrix function $\mathbf{A}(t)$ is assumed to be continuous on an interval I.

Fundamental Solution Set: $\{\mathbf{x}_1, \mathbf{x}_2, \ldots, \mathbf{x}_n\}$. The n vector solutions $\mathbf{x}_1(t)$, $\mathbf{x}_2(t)$, …, $\mathbf{x}_n(t)$ of the homogeneous system on the interval I form a **fundamental solution set**, provided they are linearly independent on I or, equivalently, their **Wronskian**

$$W[\mathbf{x}_1, \mathbf{x}_2, \ldots, \mathbf{x}_n](t):=\det[\mathbf{x}_1, \mathbf{x}_2, \ldots, \mathbf{x}_n]=\begin{vmatrix} x_{11}(t) & x_{12}(t) & \ldots & x_{1n}(t) \\ x_{21}(t) & x_{22}(t) & \ldots & x_{2n}(t) \\ \vdots & \vdots & & \vdots \\ x_{n1}(t) & x_{n2}(t) & \ldots & x_{nn}(t) \end{vmatrix}$$

is never zero on I.

Fundamental Matrix: $\mathbf{X}(t)$. An $n\times n$ matrix function $\mathbf{X}(t)$ whose column vectors form a fundamental solution set for the homogeneous system is called a **fundamental matrix.** The determinant of $\mathbf{X}(t)$ is the Wronskian of the fundamental solution set. Since the Wronskian is never zero on the interval I, then $\mathbf{X}^{-1}(t)$ exists for t in I.

General Solution to Homogeneous System: $\mathbf{X}\mathbf{c}=c_1\mathbf{x}_1+c_2\mathbf{x}_2+\ldots+c_n\mathbf{x}_n$. If $\mathbf{X}(t)$ is a fundamental matrix whose column vectors are $\mathbf{x}_1$, $\mathbf{x}_2$, …, $\mathbf{x}_n$, then a general solution to the homogeneous system is

$$\mathbf{x}(t)=\mathbf{X}(t)\mathbf{c}=c_1\mathbf{x}_1(t)+c_2\mathbf{x}_2(t)+\ldots+c_n\mathbf{x}_n(t),$$

where $\mathbf{c}=\mathrm{col}(c_1, c_2, \ldots, c_n)$ is an arbitrary constant vector.

Homogeneous Systems with Constant Coefficients. The form of a general solution for a homogeneous system with constant coefficients depends on the eigenvalues and eigenvectors of the $n\times n$ constant matrix $\mathbf{A}$. An **eigenvalue** of $\mathbf{A}$ is a number r such that the system $\mathbf{A}\mathbf{u}=r\mathbf{u}$ has a nontrivial solution $\mathbf{u}$, called an **eigenvector** of $\mathbf{A}$ associated with the eigenvalue r. Finding the eigenvalues of $\mathbf{A}$ is equivalent to finding the roots of the **characteristic equation**

$$|\mathbf{A}-r\mathbf{I}|=0.$$

The corresponding eigenvectors are found by solving the system $(\mathbf{A}-r\mathbf{I})\mathbf{u}=\mathbf{0}$.

If the matrix $\mathbf{A}$ has n linearly independent eigenvectors $\mathbf{u}_1$, $\mathbf{u}_2$, …, $\mathbf{u}_n$ and r_i is the eigenvalue corresponding to the eigenvector $\boldsymbol{u}_i$, then

$$\{e^{r_1 t}\boldsymbol{u}_1, e^{r_2 t}\boldsymbol{u}_2, \ldots, e^{r_n t}\boldsymbol{u}_n\}$$

is a fundamental solution set for the homogeneous system. A class of matrices that always has n linearly independent eigenvectors is the set of **symmetric** matrices—that is, matrices that satisfy $\boldsymbol{A}=\boldsymbol{A}^{\mathrm{T}}$.

If $\boldsymbol{A}$ has complex conjugate eigenvalues $\alpha \pm i\beta$ and associated eigenvectors $\boldsymbol{z}=\boldsymbol{a}\pm i\boldsymbol{b}$, where $\boldsymbol{a}$ and $\boldsymbol{b}$ are real vectors, then two linearly independent real vector solutions to the homogeneous system are

$$e^{\alpha t}\cos\beta t\boldsymbol{a}-e^{\alpha t}\sin\beta t\boldsymbol{b},\ e^{\alpha t}\sin\beta t\boldsymbol{a}+e^{\alpha t}\cos\beta t\boldsymbol{b}.$$

When $\boldsymbol{A}$ has a repeated eigenvalue r of multiplicity m, then it is possible that $\boldsymbol{A}$ does not have n linearly independent eigenvectors. However, associated with r are m linearly independent **generalized eigenvectors** that can be used to generate m linearly independent solutions to the homogeneous system (see Section 5.6 under "Generalized Eigenvectors").

Nonhomogeneous Normal Systems

$$\boldsymbol{x}'(t)=\boldsymbol{A}(t)\boldsymbol{x}(t)+\boldsymbol{f}(t).$$

The $n\times n$ matrix function $\boldsymbol{A}(t)$ and the vector function $\boldsymbol{f}(t)$ are assumed continuous on an interval I.

General Solution to Nonhomogeneous System: $\boldsymbol{x}_p+\boldsymbol{X}(t)\boldsymbol{c}$. If $\boldsymbol{x}_p(t)$ is any particular solution for the nonhomogeneous system and $\boldsymbol{X}(t)$ is a fundamental matrix for the associated homogeneous system, then a general solution for the nonhomogeneous system is

$$\boldsymbol{x}(t)=\boldsymbol{x}_p(t)+\boldsymbol{X}(t)\boldsymbol{c}=\boldsymbol{x}_p(t)+c_1\boldsymbol{x}_1(t)+c_2 x_2(t)+\ldots+c_n\boldsymbol{x}_n(t),$$

where $\boldsymbol{x}_1(t)$, $\boldsymbol{x}_2(t)$, $\ldots$, $\boldsymbol{x}_n(t)$ are the column vectors of $\boldsymbol{X}(t)$ and $\boldsymbol{c}=\mathrm{col}(c_1, c_2, \ldots, c_n)$ is an arbitrary constant vector.

Undetermined Coefficients. If the nonhomogeneous term $\boldsymbol{f}(t)$ is a vector whose components are polynomials, exponential or sinusoidal functions, and $\boldsymbol{A}$ is a constant matrix, then one can use an extension of the method of undetermined coefficients to decide the form of a particular solution to the nonhomogeneous system.

Variation of Parameters: $\boldsymbol{X}(t)\boldsymbol{v}(t)$. Let $\boldsymbol{X}(t)$ be a fundamental matrix for the homogeneous system. A particular solution to the nonhomogeneous system is given by the **variation of parameters** formula

$$x_p(t)=X(t)v(t)=X(t)\int X^{-1}(t)f(t)dt.$$

Matrix Exponential Function

If A is an $n\times n$ constant matrix, then the matrix exponential function

$$e^{At}:=I+At+A^2\frac{t^2}{2!}+\ldots+A^n\frac{t^n}{n!}+\ldots$$

is a fundamental matrix for the homogeneous system $x'(t)=Ax(t)$. The matrix exponential has some of the same properties as the scalar exponential e^{at}. In particular,

$$e^0=I,\quad e^{A(t+s)}=e^{At}e^{As},\ (e^{At})^{-1}=e^{-At}.$$

If $(A-rI)^k=0$ for some r and k, then the series for e^{At} has only a finite number of terms:

$$e^{At}=e^{rt}\left[I+(A-rI)t+\ldots+(A-rI)^{k-1}\frac{t^{k-1}}{(k-1)!}\right].$$

The matrix exponential function e^{At} can also be computed from any fundamental matrix $X(t)$ via the formula

$$e^{At}=X(t)X^{-1}(0).$$

Generalized Eigenvectors

If an eigenvalue r_i of an $n\times n$ constant matrix A has multiplicity m_i there exist m_i linearly independent generalized eigenvectors u satisfying $(A-r_iI)^{m_i}u=0$. Each such u determines a solution to the system $x'=Ax$ of the form

$$x(t)=e^{r_it}\left[I+(A-r_iI)t+\ldots+(A-r_iI)^{m_i-1}\frac{t^{m_i-1}}{(m_i-1)!}\right]u$$

and the totality of all such solutions is linearly independent on $(-\infty,+\infty)$ and forms a fundamental solution set for the system.

Review Problems

In Problems 1-4, find a general solution for the system $x'(t)=Ax(t)$, *where* A *is given.*

1. $A=\begin{bmatrix}6 & -3\\ 2 & 1\end{bmatrix}$.

2. $A=\begin{bmatrix}3 & 2\\ -5 & 1\end{bmatrix}$.

3. $A=\begin{bmatrix}1 & 2 & 0 & 0\\ 2 & 1 & 0 & 0\\ 0 & 0 & 1 & 2\\ 0 & 0 & 2 & 1\end{bmatrix}$.

4. $\mathbf{A}=\begin{bmatrix}1 & 1 & 0\\ 0 & 1 & 0\\ 0 & 0 & 2\end{bmatrix}$.

In Problems 5 and 6, find a fundamental matrix for the system $\boldsymbol{x}'(t)=\mathbf{A}\boldsymbol{x}(t)$, where $\mathbf{A}$ is given.

5. $\mathbf{A}=\begin{bmatrix}1 & -1\\ 2 & 4\end{bmatrix}$.

6. $\mathbf{A}=\begin{bmatrix}5 & 0 & 0\\ 0 & -4 & 3\\ 0 & 3 & 4\end{bmatrix}$.

In Problems 7-10, find a general solution for the system $\boldsymbol{x}'(t)=\mathbf{A}\boldsymbol{x}(t)+\boldsymbol{f}(t)$, where $\mathbf{A}$ and $\boldsymbol{f}(t)$ are given.

7. $\mathbf{A}=\begin{bmatrix}1 & 1\\ 4 & 1\end{bmatrix}$, $\boldsymbol{f}(t)=\begin{bmatrix}5\\ 6\end{bmatrix}$.

8. $\mathbf{A}=\begin{bmatrix}-4 & 2\\ 2 & -1\end{bmatrix}$, $\boldsymbol{f}(t)=\begin{bmatrix}e^{4t}\\ 3e^{4t}\end{bmatrix}$.

9. $\mathbf{A}=\begin{bmatrix}2 & 1 & -1\\ -3 & -1 & 1\\ 9 & 3 & -4\end{bmatrix}$, $\boldsymbol{f}(t)=\begin{bmatrix}t\\ 0\\ 1\end{bmatrix}$.

10. $\mathbf{A}=\begin{bmatrix}2 & -2 & 3\\ 0 & 3 & 2\\ 0 & -1 & 2\end{bmatrix}$, $\boldsymbol{f}(t)=\begin{bmatrix}e^{-t}\\ 2\\ 1\end{bmatrix}$.

In Problems 11 and 12, solve the given initial value problem.

11. $\boldsymbol{x}'(t)=\begin{bmatrix}0 & 1\\ -2 & 3\end{bmatrix}\boldsymbol{x}(t)$, $\boldsymbol{x}(0)=\begin{bmatrix}1\\ -1\end{bmatrix}$.

12. $\boldsymbol{x}'(t)=\begin{bmatrix}2 & 1\\ -4 & 2\end{bmatrix}\boldsymbol{x}(t)+\begin{bmatrix}te^{2t}\\ e^{2t}\end{bmatrix}$, $\boldsymbol{x}(0)=\begin{bmatrix}2\\ 2\end{bmatrix}$.

In Problems 13 and 14, find a general solution for the Cauchy-Euler system $t\mathbf{X}'(t)=\mathbf{A}\boldsymbol{x}(t)$, where $\mathbf{A}$ is given.

13. $\mathbf{A}=\begin{bmatrix}0 & 3 & 1\\ 1 & 2 & 1\\ 1 & 3 & 0\end{bmatrix}$.

14. $\mathbf{A}=\begin{bmatrix} 1 & 2 & -1 \\ 2 & 1 & 1 \\ -1 & 1 & 0 \end{bmatrix}$.

In Problems 15 and 16, find the fundamental matrix $e^{\mathbf{A}t}$ *for the system* $\mathbf{x}'(t)=\mathbf{A}\mathbf{x}(t)$, *where* $\mathbf{A}$ *is given.*

15. $\mathbf{A}=\begin{bmatrix} 4 & 2 & 3 \\ 2 & 1 & 2 \\ -1 & 2 & 0 \end{bmatrix}$.

16. $\mathbf{A}=\begin{bmatrix} 0 & 1 & 4 \\ 0 & 0 & 2 \\ 0 & 0 & 0 \end{bmatrix}$.

Technical Writing Exercises

1. Explain how the theory of homogeneous linear differential equations (as described in Chapter 4) follows from the theory of linear systems in normal form (as described in Chapter 5).

2. Discuss the similarities and differences between the method for finding solutions to a linear constant coefficient differential equation (see Chapter 4) and the method for finding solutions to a linear system in normal form that has constant coefficients (see Chapter 5).

Chapter 6

Stability

6.1 Introduction

The populations $x(t)$ and $y(t)$ at time t of two species that compete for the same food supply are governed by the system

$$\begin{cases} \frac{dx}{dt}=x(a_1-b_1x-c_1y), \\ \frac{dy}{dt}=y(a_2-b_2y-c_2x), \end{cases} \tag{6.1}$$

where a_1, a_2, b_1, b_2, c_1, and c_2 are positive constants that depend on the species and environment under investigation. Describe the long-term behavior of these populations.

That system (6.1) is a plausible model for competing species can be seen as follows. Using the logistic model for the growth of a single species, let's assume that in the absence of species y, population x is governed by the equation of the form

$$\frac{dx}{dt}=a_1x-b_1x^2=x(a_1-b_1x),$$

and, similarly, that in the absence of species x, population y is governed by

$$\frac{dy}{dt}=a_2y-b_2y^2=y(a_2-b_2y),$$

where a_1, a_2, b_1, and b_2 are positive constants. When both species are present and competing for the same food, the effect is to decrease their rate of growth. Because this effect depends on the size of both populations, it is reasonable to assume it is proportional to the product xy. Consequently, we replace the growth rate for species x by $a_1x-b_1x^2-c_1xy$ and for species y by $a_2y-b_2y^2-c_2xy$, which yields the model (6.1).

Generally, we cannot explicitly solve the nonlinear system (6.1) for x and y in terms of t. Nor can we find explicit analytic formulas for the trajectories in the phase plane. However, in this chapter, we will extend the qualitative methods to obtain useful information about certain trajectories—specifically, we will try to determine how the solutions behave near a critical point. A critical point of system (6.1) is a point where both right-hand sides are zero. If (x_0, y_0) is a critical point, then the constant functions $x(t)\equiv x_0$, $y(t)\equiv y_0$ form a solution to system (6.1). This is called an **equilibrium solution**, since the functions do not change with t.

Let's first determine the critical points for the system (6.1). Setting the right-hand sides equal to zero, we have

$$\begin{cases} x(a_1-b_1x-c_1y)=0, \\ y(a_2-b_2y-c_2x)=0. \end{cases}$$

This system has (at most) four solutions: $(0, 0)$, $(0, a_2/b_2)$, $(a_1/b_1, 0)$, and the point of intersection of the two lines $a_1-b_1x-c_1y=0$, $a_2-b_2y-c_2x=0$, which, if it exists, is the point

$$\boldsymbol{P}=\left(\frac{a_1b_2-a_2c_1}{b_1b_2-c_1c_2}, \frac{a_2b_1-a_1c_2}{b_1b_2-c_1c_2}\right). \tag{6.2}$$

Notice that the first three equilibrium solutions lie on the nonnegative axes that border the first quadrant in the xy-plane. Because populations must be nonnegative, we will confine the rest of our analysis to the first quadrant.

To determine the flow of trajectories in the first quadrant, let's start with equation:

$$\frac{dx}{dt}=x(a_1-b_1x-c_1y).$$

From this equation, we see that in the first quadrant, the sign of dx/dt is positive when $a_1-b_1x-c_1y>0$ and negative when $a_1-b_1x-c_1y<0$. Consequently, $x(t)$ increases along a trajectory that lies below the line $a_1-b_1x-c_1y=0$ and decreases along trajectories above this line [see Figure 6.1(a)]. Similarly, $y(t)$ increases along a trajectory that lies below the line $a_2-b_2y-c_2x=0$ and decreases when it is above this line [see Figure 6.1(b)].

It turns out that there are four cases to analyze, depending on the relative positions of the two lines and whether they intersect in the first quadrant. These four cases are illustrated in Figure 6.2. The arrow systems (e. g. ,↗↘) indicate whether the variables x and y are increasing or decreasing and are based on the analysis illustrated in Figure 6.1. The grey dots are the critical points in the first quadrant. The black lines denote where $dx/dt=0$ and the grey lines where $dy/dt=0$. Typical trajectories are also shown in grey.

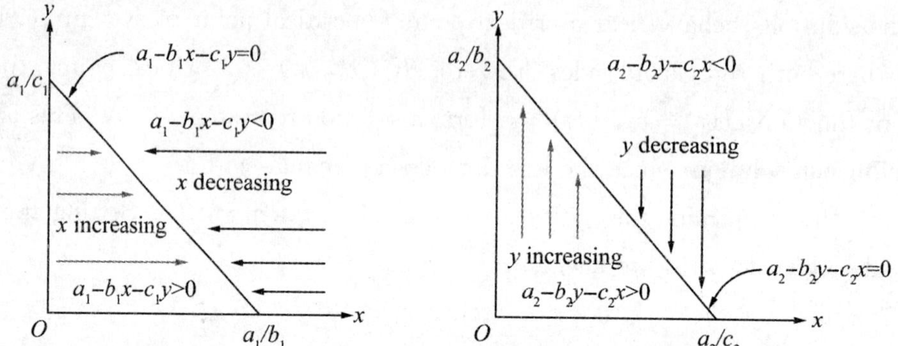

(a) Change in x (b) Change in y

Figure 6.1 Regions where the populations are increasing or decreasing

Let's begin with the case illustrated in Figure 6.2(a). If we start in region I, that is, $(x(0), y(0))=(x_0, y_0)$ lies in region I, then populations x and y are increasing (↗). At some point in time, the trajectory $(x(t), y(t))$ will move into region II. In region II, the population $x(t)$ decreases while $y(t)$ continues to increase. Observe that the trajectory cannot cross the grey line bounding region II because this would require either $x(t)$ to be increasing, which is not in region II, or $y(t)$ to increase across the grey line, which is impossible because $dy/dt=0$ on that line. Consequently, the trajectories in region II must move up and to the left (↖). Since $dx/dt=0$ on the y-axis, the trajectory cannot cross it and so must approach the critical point $(0, a_2/b_2)$ in the upper corner of region II. When the populations start in region III, both $x(t)$ and $y(t)$ are decreasing (↙). Since the trajectory cannot leave the first quadrant, the trajectory must cross the grey line and enter region II, where it then moves up toward the critical point $(0, a_2/b_2)$. To summarize, we have shown that when the initial populations $x(0)$ and $y(0)$ are positive, the species x dies off $[x(t)\to 0]$, while the population of species y approaches the limiting population a_2/b_2. Graphically, all trajectories beginning in the first quadrant approach the critical point $(0, a_2/b_2)$ as $t\to+\infty$. Typical trajectories are illustrated by grey curves in Figure 6.2(a).

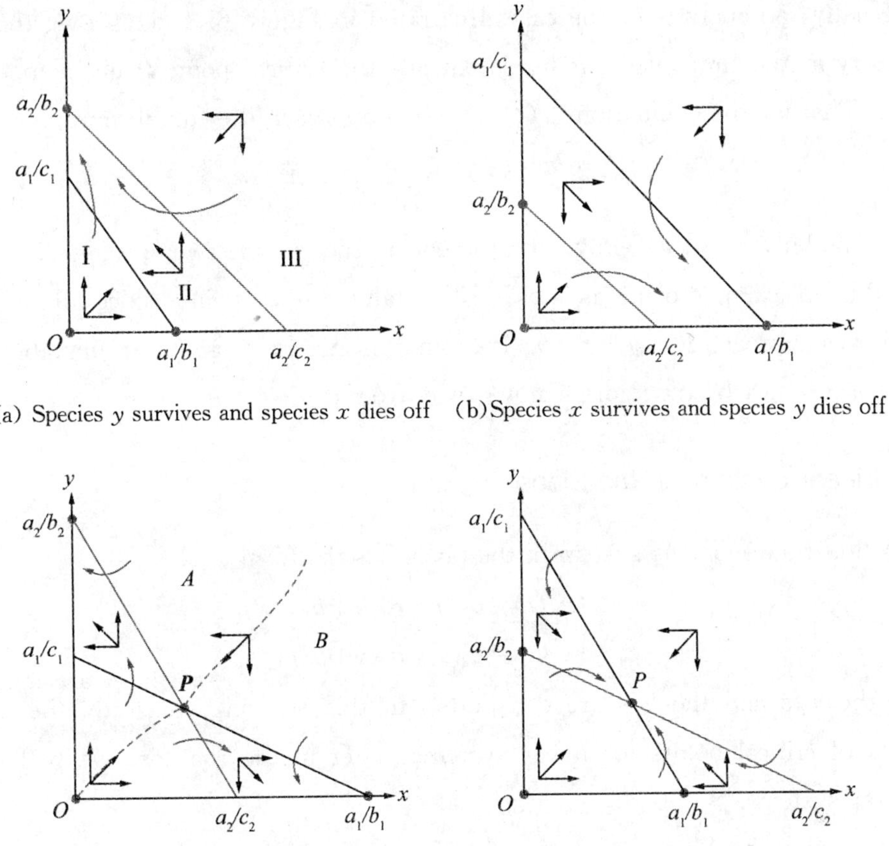

(a) Species y survives and species x dies off (b) Species x survives and species y dies off

(c) Either species x or y, but not both, survives (d) Coexistence

Figure 6.2 Phase plane diagrams for competing species

A similar analysis of the case illustrated in Figure 6.2(b) shows that species y dies off, while the population of species x approaches the limiting population a_1/b_1. Graphically, all trajectories beginning in the first quadrant approach the critical point $(a_1/b_1, 0)$ as $t \to +\infty$. Typical trajectories are again illustrated by grey curves.

The analysis of the case illustrated in Figure 6.2(c) yields a different result. There exists a dividing curve called a **separatrix** (illustrated by a dashed grey curve) that separates the first quadrant into two regions A and B. Trajectories beginning in region A approach the critical point $(0, a_2/b_2)$ as $t \to +\infty$, and those beginning in region B approach the critical point $(a_1/b_1, 0)$. In other words, for some initial populations, species x dies off, while the population of species y approaches a_2/b_2. For other initial populations, species y dies off, while the population of species x approaches a_1/b_1. Unfortunately, the separatrix is usually difficult to determine.

Finally, an analysis of the case illustrated in Figure 6. 2(d) shows that every trajectory in the first quadrant heads toward the critical point $\boldsymbol{P}$ given in equation (6. 2). That is, the populations $x(t)$ and $y(t)$ approach the equilibrium populations

$$x=\frac{a_1b_2-a_2c_1}{b_1b_2-c_1c_2},\quad y=\frac{a_2b_1-a_1c_2}{b_1b_2-c_1c_2}.$$

This is the only case that allows for the coexistence of the two species. The critical point $\boldsymbol{P}$ is an example of an asymptotically stable critical point, since all the nearby trajectories approach $\boldsymbol{P}$ as $t\to+\infty$. By comparison, the origin is an unstable critical point because nearby trajectories move away from it.

6.2 Linear Systems in the Plane

A linear autonomous system in the plane has the form

$$\begin{cases}x'(t)=a_{11}x+a_{12}y+b_1,\\ y'(t)=a_{21}x+a_{22}y+b_2,\end{cases}\tag{6.3}$$

where the a_{ij}'s and the b_i's are constants. In this section, we study the issue of stability of critical points for linear systems; later in the chapter, we will turn to nonlinear systems.

By a simple translation, we can always transform a given linear system into the form

$$\begin{cases}x'(t)=ax+by,\\ y'(t)=cx+dy,\end{cases}\tag{6.4}$$

where the origin (0,0) is now the critical point. So, without loss of generality, we analyze the linear system (6. 3), but under the additional assumption that $ad-bc\neq0$. This last condition makes (0,0) an isolated critical point.

System (6. 3) has solutions of the form $x(t)=ue^{rt}$, $y(t)=ve^{rt}$, where u, v, and r are constants to be determined. Substituting these expressions for x and y into system (6. 3), we have

$$\begin{cases}rue^{rt}=aue^{rt}+bve^{rt},\\ rve^{rt}=cue^{rt}+dve^{rt},\end{cases}$$

which simplifies to

$$\begin{cases}(r-a)u-bv=0,\\ -cu+(r-\mathrm{d})v=0.\end{cases}$$

This pair of equations has a nontrivial solution for u, v if and only if the determinant of the coefficient matrix is zero; that is,

$$(r-a)(r-d)-bc=r^2-(a+d)r+(ad-bc)=0. \tag{6.5}$$

This is called the **characteristic equation** for system (6.3).

The asymptotic (long-term) behavior of the solutions to system (6.3) is linked to the nature of the roots r_1 and r_2 of the characteristic equation—namely, whether these roots are both real or both complex, both positive, both negative, of opposite signs, etc. We consider each case in detail by first discussing a special example and then indicating what happens in general. These general results (whose proofs we omit) can be obtained using a linear change of variables.

Case 6.1 $\boldsymbol{r_1}$, $\boldsymbol{r_2}$ **are real, distinct, and positive.** In this case, the general solution to system (6.3) has the form

$$\begin{cases} x(t)=A_1 e^{r_1 t}+A_2 e^{r_2 t}, \\ y(t)=B_1 e^{r_1 t}+B_2 e^{r_2 t}, \end{cases} \tag{6.6}$$

where the A_i's and B_i's are interrelated. As a simple example, we discuss the system

$$\frac{dx}{dt}=x, \quad \frac{dy}{dt}=3y.$$

Here $r_1=1$ and $r_2=3$ and a general solution is

$$x(t)=Ae^t, \quad y(t)=Be^{3t}.$$

Consequently, the trajectories of the system lie on the curves $y=B(e^t)^3=(B/A^3)x^3$ in the phase plane. A phase plane diagram for this system is given in Figure 6.3(a). Notice that the trajectories move away from the origin and are tangent to the x-axis at the origin, except for the two along the y-axis. In this case, we say that the origin is an **unstable improper node.** It is unstable because the trajectories move away from the origin, and it is an improper node because (almost all) trajectories have the same tangent line at the origin. (A proper node occurs in Case 6.4; see Figures 6.3 and 6.8 for a comparison.)

In general, when the roots are distinct, real, and positive, the origin is an unstable improper node whose phase plane diagram near the origin resembles Figure 6.3(b). The grey lines labeled $\hat{x}$ and $\hat{y}$ represent the images of the x- and y-axes under a change of variables.

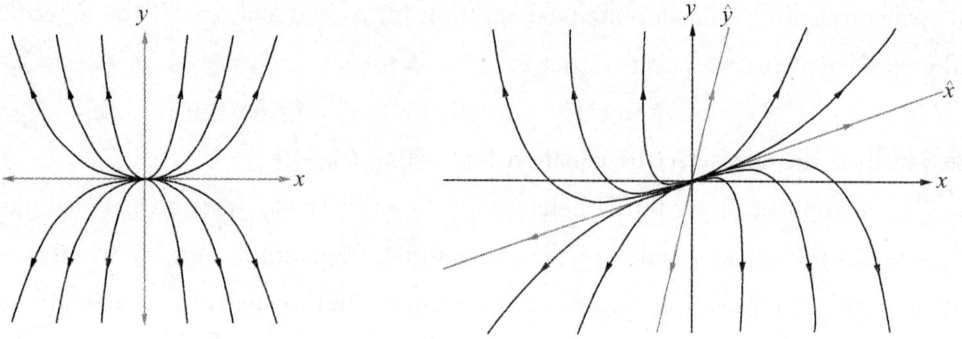

(a) Specific example (b) Typical trajectories

Figure 6.3 Unstable improper node; roots are real, distinct, and positive

We make two key observations. First, (most) trajectories in Figures 6.3(a) and 6.3(b) are becoming parallel to the y- and $\hat{y}$-axes, respectively, as they move away from the origin. Second, (most) trajectories are leaving the origin along directions that are tangent to the x- or $\hat{x}$-axes. We refer to the $\hat{x}$- and $\hat{y}$-axes as the **transformed axes.** Observe that these transformed axes also contain trajectories.

Case 6.2 r_1, r_2 are real, distinct, and negative. The general solutions to system (6.3) have the same form as in (6.6). A simple example is the system

$$\frac{dx}{dt}=-2x, \quad \frac{dy}{dt}=-y,$$

for which $r_1=-2$, $r_2=-1$ and

$$x(t)=Ae^{-2t}, \quad y(t)=Be^{-t}$$

is a general solution. The reader should verify that the integral curves for this example have the form $x=(A/B^2)y^2$. A phase plane diagram for this system is given in Figure 6.4(a).

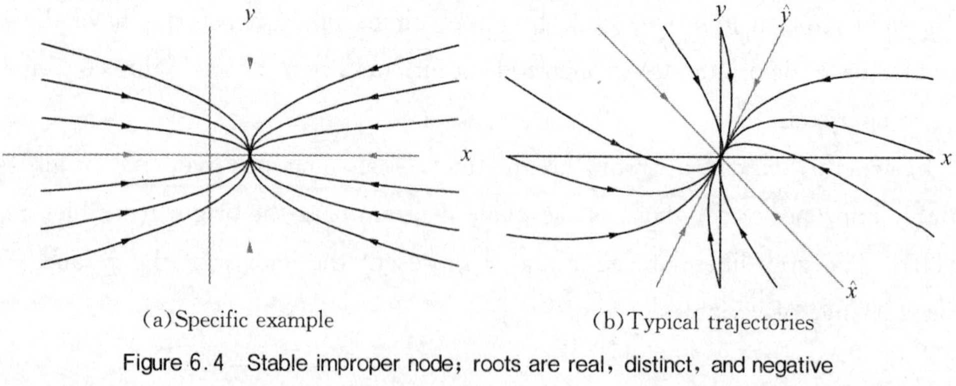

(a) Specific example (b) Typical trajectories

Figure 6.4 Stable improper node; roots are real, distinct, and negative

Notice that in this example, the trajectories are moving toward the origin. For this reason, the origin is called an **asymptotically stable** critical point. Again, since (most) trajectories approach the origin along curves that are tangent to the y-axis, the origin is an improper node. Thus, the critical point at the origin is an asymptotically stable improper node.

In general, when the roots are distinct, real, and negative, the origin is an asymptotically stable improper node whose phase plane diagram near the origin resembles that in Figure 6.4(b). Cases 6.1 and 6.2 are quite similar. They differ mainly in that when the roots are both positive, the trajectories move away from the origin (the origin is unstable), while when the roots are both negative, the trajectories approach the origin (the origin is stable).

Example 6.1 Classify the critical point at the origin and sketch a phase diagram for the system

$$\begin{cases} \dfrac{dx}{dt}=-5x+2y, \\ \dfrac{dy}{dt}=x-4y. \end{cases} \tag{6.7}$$

Solution The characteristic equation for this system is $r^2+9r+18=(r+6)(r+3)=0$, which has roots $r_1=-6$, $r_2=-3$. Thus, the origin is an asymptotically stable improper node. To sketch the phase plane diagram, we must first determine the two lines passing through the origin that correspond to the transformed axes. These lines have the form of $y=mx$ and must be trajectories for the given system. To determine the values of m, we first divide system (6.7) to obtain the phase plane equation

$$\frac{dy}{dx}=\frac{dy/dt}{dx/dt}=\frac{x-4y}{-5x+2y}. \tag{6.8}$$

Substituting in $y=mx$ gives

$$m=\frac{x-4mx}{-5x+2mx}=\frac{1-4m}{-5+2m}.$$

Solving for m yields $2m^2-m-1=(m-1)(2m+1)=0$, so $m=1$, $-1/2$. Hence, the two axes are $y=x$ and $y=-(1/2)x$. As we know, the trajectories approach one of these lines tangentially. To determine which, let's graph a few of the isoclines for equation (6.8). For example, when $dy/dx=0$, we get $x-4y=0$ or $y=(1/4)x$. So, when a trajectory crosses the line $y=(1/4)x$, its slope is zero. Moreover, when

dy/dx is infinite or undefined, then $-5x+2y=0$. Hence, when a trajectory crosses the line $y=(5/2)x$, the tangent line to the trajectory must be vertical. These isoclines are represented in Figure 6.5 by horizontal and vertical hash marks, respectively. It is now clear that the trajectories must be tangent to the line $y=x$ as they approach the origin. ◆

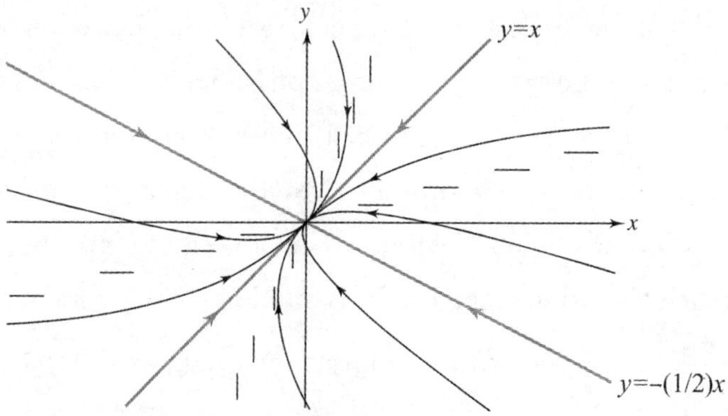

Figure 6.5 Phase plane diagram for $dx/dt=-5x+2y$, $dy/dt=x-4y$

Case 6.3 $\boldsymbol{r_1}$, $\boldsymbol{r_2}$ **are real and of opposite signs.** Again, a general solution to system (6.5) has the same form as in (6.6). A simple example is the system

$$\frac{dx}{dt}=x, \quad \frac{dy}{dt}=-y,$$

which has a general solution of the form

$$x(t)=Ae^t, \quad y(t)=Be^{-t},$$

with $r_1=1$, $r_2=-1$. The trajectories lie on the curves $y=ABx^{-1}$ [see Figure 6.6(a)].

In this case the trajectories come in along the positive and negative y-axis and go out along the positive and negative x-axis. The origin is unstable, since there are trajectories that pass arbitrarily near to the origin, but then eventually move away. (The two trajectories on the y-axis do approach the origin, but they are the only ones.) For apt reasons, the origin is said to be a **saddle point.** Hence, when the roots are real and of opposite signs, the origin is an unstable saddle point. In general, the trajectories come in alongside one of the transformed axes and go out alongside the other. This is illustrated in Figure 6.6(b). Observe that the trajectories have both of these lines as asymptotes and that both lines are themselves trajectories for the system.

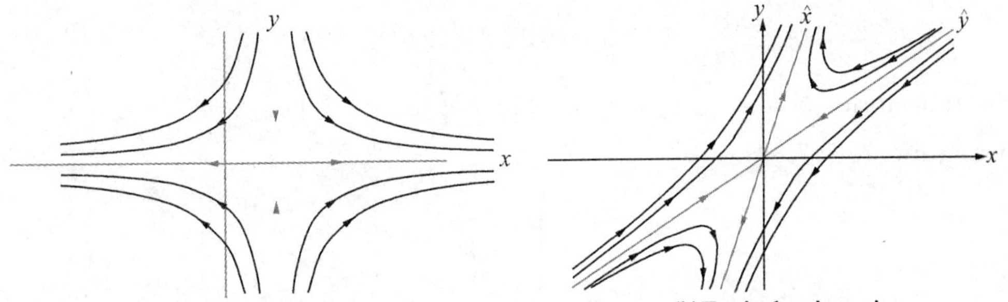

(a)Specific example (b)Typical trajectories

Figure 6.6 Unstable saddle point; roots are real and of opposite signs

Example 6.2 Classify the critical point at the origin and sketch a phase plane diagram for the system

$$\begin{cases} \frac{dx}{dt}=5x-3y, \\ \frac{dy}{dt}=4x-3y. \end{cases} \tag{6.9}$$

Solution The characteristic equation for this system is $r^2-2r-3=(r+1)(r-3)=0$, which has roots $r_1=-1$, $r_2=3$. Thus, the origin is an unstable saddle point. To sketch the phase plane diagram, we must first determine the two lines corresponding to the transformed axes. Solving for dy/dx, we have

$$\frac{dy}{dx}=\frac{dy/dt}{dx/dt}=\frac{4x-3y}{5x-3y}.$$

Next we substitute $y=mx$ and find

$$m=\frac{4x-3mx}{5x-3mx}=\frac{4-3m}{5-3m}.$$

Solving for m gives $3m^2-8m+4=(3m-2)(m-2)=0$, so $m=2$, $2/3$. Hence, the two transformed axes are $y=2x$ and $y=(2/3)x$. Since the origin is a saddle point, the trajectories are moving in toward the origin alongside one of the transformed axes and they are moving out alongside the other. On the line $y=2x$ itself, we find that

$$\frac{dx}{dt}=5x-3y=5x-3\cdot 2x=-x.$$

So, when $x>0$, this trajectory moves toward the origin, since $dx/dt<0$. The same is true for $x<0$, since $dx/dt>0$. Consequently, on the other line, $y=(2/3)x$, the trajectories must be moving away from the origin. Indeed, a quick check shows that on $y=(2/3)x$,

$$\frac{dx}{dt}=5x-3y=5x-3\cdot(2/3)x=3x,$$

so the trajectories do, in fact, recede from the origin. The phase plane diagram is shown in Figure 6.7. ◆

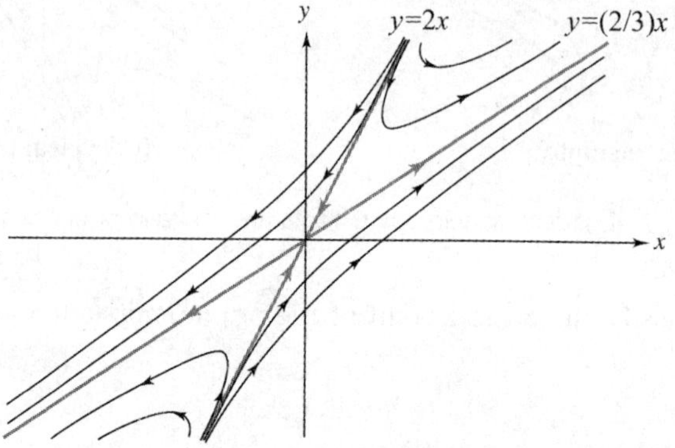

Figure 6.7 Phase plane diagram for $dx/dt=5x-3y$, $dy/dt=4x-3y$

Case 6.4 Equal roots, $r_1=r_2$. In this case, a general solution may or may not involve a factor of t times the exponential e^{rt}. As we shall see, the presence of the factor of t alters the phase plane diagram, but not the stability. We will discuss the two possibilities.

A simple example where the factor of t is not present is

$$\frac{dx}{dt}=rx, \quad \frac{dy}{dt}=ry,$$

which has the general solution

$$x(t)=Ae^{rt}, \quad y(t)=Be^{rt}.$$

Here the trajectories lie on the integral curves $y=(B/A)x$. When $r>0$, these trajectories move away from the origin, so the origin is unstable [see Figure 6.8(a)]. When $r<0$, the trajectories approach the origin and the origin is asymptotically stable [see Figure 6.8(b)]. In either case all the trajectories lie on lines passing through the origin. Because every direction through the origin defines a trajectory, the origin is called a **proper node.**

In general, when the roots are equal and a factor of t is not present, the origin is a proper node that is unstable when the root $r>0$ and asymptotically stable when $r<0$.

An example where the factor of t arises is

$$\frac{dx}{dt}=rx, \quad \frac{dy}{dt}=x+ry,$$

which has the general solution

$$x(t)=Ae^{rt}, \quad y(t)=(At+B)e^{rt}.$$

To find the trajectories, we first solve for t in terms of x to obtain

$$t=\frac{1}{r}\ln(x/A).$$

Substituting for t in the expression for $y(t)$ and simplifying, we find

$$y=\frac{B}{A}x+\frac{x}{r}\ln(x/A).$$

(Alternatively, we can deduce this formula by solving the phase plane equation.) The integral curves are sketched in Figure 6.9(a). Observe that dy/dx is undefined for $x=0$. When $r>0$, the trajectories move away from the origin, so it is unstable. Since the trajectories lie only on curves that are tangent to the y-axis at the origin, we have an improper node [see Figure 6.9(a)]. When $r<0$, the trajectories approach the origin, so the origin is an asymptotically stable point.

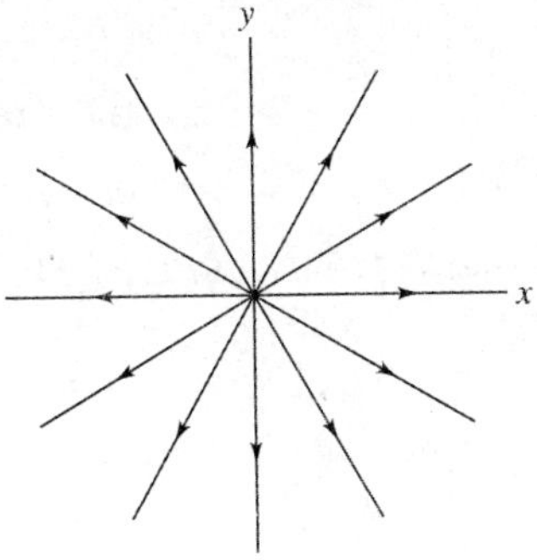

(a) Equal positive roots, unstable

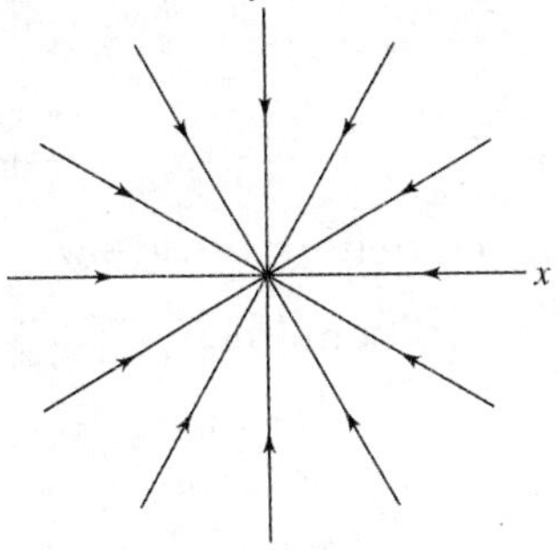

(b) Equal negative roots, stable

Figure 6.8 Proper nodes; equal roots with no factor of t

In summary, when the roots are equal and a factor of t is present in the general solution, the origin is an improper node that is unstable when $r>0$ [see Figure 6.9(b)] and asymptotically stable when $r<0$ [see Figure 6.9(c)].

Case 6.5 Complex roots, $r=\alpha\pm i\beta$. We now consider the case of complex roots $r=\alpha\pm i\beta$, where $\alpha\neq 0$ and $\beta\neq 0$. (Pure imaginary roots, $r=\pm i\beta$, will be discussed in Case 6.6.) For complex roots, a general solution to system (6.3) is given by

$$\begin{cases} x(t)=e^{\alpha t}[A_1\cos\beta t+B_1\sin\beta t], \\ y(t)=e^{\alpha t}[A_2\cos\beta t+B_2\sin\beta t]. \end{cases}$$

It is clear that when $\alpha>0$, the trajectories will travel far away from the origin.

Hence, the origin is unstable. But, when $\alpha<0$, the trajectories approach the origin and the origin is asymptotically stable.

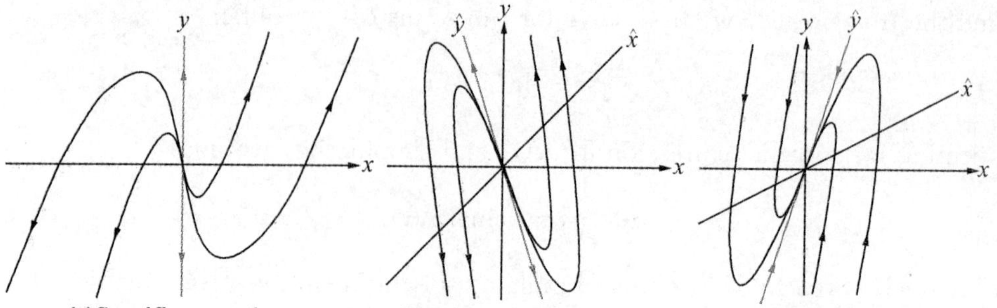

(a)Specific example; $r>0$, unstable (b)Typical trajectory; $r>0$, unstable (c)Typical trajectory; $r<0$, asymptotically stable

Figure 6.9 Improper node; equal roots with a factor of t

To better describe the behavior of trajectories near the origin, we analyze the specific system

$$\begin{cases}\dfrac{dx}{dt}=\alpha x-\beta y,\\[2mm]\dfrac{dy}{dt}=\beta x+\alpha y,\end{cases}\tag{6.10}$$

which has characteristic roots $r=\alpha\pm i\beta$. If we set $z(t)=x(t)+iy(t)$ to form a complex-valued function, then

$$\begin{aligned}\frac{dz}{dt}&=\frac{dx}{dt}+i\frac{dy}{dt}=(\alpha x-\beta y)+i(\beta x+\alpha y)\\&=\alpha(x+iy)+i\beta(x+iy)=(\alpha+i\beta)z.\end{aligned}$$

Using the polar coordinate representation $z(t)=\rho(t)e^{i\theta(t)}$, we obtain

$$\frac{dz}{dt}=\frac{d\rho}{dt}e^{i\theta}+i\rho e^{i\theta}\frac{d\theta}{dt}.$$

Setting the two expressions for dz/dt equal yields

$$\frac{d\rho}{dt}e^{i\theta}+i\rho e^{i\theta}\frac{d\theta}{dt}=(\alpha+i\beta)\rho e^{i\theta},$$

and simplifying, we have

$$\frac{d\rho}{dt}+i\rho\frac{d\theta}{dt}=\alpha\rho+i\beta\rho.$$

By equating the real and imaginary parts, we obtain the simple system

$$\frac{d\rho}{dt}=\rho\alpha,\quad\frac{d\theta}{dt}=\beta,$$

which has the general solution

$$\rho(t)=\rho_0 e^{\alpha t}, \quad \theta(t)=\beta t+\theta_0,$$

where ρ_0 and θ_0 are real-valued constants. Again, we see that if $\alpha>0$, the solution moves away from the origin, since $\rho\to+\infty$ as $t\to+\infty$. But, because θ is a linear function of t, the trajectories spiral around the origin. Similarly, when $\alpha<0$, the solutions spiral in toward the origin. We therefore refer to the origin as a **spiral point.** We can determine the direction of rotation of the spiral from the original system. For example, setting $y=0$ in equation $y'(t)=cx+dy$ in system (6.4), we find that $dy/dt=cx$. Now, if $c<0$, then as the trajectory crosses the positive x-axis ($y=0$), we have $dy/dt<0$. Consequently, the trajectory is spiraling clockwise about the origin [see Figure 6.10(a) and (b)].

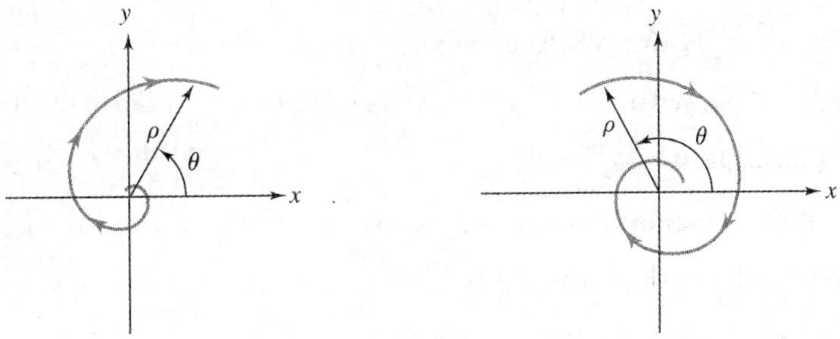

(a)$\alpha>0$, unstable (b)$\alpha<0$, asymptotically stable

Figure 6.10 Spiral point; $r=\alpha\pm i\beta$ with clockwise rotation

In the general case, the spirals may be distorted, but the stability criterion persists; when the roots $r=\alpha\pm i\beta$ are complex numbers, the origin is a spiral point that is unstable when $\alpha>0$ [see Figure 6.11(a)] and asymptotically stable when $\alpha<0$ [see Figure 6.11(b)].

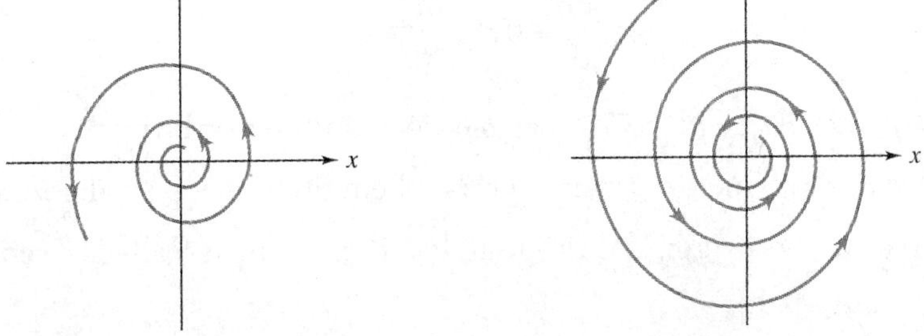

(a)$\alpha>0$, unstable (b)$\alpha<0$, asymptotically stable

Figure 6.11 General spiral point; $r=\alpha\pm i\beta$ with counterclockwise rotation

Example 6.3 Classify the critical point at the origin and sketch the phase plane diagram for the system

$$\begin{cases}\dfrac{dx}{dt}=x-4y,\\[2ex]\dfrac{dy}{dt}=4x+y.\end{cases} \tag{6.11}$$

Solution The characteristic equation for this system is $(r-1)^2+16=0$, which has complex roots $r=1\pm 4i$. Thus, the origin is an unstable spiral point, since $\alpha=1>0$. Setting $y=0$ in the equation $\dfrac{dy}{dt}=4x+y$ in system (6.11), we obtain $dy/dt=4x$. Hence, $dy/dt>0$ when a trajectory crosses the positive x-axis. Thus, the trajectories are spiraling with a counterclockwise rotation about the origin(see Figure 6.12). ◆

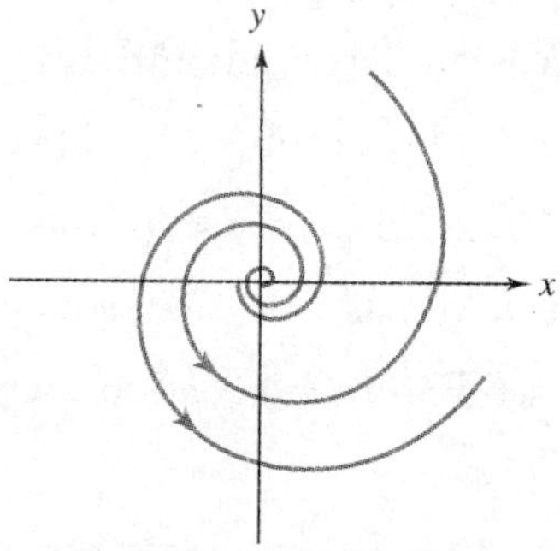

Figure 6.12 Trajectories for $dx/dt=x-4y, dy/dt=4x+y$

Case 6.6 Pure imaginary roots, $r=\pm i\beta$. When the roots are pure imaginary, $r=\pm i\beta$, a general solution has the form

$$\begin{cases}x(t)=A_1\cos\beta t+B_1\sin\beta t,\\ y(t)=A_2\cos\beta t+B_2\sin\beta t.\end{cases}$$

These solutions all have period $2\pi/\beta$ and remain bounded. A simple example is the system

$$\frac{dx}{dt}=-\beta y,\quad \frac{dy}{dt}=\beta x,$$

which is just system (6.10) with $\alpha=0$. Hence, taking $\alpha=0$ in the analysis of Case 6.5, we find that, in polar coordinates,

$$\frac{d\rho}{dt}=0,\quad \frac{d\theta}{dt}=\beta.$$

Hence, $\rho(t)\equiv\rho_0$ and $\theta(t)=\beta t+\theta_0$. Since ρ is constant (but arbitrary), the trajectories are concentric circles about the origin. Thus, the motion is a periodic rotation around a circle centered at the origin. Appropriately, the origin is called a **center** and is a stable critical point.

In general, when the roots are pure imaginary, the origin is a stable center. The trajectories are ellipses centered at the origin. A typical phase diagram is shown in

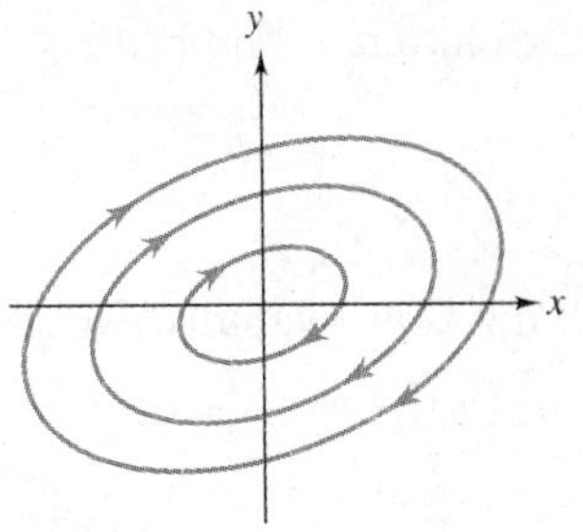

Figure 6.13 Stable center; pure imaginary roots, clockwise rotation

Figure 6.13.

We now summarize the analysis of the behavior of trajectories of a linear system near an isolated critical point at the origin.

Stability of Linear Systems

Theorem 6.1 Assume the origin (0, 0) is an isolated critical point for the linear system

$$\begin{cases} x'(t)=ax+by, \\ y'(t)=cx+dy, \end{cases}$$

where a, b, c, and d are real and $ad-bc\neq 0$. Let r_1 and r_2 be the roots of the characteristic equation

$$r^2+(a+d)r+(ad-bc)=0.$$

The stability of the origin and the classification of the origin as a critical point depends on the roots r_1 and r_2, see Table 6.1.

Table 6.1 The stability of the origin and type of critical point

Roots (r_1 and r_2)	Type of Critical Point	Stability
distinct, positive	improper node	unstable
distinct, negative	improper node	asymptotically stable
opposite signs	saddle point	unstable
equal, positive	proper or improper node	unstable
equal, negative	proper or improper node	asymptotically stable
complex-valued		
positive real part	spiral point	unstable
negative real part	spiral point	asymptotically stable
pure imaginary	center	stable

Example 6.4 Find and classify the critical point of the linear system

$$\begin{cases} \dfrac{dx}{dt}=2x+y+3, \\ \dfrac{dy}{dt}=-3x-2y-4. \end{cases} \tag{6.12}$$

Solution The critical point is the solution to the system

$$\begin{cases}2x+y+3=0,\\ -3x-2y-4=0.\end{cases}$$

Solving this system, we obtain the critical point $(-2, 1)$. Now we will use the change of variables

$$x=u-2,\quad y=v+1,$$

to translate the critical point $(-2, 1)$ to the origin $(0, 0)$. Substituting into system (6.12) and simplifying, we obtain a system of differential equations in u and v:

$$\begin{cases}\dfrac{du}{dt}=\dfrac{dx}{dt}=2(u-2)+(v+1)+3=2u+v,\\[2ex] \dfrac{dv}{dt}=\dfrac{dy}{dt}=-3(u-2)-2(v+1)-4=-3u-2v.\end{cases}$$

The characteristic equation for this linear system is $r^2-1=0$, which has roots $r=\pm 1$. Hence, the origin is an unstable saddle point for the new system. Therefore, the critical point $(-2, 1)$ is an unstable saddle point for the original system (6.12). ◆

6.3 Almost Linear Systems

Bearing in mind the discussion of the preceding section concerning stability for linear autonomous systems, we now turn to the general situation of a nonlinear system and ask: Exactly what is meant by stability of a critical point? Let's say we are dealing with the autonomous system

$$\begin{cases}\dfrac{dx}{dt}=f(x,y),\\[2ex] \dfrac{dy}{dt}=g(x,y),\end{cases}\tag{6.13}$$

where f and g are real-valued functions that do not depend explicitly on t.

Stability of a Critical Point

Definition 6.1 A critical point (x_0, y_0) of the autonomous system (6.13) is **stable** if, given any $\varepsilon>0$, there exists a $\delta>0$ such that every solution $x=\varphi(t)$, $y=\psi(t)$ of the system that satisfies

$$\sqrt{[\varphi(0)-x_0]^2+[\psi(0)-y_0]^2}<\delta$$

at $t=0$ also satisfies

$$\sqrt{[\varphi(t)-x_0]^2+[\psi(t)-y_0]^2}<\varepsilon$$

for all $t \geqslant 0$.

If (x_0, y_0) is stable and there exists an $\eta > 0$ such that every solution $x = \varphi(t)$, $y = \psi(t)$ that satisfies

$$\sqrt{[\varphi(0) - x_0]^2 + [\psi(0) - y_0]^2} < \eta$$

at $t = 0$ also satisfies

$$\lim_{t \to +\infty} \varphi(t) = x_0, \quad \lim_{t \to +\infty} \psi(t) = y_0,$$

then the critical point is **asymptotically stable.**

A critical point that is not stable is called **unstable.**

Our intuition leads us to believe that if we were to make a small perturbation of a linear system near a critical point, then the stability properties of the critical point for the perturbed system would be the same as those for the original linear system. In many cases this is true, but there are certain exceptions. These exceptions can be observed even when the perturbed system is linear.

The first is the case of pure imaginary roots. A small perturbation of such a system could result in any of three possible situations: pure imaginary roots, complex conjugate roots with positive real parts ($\alpha > 0$), or complex conjugate roots with negative real parts ($\alpha < 0$).

A second exceptional situation is that of (real) equal roots. Here an arbitrarily small perturbation may result in distinct real roots or complex conjugate roots (depending on whether the double root separates horizontally or vertically in the plane). If the equal roots are positive (negative), then under small perturbations, we would have real positive (negative) roots or complex conjugate roots whose real parts are positive (negative). In either case the stability is not affected.

The situation is similar when the perturbed system is "almost linear".

Almost Linear Systems

Definition 6.2 Let the origin (0, 0) be a critical point of the autonomous system

$$\begin{cases} x'(t) = ax + by + F(x, y), \\ y'(t) = cx + dy + G(x, y), \end{cases} \tag{6.14}$$

where a, b, c, d are constants and F, G are continuous in a neighborhood (disk) about the origin. Assume that $ad - bc \neq 0$ so that the origin is an isolated critical point for the corresponding linear system, which we obtain by setting $F = G = 0$ in system (6.14). Then, system (6.14) is said to be **almost linear** near the origin if

$$\frac{F(x,y)}{\sqrt{x^2+y^2}} \to 0 \text{ and } \frac{G(x,y)}{\sqrt{x^2+y^2}} \to 0, \text{ as } \sqrt{x^2+y^2} \to 0.$$

Stability of Almost Linear Systems

Theorem 6.2 Let r_1 and r_2 be the roots of the characteristic equation

$$r^2-(a+d)r+(ad-bc)=0 \tag{6.15}$$

for the linear system corresponding to the almost linear system (6.14). Then, the stability properties of the critical point at the origin for the almost linear system are the same as the stability properties of the origin for the corresponding linear system with one exception: When the roots are pure imaginary, the stability properties for the almost linear system cannot be deduced from the linear system (a different approach is required).

In addition to the stability, the behavior of the trajectories near the origin is essentially the same for the almost linear system and its corresponding linear system. The exceptions discussed for linear perturbations are the same as in the almost linear case. These are summarized in Table 6.2.

Table 6.2 **Stability of almost linear and linear systems**

Roots	Linear Systems		Almost Linear Systems	
r_1 and r_2	Type	Stability	Type	Stability
distinct, positive	improper node	unstable	improper node	unstable
distinct, negative	improper node	asymptotically stable	improper node	asymptotically stable
opposite signs	saddle point	unstable	saddle point	unstable
equal, positive	improper or proper node	unstable	improper or proper node or spiral point	unstable
equal, negative	improper or proper node	asymptotically stable	improper or proper node or spiral point	asymptotically stable
complex, real part				
positive	spiral point	unstable	spiral point	unstable
negative	spiral point	asymptotically stable	spiral point	asymptotically stable
pure imaginary	center	stable	center or spiral point	indeterminate

Note: Here r_1 and r_2 denote the roots of the characteristic equation (6.15) for the linear system corresponding to the almost linear system (6.14).

Now we analyze the competing species system (6.1). Using Theorem 6.2 and

Table 6. 2, we present a detailed analysis of the phase plane diagram and discuss the stability for a particular competing species problem.

Assume the populations satisfy the almost linear system

$$\begin{cases} \dfrac{dx}{dt}=x(7-x-2y)=7x-x^2-2xy, \\ \dfrac{dy}{dt}=y(5-y-x)=5y-y^2-xy. \end{cases} \tag{6. 16}$$

Then the critical points for this system are solutions to the pair of equations

$$\begin{cases} x(7-x-2y)=0, \\ y(5-y-x)=0. \end{cases}$$

One solution is (0, 0). Setting $x=0$, we also obtain $y=5$ and setting $y=0$, we get $x=7$. So three of the critical points are (0, 0), (0, 5), and (7, 0). While these points are not strictly inside the first quadrant, they do lie on its border. Therefore, knowing the behavior of trajectories near these points will help us to obtain a phase diagram for the first quadrant. A fourth critical point occurs at the intersection of the lines

$$\begin{cases} 7-x-2y=0, \\ 5-y-x=0, \end{cases}$$

which is the point (3, 2). We begin with an analysis of the trajectories near the origin (0, 0).

Critical Point (0, 0). We get the corresponding linear system by deleting the second-order terms in system (6. 16). This yields

$$\frac{dx}{dt}=7x, \quad \frac{dy}{dt}=5y.$$

The characteristic equation for this system is $(r-7)(r-5)=0$, which has roots $r_1=7$, $r_2=5$. Thus, the origin is an unstable improper node. Furthermore, this system is easy to solve (the variables are uncoupled!). We find that $x(t)=Ae^{7t}$ and $y(t)=Be^{5t}$, so we can directly compute the trajectories. They are $y=B(e^t)^5=(B/A^{5/7})x^{5/7}$. Observe that the trajectories are tangent to the y-axis and move away from the origin. In Figure 6. 14, we have sketched the phase plane diagram in the first quadrant near the

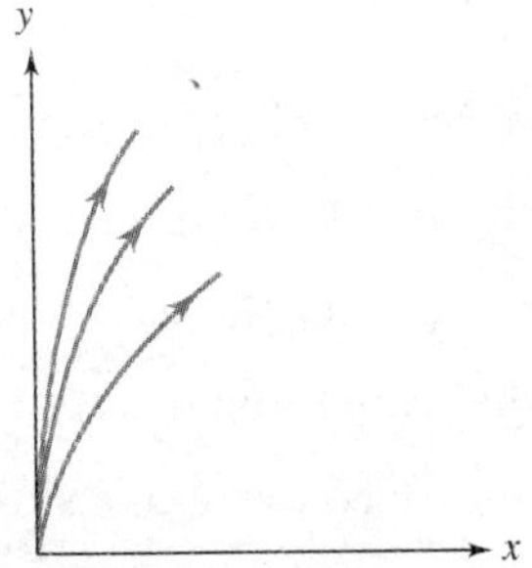

Figure 6. 14 Phase plane diagram near (0,0)

origin.

Critical Point (0, 5). We begin by translating the point (0, 5) to the origin by making the change of variables $x=u$, $y=v+5$. Substituting into system (6.16) and simplifying, we obtain the almost linear system

$$\begin{cases}\dfrac{du}{dt}=-3u-u^2-2uv,\\ \dfrac{dv}{dt}=-5u-5v-v^2-uv.\end{cases}$$

The corresponding linear system is

$$\frac{du}{dt}=-3u,\quad \frac{dv}{dt}=-5u-5v,$$

which has the characteristic equation $(r+3)(r+5)=0$ with roots $r_1=-3$, $r_2=-5$. Hence, the point (0, 5) is an asymptotically stable improper node. To find the transformed axes $\hat{u}$ and $\hat{y}$, we set $v=mu$. Substituting into

$$\frac{dv}{du}=\frac{dv/dt}{du/dt}=\frac{-5u-5v}{-3u},$$

we obtain

$$m=\frac{-5-5m}{-3},$$

and solving for m, we find $m=-5/2$. Thus, one transformed axis is $v=-(5/2)u$. What about the other axis? By substituting $v=mu$, we allow for all lines through the origin except the vertical line $u=0$. Thus, we suspect that the v-axis ($u=0$) is the missing transformed axis.

We can confirm this by solving for du/dv:

$$\frac{du}{dv}=\frac{-3u}{-5u-5v},$$

which is indeed satisfied by $u=0$. Therefore, the transformed axes are $v=-(5/2)u$ and $u=0$. If we check the direction field along the u-axis, we see that $dy/du=5/3$. Moreover, $dv/du=0$ along the line $v=-u$. Consequently, the trajectories approach the transformed axis $v=-(5/2)u$ tangentially. The phase plane diagram for the linearized system is sketched in Figure 6.15(a), and the diagram near the critical point (0,5) of the original system is shown in Figure 6.15(b).

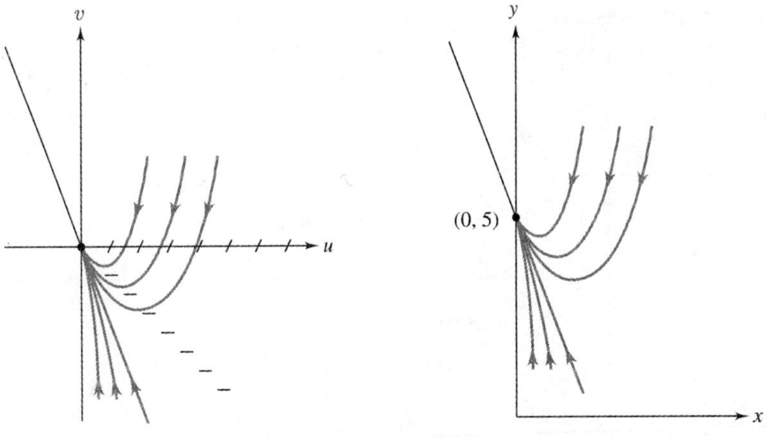

(a) About (0, 0) in uv-plane (b) About (0, 5) in xy-plane

Figure 6.15 Phase plane diagram near the critical point (0, 5)

Critical Point (7, 0). To translate the point (7, 0) to the origin, we set $x=u+7, y=v$. Substituting into system (6.16) and simplifying yields the almost linear system

$$\begin{cases}\dfrac{du}{dt}=-7u-14v-u^2-2uv,\\[2ex] \dfrac{dv}{dt}=-2v-v^2-uv.\end{cases}$$

The corresponding linear system

$$\frac{du}{dt}=-7u-14v,\quad \frac{dv}{dt}=-2v$$

has the characteristic equation $(r+7)(r+2)=0$ with roots $r_1=-7$, $r_2=-2$. Thus, the point (7, 0) is an asymptotically stable improper node. To get the transformed axes, we set $v=mu$. For the linear system, substituting into

$$\frac{dv}{du}=\frac{dv/dt}{du/dt}=\frac{-2v}{-7u-14v},$$

we obtain

$$m=\frac{-2m}{-7-14m},$$

which yields $m=0, -5/14$. Therefore, the two transformed axes are $v=0$ and $v=-(5/14)u$. We now check the direction field along a few lines. When $u=0$, $dy/du=1/7$, and when $v=-u$, $dy/du=2/7$. Moreover, dy/du is undefined when $v=-(1/2)u$. Consequently, the trajectories approach the transformed axis $v=-(5/14)u$ tangentially. The phase plane diagrams in the uv- and xy-planes are sketched in Figure 6.16(a) and (b).

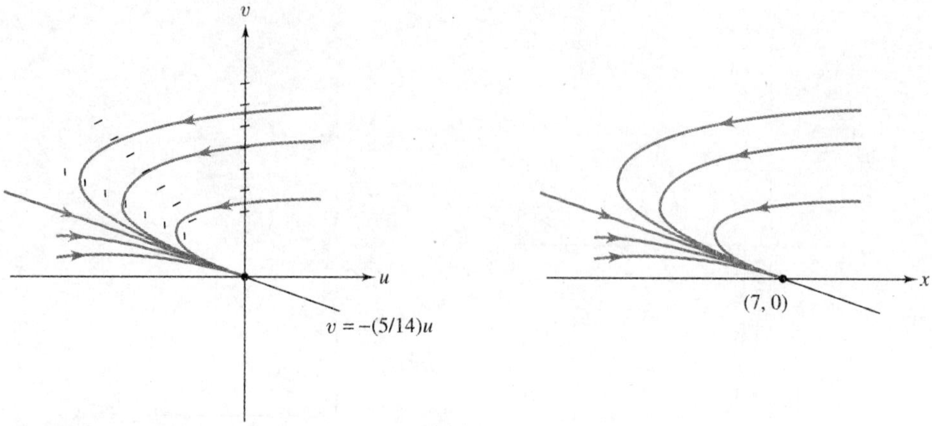

(a) About (0, 0) in uv-plane (b) About (7, 0) in xy-plane

Figure 6.16 Phase plane diagram near the critical point (7, 0)

Critical Point (3, 2). Setting $x=u+3$, $y=v+2$ and substituting into system (6.16), we obtain the almost linear system

$$\begin{cases}\dfrac{\mathrm{d}u}{\mathrm{d}t}=-3u-6v-u^2-2uv,\\[2mm]\dfrac{\mathrm{d}v}{\mathrm{d}t}=-2u-2v-v^2-uv.\end{cases}$$

The corresponding linear system has the characteristic equation $r^2+5r-6=(r+6)(r-1)=0$ with roots $r=-6, 1$. Hence, the point (3, 2) is an unstable saddle point. To obtain the transformed axes, we set $v=mu$. For the linear system, substituting into

$$\frac{\mathrm{d}v}{\mathrm{d}u}=\frac{\mathrm{d}v/\mathrm{d}t}{\mathrm{d}u/\mathrm{d}t}=\frac{-2u-2v}{-3u-6v},$$

we obtain

$$m=\frac{-2-2m}{-3-6m}.$$

Solving for m yields $m=1/2, -2/3$. Thus, the two transformed axes are $v=(1/2)u$ and $v=-(2/3)u$. For a saddle point, the trajectories move toward the critical point alongside one transformed axis and away alongside the other. If we set $v=-(2/3)u$, then for the linear system, we have $\mathrm{d}u/\mathrm{d}t=-3u-6(-2/3)u=u$. The trajectories along this transformed axis are moving away from the origin because $\mathrm{d}u/\mathrm{d}t>0$ if $u>0$ and $\mathrm{d}u/\mathrm{d}t<0$ for $u<0$. Checking the line $v=(1/2)u$, we see that $\mathrm{d}u/\mathrm{d}t=-3u-6(1/2)u=-6u$, so the trajectories are moving toward the origin. The phase plane diagrams in the uv- and xy-planes are sketched in Figure 6.17.

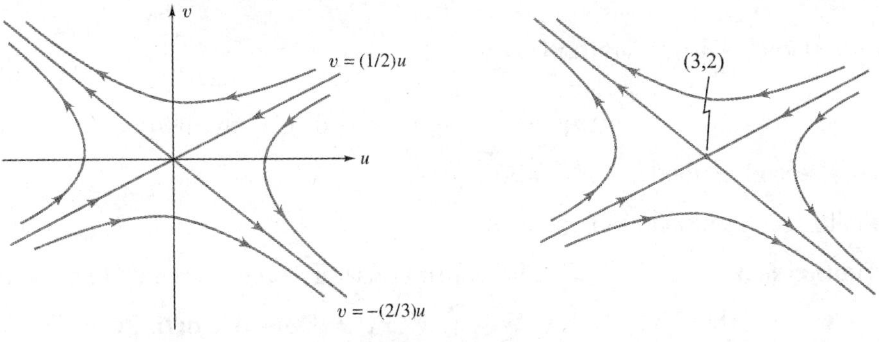

(a) About (0, 0) in uv-plane (b) About (3, 2) in xy-plane

Figure 6.17 Phase plane diagram near the critical point (3, 2)

In Figure 6.18, we have combined the results of each of the four critical points and sketched the phase plane diagram in the first quadrant for the system (6.16). We have included four trajectories that separate the first quadrant into four regions labeled I, II, III, and IV. (These trajectories are difficult to find and the curves sketched represent only rough approximations of the true curves.) These curves are called separatrices. Observe that a trajectory in region I remains in that region and approaches the critical point (0, 5) as $t \to +\infty$. Similarly, trajectories in region II also approach (0, 5) as $t \to +\infty$. The trajectories in regions III and IV approach the critical point (7, 0) as $t \to +\infty$. Regions I and II form part of the regions of asymptotic stability for the critical point (0, 5); that is, a trajectory beginning in that region is attracted to (0, 5). Similarly, regions III and IV are part of the regions of asymptotic stability for the critical point (7, 0).

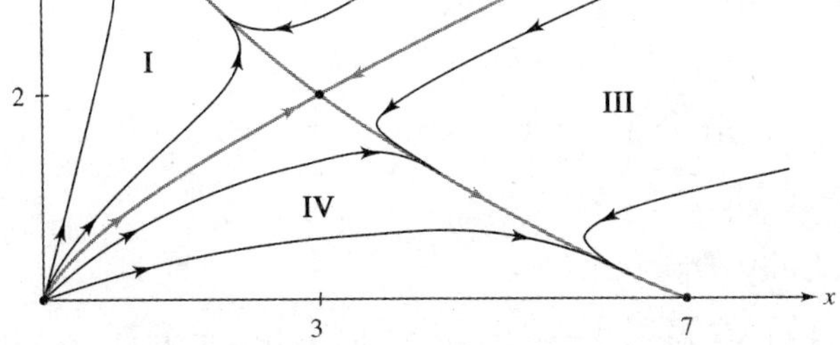

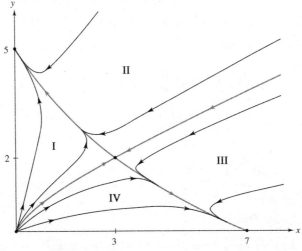

Figure 6.18 Phase plane diagram for the competing species system (6.16)

6.4 Lyapunov's Direct Method

In this section, we introduce Lyapunov's direct method (also known as **Lyapunov's second method**).

Definite and Semidefinite Functions

Definition 6.3 Let $V(x, y)$ be continuous on some domain D containing the origin and assume that $V(0,0)=0$. We say $V(x,y)$ is **positive definite** on D if $V(x,y)>0$ for all (x,y) in D except the origin.

If $V(x,y)\geqslant 0$ for all (x,y) in D, we say that $V(x,y)$ is **positive semidefinite** on D.

Similarly, we define **negative definite** and **negative semidefinite** by reversing the appropriate signs in the above inequalities.

The Derivative $\dot{V}(x,y)$

Definition 6.4 Suppose the first-order partial derivatives of $V(x, y)$ are continuous in a domain D containing the origin. The derivative of V with respect to the system

$$\frac{\mathrm{d}x}{\mathrm{d}t}=f(x,y), \quad \frac{\mathrm{d}y}{\mathrm{d}t}=g(x,y)$$

is the function

$$\dot{V}(x,y):=V_x(x,y)f(x,y)+V_y(x,y)g(x,y).$$

Remark As we saw in the preceding example, the derivative of $V(x,y)$ with respect to the system, when evaluated at a solution $x=\varphi(t)$, $y=\psi(t)$ of the system, is just the total derivative of $V(\varphi(t),\psi(t))$ with respect to t; that is,

$$\begin{aligned}\frac{\mathrm{d}V}{\mathrm{d}t}(\varphi,\psi)&=V_x(\varphi,\psi)\frac{\mathrm{d}\varphi}{\mathrm{d}t}+V_y(\varphi,\psi)\frac{\mathrm{d}\psi}{\mathrm{d}t}\\&=V_x(\varphi,\psi)f(\varphi,\psi)+V_y(\varphi,\psi)g(\varphi,\psi)\\&=\dot{V}(\varphi,\psi).\end{aligned}$$

We now state and prove Lyapunov's theorem, which gives sufficient conditions for the origin to be stable.

Lyapunov's Stability Theorem

Theorem 6.3 Let $V(x,y)$ be a positive definite function on a domain D containing the origin. Suppose the origin is an isolated critical point for the system

$$\begin{cases}\dfrac{dx}{dt}=f(x,y),\\ \dfrac{dy}{dt}=g(x,y)\end{cases} \tag{6.17}$$

and let $\dot{V}(x,y)$ be the derivative of $V(x,y)$ with respect to this system.

(i) If $\dot{V}(x,y)$ is negative semidefinite on D, then the origin is stable.

(ii) If $\dot{V}(x,y)$ is negative definite on D, then the origin is asymptotically stable.

Proof (i) Let B_r denote the open disk

$$B_r := \{(x,y): x^2+y^2<r^2\}$$

and $\overline{B}_r$ be the closed disk of radius r. For a given $\varepsilon>0$, we need to determine $\delta>0$ such that for any solution $x=\varphi(t)$, $y=\psi(t)$ of system (6.17) satisfying $(\varphi(t_0), \psi(t_0))\in B_\delta$, we have $(\varphi(t),\psi(t))\in B_\varepsilon$ for all $t\geqslant t_0$. To specify δ, we proceed as follows.

Choose $\alpha>0$ so that $\overline{B}_\alpha\subset D$ and, for $\varepsilon<\alpha$, let $\overline{A}_{\varepsilon,\alpha}$ denote the closed annulus

$$\overline{A}_{\varepsilon,\alpha} := \{(x,y): \varepsilon^2\leqslant x^2+y^2\leqslant \alpha^2\}$$

with inner radius ε and outer radius α. Let μ denote the minimum value of V on $\overline{A}_{\varepsilon,\alpha}$. Since V is continuous and is positive everywhere except at the origin, we have $\mu>0$. Also, since V is zero at the origin, we can select a $\delta>0$ such that $\delta<\varepsilon$ and $V(x,y)<\mu$ for all (x,y) in B_δ. We now show that for this δ, the stability condition is satisfied.

Let $0<\varepsilon<\alpha$ and $x=\varphi(t)$, $y=\psi(t)$ be a nontrivial solution to system (6.17) such that $(\varphi(t_0),\psi(t_0))\in B_\delta$. If $(\varphi(t),\psi(t))\in B_\varepsilon$ for all $t\geqslant t_0$, then we are done. If this is not the case, then there is a first value of t, say $t_1>t_0$, where the trajectory touches the boundary of B_ε. Then on the interval $[t_0, t_1]$, we have $(\varphi(t),\psi(t))\in \overline{B}_\varepsilon\subset B_\alpha$.

Since $\dot{V}$ is negative semidefinite, we have for $t\in[t_0, t_1]$,

$$\begin{aligned} V(\varphi(t),\psi(t))-V(\varphi(t_0),\psi(t_0)) &= \int_{t_0}^{t}\frac{dV}{ds}(\varphi(s),\psi(s))ds \\ &= \int_{t_0}^{t}\dot{V}(\varphi(s),\psi(s))ds \\ &\leqslant 0. \end{aligned}$$

Thus,

$$V(\varphi(t),\psi(t))\leqslant V(\varphi(t_0),\psi(t_0)) \quad \text{for } t\in[t_0,t_1]. \tag{6.18}$$

But $V(\varphi(t_0),\psi(t_0))<\mu$ because $(\varphi(t_0),\psi(t_0))\in B_\delta$, and $V(\varphi(t_1),\psi(t_1))\geqslant\mu$, since

$(\varphi(t_1),\psi(t_1))\in\overline{A}_{\varepsilon,\alpha}$. Consequently, from (6.18),

$$\mu\leqslant V(\varphi(t_1),\psi(t_1))\leqslant V(\varphi(t_0),\psi(t_0))<\mu,$$

which is a contradiction. Therefore $(\varphi(t),\psi(t))\in B_\varepsilon$ for all $t\geqslant t_0$. Hence, the stability is established.

(ii) Here $\dot{V}$ is negative definite on D, so we already know that the origin is stable from part (i). It remains to be shown that there exists an $\eta>0$ such that if $x=\varphi(t)$, $y=\psi(t)$ is a solution to system (6.17) for which $(\varphi(t_0),\psi(t_0))$ lies in B_η, then

$$\lim_{t\to+\infty}\varphi(t)=0, \quad \lim_{t\to+\infty}\psi(t)=0.$$

Now if we can prove that $V(\varphi(t),\psi(t))\to 0$ as $t\to+\infty$, then it will follow from the positive definiteness of V that $\varphi(t)\to 0$ and $\psi(t)\to 0$ as $t\to+\infty$.

To specify the constant η, take $\varepsilon=\alpha/2$ and for this ε choose δ as in the proof of part (i). We will show that with $\eta=\delta$ the asymptotic stability criterion is satisfied.

Suppose to the contrary that there exists a solution $x=\varphi(t)$, $y=\psi(t)$ to system (6.17) such that $(\varphi(t_0),\psi(t_0))$ lies in B_η but $(\varphi(t),\psi(t))$ does not approach the origin as $t\to+\infty$. Since $V(\varphi(t),\psi(t))$ is decreasing for $t\geqslant t_0$ [see (6.18)], it follows that there is an $L>0$ such that

$$V(\varphi(t),\psi(t))\geqslant L>0 \quad \text{for all } t\geqslant t_0.$$

Also, since V is continuous and $V(0,0)=0$, there exists an $h>0$ such that

$$V(x,y)<L \quad \text{for all } (x,y)\in B_h.$$

Hence, the trajectory $(\varphi(t),\psi(t))$ must lie in the annulus $\overline{A}_{h,\alpha/2}$ for $t\geqslant t_0$.

Let $\gamma(<0)$ be the maximum value of $\dot{V}$ on the annulus $\overline{A}_{h,\alpha/2}$. Then, since the trajectory $(\varphi(t),\psi(t))$ lies in $\overline{A}_{h,\alpha/2}$, we have

$$\frac{\mathrm{d}V}{\mathrm{d}t}(\varphi(t),\psi(t))\leqslant\gamma.$$

So, integrating we find

$$V(\varphi(t),\psi(t))\leqslant V(\varphi(t_0),\psi(t_0))+\gamma(t-t_0) \quad \text{for all } t\geqslant t_0.$$

Since $\gamma<0$, it now follows that, for t sufficiently large, $V(\varphi(t),\psi(t))<0$, which contradicts the fact that V is positive definite. Hence, we must have $V(\varphi(t),\psi(t))\to 0$ as $t\to+\infty$. ◆

In the next example, we use Lyapunov's direct method to determine the stability of the origin for an autonomous system whose corresponding linear system does not have the origin as an isolated critical point; such a system cannot be analyzed using the results of Section 6.3.

Example 6.5 Use Lyapunov's direct method to determine the stability of the origin for the system

$$\begin{cases} \dfrac{dx}{dt} = -2y^3, \\ \dfrac{dy}{dt} = x - 3y^3. \end{cases} \tag{6.19}$$

Solution The origin is an isolated critical point for this system, since the pair of equations

$$-2y^3 = 0, \quad x - 3y^3 = 0$$

has only $x = y = 0$ as a solution. [Note, however, that for the corresponding linear system $dx/dt = 0$, $dy/dt = x$, the origin is not an isolated critical point ($ad - bc = 0$). Hence, system (6.19) is not an almost linear system.]

Since $f(x, y) = -2y^3$ and $g(x, y) = x - 3y^3$, let's try to find a Lyapunov function of the form $V(x, y) = ax^2 + by^4$, with a, $b > 0$. Clearly, V is positive definite on any domain containing the origin.

Next, computing $\dot{V}(x, y)$, we find

$$\begin{aligned} \dot{V}(x, y) &= V_x(x, y) f(x, y) + V_y(x, y) g(x, y) \\ &= (2ax)(-2y^3) + (4by^3)(x - 3y^3) \\ &= 4(b - a)xy^3 - 12by^6. \end{aligned}$$

So, if we take $a = b = 1$, then

$$\dot{V}(x, y) = -12y^6 \leqslant 0,$$

and we get that $V(x, y)$ is negative semidefinite. It now follows from Theorem 6.3 that the origin is a stable critical point for the system (6.19). ◆

The next result will give conditions for the instability of a system.

Lyapunov's Instability Theorem

Theorem 6.4 Suppose the origin is an isolated critical point for the autonomous system (6.17) and let $V(x, y)$ with $V(0, 0) = 0$ be defined and continuous on a domain D containing the origin. Suppose further that $V(x, y)$ is positive definite on D. If, in every disk centered at the origin, there exists some point (a, b) such that $V(a, b)$ is positive, then the origin is unstable.

Proof Fix $t_0 > 0$ and let $\alpha > 0$ be chosen so that $\overline{B}_\alpha \subset D$, where $\overline{B}_\alpha$ is the closed disk with center (0, 0) and radius α. Since V is continuous, there exists an $L>0$ such that $|V(x,y)|\leqslant L$ for all (x,y) in $\overline{B}_\alpha$. We will show that no matter how small we take $\varepsilon>0$, there is a solution $x=\varphi(t)$, $y=\psi(t)$ with $(\varphi(t_0),\psi(t_0))$ in B_ε such that the trajectory $(\varphi(t),\psi(t))$ eventually leaves the disk B_α. Thereby, it follows that the origin is unstable.

Now, by hypothesis, given ε, with $0<\varepsilon<\alpha$, there exists a point (a, b) in B_ε for which $V(a,b)>0$. Let $x=\varphi(t)$, $y=\psi(t)$ be the solution to system (6.17) satisfying the initial condition

$$\varphi(t_0)=a,\quad \psi(t_0)=b.$$

Also, let t_1 be the first value of $t>t_0$ for which the trajectory $(\varphi(t),\psi(t))$ leaves B_α [$t_1=+\infty$ means $(\varphi(t),\psi(t))$ remains in B_α]. Since $\dot{V}$ is positive definite on B_α, we have

$$\begin{aligned}V(\varphi(t),\psi(t))-V(\varphi(t_0),\psi(t_0))&=\int_{t_0}^{t}\frac{\mathrm{d}V}{\mathrm{d}s}(\varphi(s),\psi(s))\mathrm{d}s\\&\geqslant 0\end{aligned}$$

for all t in $[t_0, t_1)$. If we let V_0 denote $V(\varphi(t_0),\psi(t_0))$, then $V_0>0$ and, for t in $[t_0, t_1)$,

$$V(\varphi(t),\psi(t))\geqslant V_0.$$

Now since V is continuous, there exists a $\lambda>0$ such that $|V(x,y)|<V_0$ for (x,y) in B_λ. Therefore, the trajectory $(\varphi(t),\psi(t))$ must lie in the annulus $\overline{A}_{\lambda,\alpha}$ for $t\in[t_0, t_1)$. Let $\mu>0$ be the minimum value of $\dot{V}$ on $\overline{A}_{\lambda,\alpha}$. Then, for t in $[t_0,t_1)$, we have

$$\begin{aligned}V(\varphi(t),\psi(t))-V(\varphi(t_0),\psi(t_0))&=\int_{t_0}^{t}\frac{\mathrm{d}V}{\mathrm{d}s}(\varphi(s),\psi(s))\mathrm{d}s\\&\geqslant \mu(t-t_0).\end{aligned}$$

That is,

$$V(\varphi(t),\psi(t))\geqslant V_0+\mu(t-t_0)\quad \text{for } t\in[t_0,t_1).$$

If $t_1=+\infty$, then it follows that $V(\varphi(t),\psi(t))\to+\infty$ as $t\to+\infty$. But this is absurd, since $|V(x,y)|\leqslant L$ for (x,y) in B_α. Hence, $t_1<\infty$; that is, the trajectory $(\varphi(t),\psi(t))$ leaves the disk B_α. ◆

Example 6.6 Determine the stability of the critical point at the origin for the system

$$\begin{cases}\dfrac{\mathrm{d}x}{\mathrm{d}t}=-y^3,\\[2mm]\dfrac{\mathrm{d}y}{\mathrm{d}t}=-x^3.\end{cases}\tag{6.20}$$

Solution The origin is not an isolated critical point for the corresponding linear system, so we cannot use the comparison approach of Section 6.3. Suspecting that the origin is unstable, we will attempt to apply Theorem 6.4.

Notice that $f(x,y)=-y^3$ is a function only of y and $g(x,y)=-x^3$ a function only of x. So let's try $V(x,y)=-xy$. Then V is continuous, $V(0,0)=0$, and in every disk centered at the origin, V is positive at some point (namely, those points of the disk lying in the second and fourth quadrants). Moreover,

$$\dot{V}(x,y)=(-y)(-y^3)+(-x)(-x^3)=y^4+x^4,$$

which is positive definite in any domain containing the origin. Hence, by Lyapunov's instability theorem, system (6.20) is unstable.

Exercises

In Problems 1-7, use Lyapunov's direct method to determine the stability of the origin for the given system.

1. $\frac{dx}{dt}=-3x^3+2xy^2$, $\frac{dy}{dt}=-y^3$.

[Hint: Try $V(x,y)=x^2+y^2$.]

2. $\frac{dx}{dt}=-y-x^3-xy^2$, $\frac{dy}{dt}=x-x^2y-y^3$.

[Hint: Try $V(x,y)=x^2+y^2$.]

3. $\frac{dx}{dt}=x^3y+x^2y^3-x^5$, $\frac{dy}{dt}=-2x^4-6x^3y^2-2y^5$.

[Hint: Try $V(x,y)=2x^2+y^2$.]

4. $\frac{dx}{dt}=2x^3$, $\frac{dy}{dt}=2x^2y-y^3$.

[Hint: Try $V(x,y)=x^2-y^2$.]

5. $\frac{dx}{dt}=x+y+y^3$, $\frac{dy}{dt}=x+y+x^5$.

[Hint: Try $V(x,y)=xy$.]

6. $\frac{dx}{dt}=2y-x^3$, $\frac{dy}{dt}=-x^3-y^5$.

[Hint: Try $V(x,y)=ax^4+by^2$.]

7. $\frac{dx}{dt}=y^3-2x^3$, $\frac{dy}{dt}=-3x-y^3$.

[Hint: Try $V(x,y)=ax^2+by^4$.]

In Problems 8-12, use Lyapunov's direct method to determine the stability of the zero solution for the given equation.

8. $\dfrac{d^2x}{dt^2}+|x|\dfrac{dx}{dt}+x^3=0.$

9. $\dfrac{d^2x}{dt^2}+(1-x^2)\dfrac{dx}{dt}+x=0.$

10. $\dfrac{d^2x}{dt^2}+\left[1-\left(\dfrac{dx}{dt}\right)^2\right]\dfrac{dx}{dt}+x=0.$

11. $\dfrac{d^2x}{dt^2}+x^2\dfrac{dx}{dt}+x^5=0.$

12. $\dfrac{d^2x}{dt^2}+\dfrac{dx}{dt}+\sin x=0.$

[Hint: Use $V(x,y)=(1/2)(x+y)^2+x^2+(1/2)y^2$, where $y=dx/dt$.]

Chapter 7

Fractional Differential Equations

Fractional calculus deals with the study of so-called fractional-order integral and derivative operators over real or complex domains and their applications.

In the last decade, fractional calculus has been recognized as one of the best tools to describe long-memory processes. Such models are interesting for engineers and physicists but also for pure mathematicians. The most important among such models are those described by differential equations containing fractional-order derivatives. Their evolutions behave in a much more complex way than in the classical integer-order case and the study of the corresponding theory is a hugely demanding task. Although some results of qualitative analysis for fractional differential equations can be similarly obtained, many classical methods are hardly applicable directly to fractional differential equations. New theories and methods are thus required to be specifically developed, whose investigation becomes more challenging. Comparing with classical theory of differential equations, the researches on the theory of fractional differential equations are only on their initial stage of development.

7.1 Riemann-Liouville Integrals

Definition 7.1 Let $n\in\mathbf{R}_+$. The operator J_a^n, defined on $L_1[a,b]$ by

$$J_a^n f(x):=\frac{1}{\Gamma(n)}\int_a^x (x-1)^{n-1}f(t)\,\mathrm{d}t$$

for $a\leqslant x\leqslant b$, is called the **Riemann-Liouville fractional integral operator of order *n*.**

For $n=0$, we set $J_a^0:=I$, the identity operator.

The definition for $n=0$ is quite convenient for future manipulations. It is evident that the Riemann-Liouville fractional integral coincides with the classical definition of J_a^n in the case $n\in\mathbf{N}$, except for the fact that we have extended the domain from Riemann integrable functions to Lebesgue integrable functions (which will not lead to any problems in our development). Moreover, in the case $n \geqslant 1$, it is obvious that the integral $J_a^n f(x)$ exists for every $x \in [a,b]$ because the integrand is the product of an integrable function f and the continuous function $(x-\cdot)^{n-1}$. In the case $0<n<1$ though, the situation is less clear at first sight. However, the following result asserts that this definition is justified.

Theorem 7.1 Let $f \in L_1[a,b]$ and $n>0$. Then, the integral $J_n^a f(x)$ exists for almost every $x \in [a,b]$. Moreover, the function $J_a^n f$ itself is also an element of $L_1[a,b]$.

Proof We write the integral in question as

$$\int_a^x (x-t)^{n-1} f(t)\mathrm{d}t = \int_{-\infty}^{\infty} \varphi_1(x-t)\varphi_2(t)\mathrm{d}t,$$

where

$$\varphi_1(u)=\begin{cases} u^{n-1}, & \text{for } 0<u\leqslant b-a, \\ 0, & \text{else,} \end{cases}$$

and

$$\varphi_2(u)=\begin{cases} f(u), & \text{for } a\leqslant u\leqslant b, \\ 0, & \text{else.} \end{cases}$$

By construction, $\varphi_j \in L_1(\mathbf{R})$ for $j \in \{1,2\}$, and thus by a classical result on Lebesgue integration the desired result follows.

One important property of integer-order integral operators is preserved by our generalization.

Theorem 7.2 Let $m,n \geqslant 0$ and $\varphi \in L_1[a,b]$. Then,

$$J_a^m J_a^n \varphi = J_a^{m+n} \varphi$$

holds almost everywhere on $[a,b]$. If additionally $\varphi \in C[a,b]$ or $m+n \geqslant 1$, then the identity holds everywhere on $[a,b]$.

Corollary 7.1 Under the assumptions of Theorem 7.2,

$$J_a^m J_a^n \varphi = J_a^n J_a^m \varphi.$$

Example 7.1 Let $f(x)=(x-a)^\beta$ for some $\beta>-1$ and $n>0$. Then,

$$J_a^n f(x) = \frac{\Gamma(\beta+1)}{\Gamma(n+\beta+1)}(x-a)^{n+\beta}.$$

Solution In view of the well known corresponding result in the case $n \in \mathbf{N}$, this result is precisely what one would expect from a sensible generalization of the integral operator.

$$J_a^n f(x)=\frac{1}{\Gamma(n)}\int_a^x (t-a)^\beta (x-t)^{n-1}\mathrm{d}t$$
$$=\frac{1}{\Gamma(n)}(x-a)^{n+\beta}\int_0^1 s^\beta (1-s)^{n-1}\mathrm{d}s$$
$$=\frac{\Gamma(\beta+1)}{\Gamma(n+\beta+1)}(x-a)^{n+\beta}. \quad \blacklozenge$$

Next we discuss the interchange of limit operation and fractional integration.

Theorem 7.3 Let n>0. Assume that $(f_k)_{k=1}^\infty$ is a uniformly convergent sequence of continuous functions on $[a,b]$. Then we may interchange the fractional integral operator and the limit process, i. e. ,

$$(J_a^n \lim_{k\to\infty} f_k)(x)=(\lim_{k\to\infty} J_a^n f_k)(x).$$

In particular, the sequence of functions $(J_a^n f_k)_{k=1}^\infty$ is uniformly convergent.

Proof We denote the limit of the sequence (f_k) by f. It is well known that f is continuous. We then find

$$|J_a^n f_k(x)-J_a^n f(x)|\leqslant\frac{1}{\Gamma(n)}\int_a^x |f_k(t)-f(t)|(x-t)^{n-1}\mathrm{d}t$$
$$\leqslant\frac{1}{\Gamma(n)}\| f_k-f\|_\infty\int_a^x (x-t)^{n-1}\mathrm{d}t$$
$$=\frac{1}{\Gamma(n+1)}\| f_k-f\|_\infty (x-a)^n$$
$$\leqslant\frac{1}{\Gamma(n+1)}\| f_k-f\|_\infty (b-a)^n,$$

which converges to zero as $k\to\infty$ uniformly for all $x\in[a,b]$.

Corollary 7.2 Let f be analytic in $(a-h,a+h)$ for some $h>0$, and let $n>0$. Then

$$J_a^n f(x)=\sum_{k=0}^\infty \frac{(-1)^k (x-a)^{k+n}}{k!(n+k)\Gamma(n)}D^k f(x)$$

for $a\leqslant x<a+h/2$, and

$$J_a^n f(x)=\sum_{k=0}^\infty \frac{(x-a)^{k+n}}{\Gamma(k+1+n)}D^k f(a)$$

for $a\leqslant x<a+h$. In particular, $J_a^n f$ is analytic in $(a,a+h)$.

Proof For the first statement, we use the definition of the Riemann-Liouville integral operator J_a^n, viz.

$$J_a^n f(x)=\frac{1}{\Gamma(n)}\int_a^x f(t)(x-t)^{n-1}\mathrm{d}t,$$

and expand $f(t)$ into a power series about x. Since $x\in[a,a+h/2)$, the power series converges in the entire interval of integration. Thus, by Theorem 7.3, it is legal to exchange summation and integration. Then we use the explicit representation for the fractional integral of the power function that we had derived in Example 7.1 to find the final result.

For the second statement, we proceed in a similar way, but we now expand the power series at a and not at x. This allows us again to conclude the convergence of the series in the required interval.

The analyticity of $J_a^n f$ follows immediately from the second statement.

In the last two theorems of this section, we introduce another important property of fractional integral operators, namely the continuity with respect to the order of the operator. We first look at the case that we work in Lebesgue spaces. Under this assumption, the situation is very simple.

Theorem 7.4 Let $1\leqslant p<\infty$ and let $(m_k)_{k=1}^{\infty}$ be a convergent sequence of nonnegative numbers with limit m. Then, for every $f\in L_p[a,b]$,

$$\lim_{k\to\infty} J_a^{m_k} f = J_a^m f$$

where the convergence is in the sense of the $L_p[a,b]$ norm.

Theorem 7.5 Let $f\in C[a,b]$ and $m\geqslant 0$. Moreover, assume that (m_k) is a sequence of positive numbers such that $\lim_{k\to\infty} m_k=m$. Then, for every $\varepsilon>0$,

$$\lim_{k\to\infty}\sup_{x\in[a+\varepsilon,b]}|J_a^{m_k}f(x)-J_a^m f(x)|=0.$$

If additionally $m>0$ or $f(x)=O((x-a)^{\delta})$ as $x\to a$ for some $\delta>0$, then

$$\lim_{k\to\infty}\| J_a^{m_k}f-J_a^m f\|_{\infty}=0.$$

7.2 Riemann-Liouville Derivatives

Having established these fundamental properties of Riemann-Liouville integral operators, we now come to the corresponding differential operators.

Definition 7.2 Let $n\in\mathbf{R}_+$ and $m=\lceil n\rceil$. The operator D_a^n, defined by

$$D_a^n f := D^m J_a^{m-n} f$$

is called the **Riemann-Liouville fractional differential operator of order n.**

For $n=0$, we set $D_a^0 := I$, the identity operator.

Lemma 7.1 Let $n\in\mathbf{R}_+$ and let $m\in\mathbf{N}$ such that $m>n$. Then,

$$D_a^n = D^m J_a^{m-n}$$

Lemma 7.2 Let $f\in A^1[a,b]$ and $0<n<1$. Then, $D_a^n f$ exists almost everywhere in $[a,b]$. Moreover, $D_a^n f\in L_p[a,b]$ for $1\leqslant p<1/n$ and

$$D_a^n f(x)=\frac{1}{\Gamma(1-n)}\left(\frac{f(a)}{(x-a)^n}+\int_a^x f'(t)(x-t)^{-n}\mathrm{d}t\right).$$

Example 7.2 Let $f(x)=(x-a)^\beta$ for some $\beta>-1$ and $n>0$. Then, in view of Example 7.1,

$$D_a^n f(x)=D^{[n]}J_a^{[n]-n}f(x)=\frac{\Gamma(\beta+1)}{\Gamma([n]-n+\beta+1)}D^{[n]}[(\cdot-a)^{[n]-n+\beta}](x).$$

Specifically, if $n-\beta\in\mathbf{N}$, then

$$D_a^n[(\cdot-a)^{n-m}](x)=0 \text{ for all } n>0,\ m\in\{1,2,\ldots,[n]\}.$$

On the other hand, if $n-\beta\notin\mathbf{N}$, we find

$$D_a^n[(\cdot-a)^\beta](x)=\frac{\Gamma(\beta+1)}{\Gamma(\beta+1-n)}(x-a)^{\beta-n}.\ \blacklozenge$$

Theorem 7.6 Assume that $n_1,n_2\geqslant 0$. Moreover, let $\varphi\in L_1[a,b]$ and $f=J_a^{n_1+n_2}\varphi$. Then,

$$D_a^{n_1}D_a^{n_2}f=D_a^{n_1+n_2}f.$$

Proof By our assumption on f and the definition of the Riemann-Liouville differential operator,

$$D_a^{n_1}D_a^{n_2}f=D_a^{n_1}D_a^{n_2}J_a^{n_1+n_2}\varphi=D^{[n_1]}J_a^{[n_1]-n_1}D^{[n_2]}J_a^{[n_2]-n_2}J_a^{n_1+n_2}\varphi.$$

The property of the integral operators allows us to rewrite this expression as

$$\begin{aligned}D_a^{n_1}D_a^{n_2}f&=D^{[n_1]}J_a^{[n_1]-n_1}D^{[n_2]}J_a^{[n_2]+n_1}\varphi\\&=D^{[n_1]}J_a^{[n_1]-n_1}D^{[n_2]}J_a^{[n_2]}J_a^{n_1}\varphi.\end{aligned}$$

This is equivalent to

$$D_a^{n_1}D_a^{n_2}f=D^{[n_1]}J_a^{[n_1]-n_1}J_a^{n_1}\varphi=D^{[n_1]}J_a^{[n_1]}\varphi.$$

Therefore,

$$D_a^{n_1}D_a^{n_2}f=\varphi.$$

The proof that $D_a^{n_1+n_2}f=\varphi$ goes along similar lines.

Theorem 7.7 Let $n\geqslant 0$. Then, for every $f\in L_1[a,b]$,

$$D_a^n J_a^n f=f$$

almost everywhere.

Proof The case $n=0$ is trivial for then D_a^n and J_a^n are both the identity operator.

For $n>0$, we proceed as in the proof of Theorem 7.6: Let $m=[n]$. Then, by the definition of D_a^n, the semigroup property of fractional integration (which may be applied here since $m \in \mathbf{N}$),

$$D_a^n J_a^n f(x) = D^m J_a^{m-n} J_a^n f(x) = D^m J_a^m f(x) = f(x).$$

Now we come to an analogue of Theorem 7.3.

Theorem 7.8 Let $n > 0$. Assume that $(f_k)_{k=1}^{\infty}$ is a uniformly convergent sequence of continuous functions on $[a,b]$, and that $D_a^n f_k$ exists for every k. Moreover, assume that $(D_a^n f_k)_{k=1}^{\infty}$ converges uniformly on $[a+\varepsilon, b]$ for every $\varepsilon > 0$. Then, for every $x \in (a,b]$, we have

$$(\lim_{k\to\infty} D_a^n f_k)(x) = (D_a^n \lim_{k\to\infty} f_k)(x).$$

Proof We recall that $D_a^n = D^{\lceil n \rceil} J_a^{\lceil n \rceil - n}$. By Theorem 7.3, the sequence $(J_a^{\lceil n \rceil - n} f_k)_k$ is uniformly convergent, and we may interchange the limit operation and the fractional integral. By assumption, the $\lceil n \rceil$th derivative of this series converges uniformly on every compact subinterval of $(a,b]$. Thus, by a standard theorem from analysis, we may also interchange the limit operator and the differential operator whenever $a < x \leqslant b$.

We can immediately deduce an analogue of Corollary 7.2.

Corollary 7.3 Let f be analytic in $(a-h, a+h)$ for some $h > 0$, and let $n > 0$, $n \notin \mathbf{N}$. Then

$$D_a^n f(x) = \sum_{k=0}^{\infty} \binom{n}{k} \frac{(x-a)^{k-n}}{\Gamma(k+1-n)} D^k f(x)$$

for $a < x < a + h/2$, and

$$D_a^n f(x) = \sum_{k=0}^{\infty} \frac{(x-a)^{k-n}}{\Gamma(k+1-n)} D^k f(a)$$

for $a < x < a+h$. In particular, $D_a^n f$ is analytic in $(a, a+h)$.

Theorem 7.9 Let f_1 and f_2 be two functions defined on $[a,b]$ such that $D_a^n f_1$ and $D_a^n f_2$ exist almost everywhere. Moreover, let C_1, $C_2 \in \mathbf{R}$. Then, $D_a^n(c_1 f_1 + c_2 f_2)$ exists almost everywhere, and

$$D_a^n(c_1 f_1 + c_2 f_2) = c_1 D_a^n f_1 + c_2 D_a^n f_2.$$

Proof This linearity property of the fractional differential operator is an immediate consequence of the definition of D_a^n.

Theorem 7.10 (Leibniz' formula for Riemann-Liouville operators) Let $n > 0$, and assume that f and g are analytic on $(a-h, a+h)$ with some $h > 0$. Then,

$$D_a^n[fg](x) = \sum_{k=0}^{\lceil n \rceil} \binom{n}{k} (D_a^k f)(x)(D_a^{n-k} g)(x) + \sum_{k=\lceil n \rceil + 1}^{\infty} \binom{n}{k} (D_a^k f)(x)(J_a^{k-n} g)(x)$$

for $a < x < a + h/2$.

Proof In view of Corollary 7. 3, we have

$$D_a^n[fg](x)=\sum_{k=0}^{\infty}\binom{n}{k}\frac{(x-a)^{k-n}}{\Gamma(k+1-n)}D^k[fg](x).$$

Now we apply the standard Leibniz formula to $D^k[fg]$ and interchange the order of summation. This yields

$$\begin{aligned}D_a^n[fg](x)&=\sum_{k=0}^{\infty}\binom{n}{k}\frac{(x-a)^{k-n}}{\Gamma(k+1-n)}\sum_{j=0}^{k}\binom{k}{j}D^j(x)D^{k-j}g(x)\\&=\sum_{j=0}^{\infty}\sum_{k=j}^{\infty}\binom{n}{k}\frac{(x-a)^{k-n}}{\Gamma(k+1-n)}\binom{k}{j}D^jf(x)D^{k-j}g(x)\\&=\sum_{j=0}^{\infty}D^jf(x)\sum_{l=0}^{\infty}\binom{n}{l+j}\frac{(x-a)^{l+j-n}}{\Gamma(l+j+1-n)}\binom{l+j}{j}D^lg(x).\end{aligned}$$

The observation

$$\binom{n}{l+j}\binom{l+j}{j}=\binom{n}{j}\binom{n-j}{l}$$

gives us

$$\begin{aligned}D_a^n[fg](x)&=\sum_{j=0}^{\infty}D^jf(x)\binom{n}{j}\sum_{l=0}^{\infty}\binom{n-j}{l}\frac{(x-a)^{l+j-n}}{\Gamma(l+j+1-n)}D^lg(x)\\&=\sum_{j=0}^{[n]}\binom{n}{j}D^jf(x)\sum_{l=0}^{\infty}\binom{n-j}{l}\frac{(x-a)^{l+j-n}}{\Gamma(l+j+1-n)}D^lg(x)\\&\quad+\sum_{j=[n]+1}^{\infty}\binom{n}{j}D^jf(x)\sum_{l=0}^{\infty}\binom{n-j}{l}\frac{(x-a)^{l+j-n}}{\Gamma(l+j+1-n)}D^lg(x).\end{aligned}$$

By the first parts of Corollaries 7. 3 and 7. 2, respectively, we may replace the inner sums, and the desired result follows.

Theorem 7. 11 Let $f\in C^m[a,b]$ for some $m\in\mathbf{N}$. Then,

$$\lim_{n\to m_-}D_a^nf=D^mf$$

in a pointwise sense on $(a,b]$. The convergence is uniform on $[a,b]$ if additionally $f(x)=O((x-a)^{m+\delta})$ for some $\delta>0$ as $x\to a_+$.

Lemma 7. 3 Let $n>0$, $n\notin\mathbf{N}$, and $m=[n]$. Assume that $f\in C^m[a,b]$ and $x\in[a,b]$. Then,

$$D_a^nf(x)=\frac{1}{\Gamma(-n)}\int_a^x(x-t)^{-n-1}f(t)\,dt.$$

7.3 Relations between Riemann-Liouville Integrals and Derivatives

Having established a theory of Riemann-Liouville differential and integral operators separately, we now investigate how they interact.

Theorem 7.12 Let $n>0$. If there exists some $\varphi \in L_1[a,b]$ such that $f=J_a^n\varphi$, then

$$J_a^n D_a^n f = f$$

almost everywhere.

Proof This is an immediate consequence of the previous result: We have, by definition of f and Theorem 7.7, that

$$J_a^n D_a^n f = J_a^n [D_a^n J_a^n \varphi] = J_a^n \varphi = f.$$

If f is not as required in the assumptions of Theorem 7.11, then we have a different representation for $J_a^n D_a^n f$.

Theorem 7.13 Let $n>0$ and $m=[n]+1$. Assume that f is such that $J_a^{m-n} f \in A^m[a,b]$. Then,

$$J_a^n D_a^n f(x) = f(x) - \sum_{k=0}^{m-1} \frac{(x-a)^{n-k-1}}{\Gamma(n-k)} \lim_{z\to a^+} D^{m-k-1} J_a^{m-n} f(z).$$

Specifically, for $0<n<1$, we have

$$J_a^n D_a^n f(x) = f(x) - \frac{(x-a)^{n-1}}{\Gamma(n)} \lim_{z\to a^+} J_a^{1-n} f(z).$$

Theorem 7.14 (Fractional Taylor Expansion) Under the assumptions of Theorem 7.13, we have

$$f(x) = \frac{(x-a)^{n-m}}{\Gamma(n-m+1)} \lim_{z\to a_+} J_a^{m-n} f(z) + \sum_{k=1}^{m-1} \frac{(x-a)^{k+n-m}}{\Gamma(k+n-m+1)} \lim_{z\to a_+} D_a^{k+n-m} f(z) + J_a^n D_a^n f(x).$$

Note that in the case $n\in \mathbf{N}$, we have $m=n+1$ and hence the limit outside the sum vanishes. We may thus retrieve the classical result (with m replaced by $m-1$).

7.4 Caputo's Derivative

It turns out that the Riemann-Liouville derivatives have certain disadvantages when trying to model real-world phenomena with fractional differential equations. We shall therefore now discuss a modified concept of a fractional derivative. As we will see below when comparing the two ideas, this second one seems to be better suited to such tasks.

Definition 7.3 Let $n \geqslant 0$ and $m = \lceil n \rceil$. Then, we define the operator $\hat{D}_a^n$ by

$$\hat{D}_a^n f := J_a^{m-n} D^m f$$

whenever $D^m f \in L_1[a,b]$.

Theorem 7.15 Let $n \geqslant 0$ and $m = \lceil n \rceil$. Moreover assume that $f \in A^m[a,b]$. Then,

$$\hat{D}_a^n f = D_a^n [f - T_{m-1}[f;a]]$$

almost everywhere. Here, $T_{m-1}|f;a|(x) = \sum_{k=0}^{m-1} \frac{f^{(k)}(a)}{k!}(x-a)$; in the case $m=0$, we define $T_{m-1}[f;a] := 0$.

Definition 7.4 Assume that $n \geqslant 0$ and that f is such that $D_a^n[f - T_{m-1}[f;a]]$ exists, where $m = \lceil n \rceil$. Then we define the function $D_{*a}^n f$ by

$$D_{*a}^n f := D_a^n [f - T_{m-1}[f;a]].$$

The operator D_a^n is called the **Caputo differential operator of order *n*.**

Once again we note for $n \in \mathbf{N}$ that $m = n$ and hence

$$D_{*a}^n f = D_a^n [f - T_{n-1}[f;a]] = D^n f - D^n (T_{n-1}[f;a]) = D^n f$$

because $T_{n-1}[f;a]$ is a polynomial of degree $n-1$ that is annihilated by the classical operator D^n. So in this case we recover the usual differential operator as well. In particular, D_{*a}^0 is once again the identity operator.

Taking into account the definition of the Caputo operator and Lemma 7.3, we obtain a direct consequence.

Lemma 7.4 Under the assumptions of Lemma 7.3, we have

$$D_{*a}^n f(x) = \frac{1}{\Gamma(-n)} \int_a^x (x-t)^{-n-1} (f(t) - T_{m-1}[f;a](t)) \mathrm{d}t.$$

Remark 7.1 As in the case of the Riemann-Liouville operators, we see that the Caputo derivatives are not local either.

The following result is the relation between the Riemann-Liouville operator and the Caputo operator.

Lemma 7.5 Let $n \geqslant 0$ and $m = \lceil n \rceil$. Assume that f is such that both $D_{*a}^n f$ and $D_a^n f$ exist. Then,

$$D_{*a}^n f(x) = D_a^n f(x) - \sum_{k=0}^{m-1} \frac{D^k f(a)}{\Gamma(k-n+1)} (x-a)^{k-n}.$$

Proof In view of the definition of the Caputo derivative and Example 7.2,

$$D_{*a}^{n}f(x)=D_{a}^{n}f(x)-\sum_{k=0}^{m-1}\frac{D^{k}f(a)}{k!}D_{a}^{n}[(\cdot-a)^{k}](x)$$

$$=D_{a}^{n}f(x)-\sum_{k=0}^{m-1}\frac{D^{k}f(a)}{\Gamma(k-n+1)}(x-a)^{k-n}.\ \blacklozenge$$

An immediate consequence of this Lemma is Lemma 7.6.

Lemma 7.6 Assume the hypotheses of Lemma 7.5. Then,

$$D_{a}^{n}f=D_{*a}^{n}f$$

holds if and only if f has an m-fold zero at a, i. e. if and only if

$$D^{k}f(a)=0 \quad \text{for} \quad k=0,1,\ldots,m-1.$$

When it comes to the composition of Riemann-Liouville integrals and Caputo differential operators, we find that the Caputo derivative is also a left inverse of the Riemann-Liouville integral.

Theorem 7.16 If f is continuous and $n\geqslant 0$, then

$$D_{*a}^{n}J_{a}^{n}f=f.$$

Once again, we find that the Caputo derivative is not the right inverse of the Riemann-Liouville integral.

Theorem 7.17 Assume that $n\geqslant 0$, $m=[n]$, and $f\in A^{m}[a,b]$. Then

$$J_{a}^{n}D_{*a}^{n}f(x)=f(x)-\sum_{k=0}^{m-1}\frac{D^{k}f(a)}{k!}(x-a)^{k}.$$

Proof By Theorem 7.15 and Definition 7.3, we have

$$D_{*a}^{n}f=\hat{D}_{a}^{n}f=J_{a}^{m-n}D^{m}f.$$

Thus, applying the operator J_{a}^{n} to both sides of this equation and using the semigroup property of fractional integration, we obtain

$$J_{a}^{n}D_{*a}^{n}f=J_{a}^{n}J_{a}^{m-n}D^{m}f=J_{a}^{m}D^{m}f.$$

By the classical version of Taylor's theorem, we have that

$$f(x)=\sum_{k=0}^{m-1}\frac{D^{k}f(a)}{k!}(x-a)^{k}+J_{a}^{m}D^{m}f(x).$$

Combining these two equations we derive the claim. ◆

A fractional analogue of Taylor's theorem follows immediately.

Corollary 7.4 (Taylor Expansion for Caputo Derivatives) Under the assumptions of Theorem 7.17,

$$f(x)=\sum_{k=0}^{m-1}\frac{D^{k}f(a)}{k!}(x-a)^{k}+J_{a}^{n}D_{*a}^{n}f(x).$$

The previous result has dealt with the concatenation of two Caputo differential operators. In some instances, however, it may also be useful to concatenate a Caputo operator with a differential operator of Riemann-Liouville type.

Theorem 7.18 Let $f \in C^{\mu}[a,b]$ for some $\mu \in \mathbf{N}$. Moreover, let $n \in [0, \mu]$. Then,

$$D_a^{\mu-n} D_{*a}^n f = D^{\mu} f.$$

Notice that the operator D^{μ} appearing on the right-hand side of the claim is a classical (integer-order) differential operator.

Proof If n is an integer then both differential operators on the left-hand side reduce to integer-order operators and hence we obtain the desired result by an application of the definition of the iterated operators.

If n is not an integer then we may invoke Theorem 7.15 to conclude that

$$D_{*a}^n f = \hat{D}_a^n f = J_a^{\lceil n \rceil - n} D^{\lceil n \rceil} f.$$

Combining this with the definition of the Riemann-Liouville derivative and using the property of fractional integration, we find

$$\begin{aligned} D_a^{\mu-n} D_{*a}^n f &= D^{\mu-\lceil n \rceil+1} J_a^{n+1-\lceil n \rceil} J_a^{\lceil n \rceil - n} D^{\lceil n \rceil} f = D^{\mu-\lceil n \rceil+1} J_a^1 D^{\lceil n \rceil} f \\ &= D^{\mu-\lceil n \rceil} D^{\lceil n \rceil} f = D^{\mu} f. \end{aligned}$$ ◆

The main computational rules for the Caputo derivative are similar, but not identical, to those for the Riemann-Liouville derivative.

Theorem 7.19 Let $f_1, f_2: [a,b] \to \mathbf{R}$ be such that $D_{*a}^n f_1$ and $D_{*a}^n f_2$ exist almost everywhere and let $C_1, C_2 \in \mathbf{R}$. Then, $D_{*a}^n (c_1 f_1 + c_2 f_2)$ exists almost everywhere, and

$$D_{*a}^n (c_1 f_1 + c_2 f_2) = c_1 D_{*a}^n f_1 + c_2 D_{*a}^n f_2.$$

Proof This linearity property of the fractional differential operator is an immediate consequence of the definition of D_{*a}^n.

For the formula of Leibniz, we only state the case $0 < n < 1$ explicitly. ◆

Theorem 7.20 (Leibniz' Formula for Caputo Operators) Let $0 < n < 1$, and assume that f and g are analytic on $(a-h, a+h)$. Then,

$$\begin{aligned} D_{*a}^n [fg](x) &= \frac{(x-a)^{-n}}{\Gamma(1-n)} g(a)(f(x) - f(a)) + (D_{*a}^n g(x)) f(x) \\ &\quad + \sum_{k=1}^{\infty} \binom{n}{k} (J_a^{k-n} g(x)) D_{*a}^k f(x). \end{aligned}$$

Proof We apply the definition of the Caputo derivative and find

$$D_{*a}^n [fg] = D_a^n [fg - f(a) g(a)] = D_a^n [fg] - f(a) g(a) D_a^n [1].$$

Next we use Leibniz' formula for Riemann-Liouville derivatives and find

$$D^n_{*a}[fg]=f(D^n_a g)+\sum_{k=1}^{\infty}\binom{n}{k}(D^k_a f)(J^{k-n}_a g)-f(a)g(a)D^n_a[1].$$

Now we add and subtract $f\cdot g(a)(D^n_a[1])$ and rearrange to obtain

$$\begin{aligned}D^n_{*a}[fg]&=f(D^n_a[g-g(a)]+\sum_{k=1}^{\infty}\binom{n}{k}(D^k_a f)(J^{k-n}_a g)+g(a)(f-f(a))D^n_a[1]\\&=f\times(D^n_{*a}g)+\sum_{k=1}^{\infty}\binom{n}{k}(D^k_{*a}f)(J^{k-n}_a g)+g(a)(f-f(a))\times D^n_a[1],\end{aligned}$$

where we have used the fact that, for $k\in\mathbf{N}$, $D^k_a=D^k=D^k_{*a}$. To finally complete the proof, it only remains to use the explicit expression for $D^n_a[1]$ from Example 7.2. ◆

7.5 Existence and Uniqueness Results for Riemann-Liouville Fractional Differential Equations

In this part of the text, we now discuss the classical questions concerning ordinary differential equations involving fractional derivatives, i.e., the questions of existence and uniqueness of solutions. We shall mainly be interested in initial value problems (Cauchy problems), and in particular in global results.

The fundamental result is an existence and uniqueness theorem. Without loss of generality, we assume in this result and in the ensuing developments that the fractional derivatives are developed at the point 0.

Theorem 7.21 Let $n>0$, $n\notin\mathbf{N}$ and $m=[n]$. Moreover, let $K>0$, $h^*>0$, and $b_1, b_2, \ldots, b_m\in\mathbf{R}$. Define

$$G:=\{(x,y)\in\mathbf{R}^2: 0\leqslant x\leqslant h^*,\ y\in\mathbf{R}\text{ for }x=0\text{ and }$$

$$|x^{m-n}y-\sum_{k=1}^{m}b_k x^{m-k}/\Gamma(n-k+1)|<K\text{ else}\},$$

and assume that the function $f: G\to\mathbf{R}$ is continuous and bounded in G and that it fulfils a Lipschitz condition with respect to the second variable, i.e., there exists a constant $L>0$ such that, for all (x,y_1) and $(x,y_2)\in G$, we have

$$|f(x,y_1)-f(x,y_2)|<L|y_1-y_2|.$$

Then the differential equation

$$D^n_0 y(x)=f(x,y(x))$$

equipped with the initial conditions

$$D^{n-k}_0 y(0)=b_k\quad(k=1, 2, \ldots, m-1),\quad \lim_{z\to 0^+}J^{m-n}_0 y(z)=b_m$$

has a uniquely defined continuous solution $y\in C(0,h]$, where

$$h:=\min\left\{h^*,\tilde{h},\left[\frac{\Gamma(n+1)K}{M}\right]^{1/m}\right\}$$

with $M:=\sup_{(x,z)\in G}|f(x,z)|$ and $\tilde{h}$ being an arbitrary positive number satisfying the constraint

$$\tilde{h}<\left(\frac{\Gamma(2n-m+1)}{\Gamma(n-m+1)L}\right)^{1/n}.$$

The result is very similar to the known classical results for first-order equations. Therefore, it is probably not surprising to find that the proof is analogous as well. Specifically, we shall first transform the initial value problem into an equivalent Volterra integral equation (Lemma 7.7), and then we are going to prove the existence and uniqueness of the solution of this integral equation by a Picard-type iteration process (i. e., by using a variant of Banach's fixed point theorem in a suitably chosen complete metric space), cf. Lemma 7. 9. Theorem 7. 21 is thus an immediate consequence of these two lemmas.

Lemma 7.7 Assume the hypotheses of Theorem 7. 21 and let $h>0$. The function $y\in C(0,h]$ is a solution of the differential equation

$$D_0^n y(x)=f(x,y(x)),$$

equipped with the initial conditions

$$D_0^{n-k}y(0)=b_k\quad(k=1,2,\dots,m-1),\qquad \lim_{z\to 0^+}J_0^{m-n}y(z)=b_m,$$

if and only if it is a solution of the Volterra integral equation

$$y(x)=\sum_{k=1}^{m}\frac{b_k x^{n-k}}{\Gamma(n-k+1)}+\frac{1}{\Gamma(n)}\int_0^x(x-t)^{n-1}f(t,y(t))\mathrm{d}t.$$

Proof Assume first that y is a solution of the integral equation. We can rewrite this equation in the shorter form

$$y(x)=\sum_{k=1}^{m}\frac{b_k x^{n-k}}{\Gamma(n-k+1)}+J_0^n f(\cdot,y(\cdot))(x).$$

Now we apply the differential operator D_0^n to both sides of this relation and immediately obtain, in view of Example 7. 2 and Theorem 7. 7, that y also solves the differential equation. With respect to the initial conditions, we look at the case $1\leqslant k\leqslant m-1$ first and find, by an application of D_0^{n-k} to the Volterra equation, that

$$D_0^{n-k}y(x)=\sum_{j=1}^{m}\frac{b_j D^{n-k}(\cdot)^{n-j}(x)}{\Gamma(n-j+1)}+D_0^{n-k}J_0^{n-k}J_0^k f(\cdot,y(\cdot))(x)$$

in view of the semigroup property of fractional integration. By Example 7. 2, we find that the summands vanish identically for $j>k$. Moreover, by the same example, the summands for $j<k$ vanish if $x=0$. Thus, according to Theorem 7. 7,

$$D_0^{n-k}y(0)=\frac{b_k D_0^{n-k}(\cdot)^{n-k}(0)}{\Gamma(n-k+1)}+J_0^k f(\cdot,y(\cdot))(0).$$

Since $k\geqslant 1$, the integral vanishes, and once again applying Example 7.2, we find that $D_0^{n-k}(\cdot)^{n-k}(x)=\Gamma(n-k+1)$. Thus, $D_0^{n-k}y(0)=b_k$ as required by the initial condition. Finally, for $k=m$, we apply the operator J_0^{m-n} to both sides of the integral equation and find that, in the limit $z\to 0$, all the summands of the sum vanish except for the mth. The integral $J_0^{m-n}J_0^n f(\cdot,y(\cdot))(z)=J_0^m f(\cdot,y(\cdot))(z)$ also vanishes as $z\to 0$. Thus, we find

$$\lim_{z\to 0^+} J_0^{m-n}y(z)=\lim_{z\to 0^+} J_0^{m-n}\frac{b_m J_0^{m-n}(\cdot)^{n-m}(z)}{\Gamma(n-m+1)}=b_m$$

because of Example 7.1. Hence, y solves the given initial value problem.

If y is a continuous solution of the initial value problem, then we define $z(x):=f(x,y(x))$. By assumption, z is a continuous function and $z(x)=f(x,y(x))=D_0^n y(x)=D^m J_0^{m-n}y(x)$. Thus, $D^m J_0^{m-n}y$ is continuous too, i.e., $J_0^{m-n}y\in C^m(0,h]$.

We may therefore apply Theorem 7.12 to derive

$$y(x)=J_0^n D_0^n y(x)+\sum_{k=1}^{m}c_k x^{n-k}=J_0^n f(\cdot,y(\cdot))(x)+\sum_{k=1}^{m}c_k x^{n-k}$$

with certain constants $c_1, c_2, \ldots, c_m$. Introducing the initial conditions as indicated above, we can determine these constants c_k as $c_k=b_k/\Gamma(n-k+1)$. ◆

Lemma 7.8 (Weissinger's Fixed Point Theorem) Assume (U,d) to be a nonempty complete metric space, and let $\alpha_j\geqslant 0$ for every $j\in\mathbf{N}_0$ and such that $\sum_{j=0}^{\infty}\alpha_j$ converges. Furthermore, let the mapping $A:U\to U$ satisfy the inequality

$$d(A^j u, A^j v)\leqslant \alpha_j d(u,v)$$

for every $j\in\mathbf{N}$ and every $u,v\in U$. Then, A has a uniquely determined fixed point u^*. Moreover, for any $u_0\in U$, the sequence $(A^j u_0)_{j=1}^{\infty}$ converges to this fixed point u^*.

Lemma 7.9 Under the assumptions of Theorem 7.21, the Volterra equation

$$y(x)=\sum_{k=1}^{m}\frac{b_k x^{n-k}}{\Gamma(n-k+1)}+\frac{1}{\Gamma(n)}\int_0^x (x-t)^{n-1}f(t,y(t))\mathrm{d}t$$

possesses a uniquely determined solution $y\in C(0,h]$.

Proof We define the set

$$B:=\left\{y\in C(0,h]:\sup_{0<x<h}\left|x^{m-n}y(x)-\sum_{k=1}^{m}\frac{b_k x^{m-k}}{\Gamma(n-k+1)}\right|\leqslant K\right\}$$

and on this set, we define the operator A by

$$Ay(x) := \sum_{k=1}^{m} \frac{b_k x^{n-k}}{\Gamma(n-k+1)} + \frac{1}{\Gamma(n)} \int_0^x (x-t)^{n-1} f(1, y(t)) \mathrm{d}t.$$

Then we note that, for $y \in B$, Ay is also a continuous function on $(0, h]$. Moreover,

$$\begin{aligned}\left| x^{m-n} Ay(x) - \sum_{k=1}^{m} \frac{b_k x^{m-k}}{\Gamma(n-k+1)} \right| &= \left| \frac{x^{m-n}}{\Gamma(n)} \int_0^x (x-t)^{n-1} f(t, y(t)) \mathrm{d}t \right| \\ &\leqslant \frac{x^{m-n}}{\Gamma(n)} M \int_0^x (x-t)^{n-1} \mathrm{d}t \\ &\leqslant \frac{x^{m-n}}{\Gamma(n)} M \frac{x^n}{n} = \frac{x^m M}{\Gamma(n+1)} \leqslant K\end{aligned}$$

for $x \in (0, h]$, where the last inequality follows from the definition of h. This shows that $Ay \in B$ if $y \in B$, i. e. , the operator A maps the set B into itself.

Next we introduce a new set

$$\hat{B} := \{y \in C(0, h] : \sup_{0 < x \leqslant h} |x^{m-n} y(x)| < \infty\},$$

and on this set, we define a norm $\|\cdot\|_{\hat{B}}$ by

$$\|y\|_{\hat{B}} := \sup_{0 < x \leqslant h} |x^{m-n} y(x)|.$$

It is easily seen that $\hat{B}$, equipped with this norm, is a normed linear space, and that B is a complete subset of this space.

We use the definition of A to rewrite the Volterra equation more compactly as

$$y = Ay.$$

Hence, in order to prove the desired result, it is sufficient to show that the operator A has a unique fixed point. For this purpose, we shall employ Weissinger's fixed point theorem (Lemma 7. 8). In this context we prove, for $y, \tilde{y} \in B$,

$$\|A^j y - A^j \tilde{y}\|_{\hat{B}} \leqslant \left(\frac{L h^n \Gamma(n-m+1)}{\Gamma(2n-m+1)} \right)^j \|y - \tilde{y}\|_{\hat{B}}. \tag{7. 1}$$

This can be shown by induction: In the case $j=0$, the statement is trivially true. For the induction step $j-1 \to j$, we proceed as follows. We write

$$\begin{aligned}\|A^j y - A^j \tilde{y}\|_{\hat{B}} &= \sup_{0 < x \leqslant h} |x^{m-n} (A^j y(x) - A^j \tilde{y}(x))| \\ &= \sup_{0 < x \leqslant h} |x^{m-n} (A A^{j-1} y(x) - A A^{j-1} \tilde{y}(x))| \\ &= \sup_{0 < x \leqslant h} \frac{x^{m-n}}{\Gamma(n)} \left| \int_0^x (x-t)^{n-1} [f(t, A^{j-1} y(t)) - f(t, A^{j-1} \tilde{y}(t))] \mathrm{d}t \right| \\ &\leqslant \sup_{0 < x \leqslant h} \frac{x^{m-n}}{\Gamma(n)} \int_0^x (x-t)^{n-1} |f(t, A^{j-1} y(t)) - f(t, A^{j-1} \tilde{y}(t))| \mathrm{d}t \\ &\leqslant \frac{L}{\Gamma(n)} \sup_{0 < x \leqslant h} x^{m-n} \int_0^x (x-t)^{n-1} |A^{j-1} y(t) - A^{j-1} \tilde{y}(t)| \mathrm{d}t\end{aligned}$$

by definition of the operator A and the Lipschitz condition on f. In the next step, we estimate further to find

$$\begin{aligned}\| A^j y - A^j \tilde{y} \|_{\hat{B}} &\leqslant \frac{L}{\Gamma(n)} \sup_{0<x\leqslant h} x^{m-n} \int_0^x (x-t)^{n-1} |A^{j-1}y(t) - A^{j-1}\tilde{y}(t)| \, dt \\ &\leqslant \frac{L}{\Gamma(n)} \sup_{0<x\leqslant h} x^{m-n} \int_0^x (x-t)^{n-1} t^{n-m} t^{m-n} |A^{j-1}y(t) - A^{j-1}\tilde{y}(t)| \, dt \\ &\leqslant \frac{L}{\Gamma(n)} \| A^{j-1}y - A^{j-1}\tilde{y} \|_{\hat{B}} \sup_{0<x\leqslant h} x^{m-n} \int_0^x (x-t)^{n-1} t^{n-m} \, dt \\ &= \frac{L}{\Gamma(n)} \| A^{j-1}y - A^{j-1}\tilde{y} \|_{\hat{B}} \sup_{0<x\leqslant h} \frac{\Gamma(n)\Gamma(n-m+1)}{\Gamma(2n-m+1)} x^n \\ &= \frac{Lh^n \Gamma(n-m+1)}{\Gamma(2n-m+1)} \| A^{j-1}y - A^{j-1}\tilde{y} \|_{\hat{B}}.\end{aligned}$$

Now we use the induction hypothesis, proving inequality (7. 1). Therefore, we may apply Lemma 7. 8 with $\alpha_j = \gamma^j$, where $\gamma = (Lh^n \Gamma(n-m+1)/\Gamma(2n-m+1))$. It remains to prove that the series $\sum_{j=0}^{\infty} \alpha_j$ is convergent. This, however, is trivial in view of the fact that $h \leqslant \tilde{h}$ and the definition of $\tilde{h}$ that implies $\gamma < 1$. Thus, an application of the fixed point theorem yields the existence and the uniqueness of the solution of our integral equation.

Remark 7. 2 The proof of Lemma 7. 9 also gives us, at least in theory, a constructive method to find the solution of the initial value problem by means of the calculation of the sequence $(A^j y_0)_{j=0}^{\infty}$, where y_0 is an arbitrary element of B. The limit of this sequence is the desired solution. Typically, one chooses

$$y_0(x) = \sum_{k=1}^{m} \frac{b_k x^{n-k}}{\Gamma(n-k+1)}.$$

In this case, we call the sequence $(A^j y_0)_{j=1}^{\infty}$ the **Picard iteration sequence** corresponding to the given initial value problem.

Remark 7. 3 Theorem 7. 21 can be interpreted as an analogue of the Picard-Lindelöf theorem for first-order differential equations. We may ask ourselves whether the conditions are too sharp in the fractional setting. It turns out that it is possible to prove a weaker result under weaker assumptions. If we drop the Lipschitz condition on f, the existence of the solution can still be shown. This corresponds to Peano's existence theorem in the classical theory. The proof is essentially similar, just replacing Weissinger's theorem by Schauder's fixed point theorem.

For a further illustration of this remark, we discuss a very simple example of a fractional differential equation with a right-hand side that does not fulfil a Lipschitz condition.

Example 7.3 Consider the differential equation

$$D_0^n y(x)=[y(x)]^\mu,$$

where $0<\mu<1$. In this case, the right-hand side of the equation is continuous but the Lipschitz condition is violated. If we select the initial condition corresponding to this differential equation as

$$\lim_{z\to 0^+} J_0^{1-n} y(z)=0 \text{ and } D_0^{n-k} y(0)=0 \quad (k=1, 2, \ldots, [n]-1),$$

we easily see that one solution is $y\equiv 0$. However, an explicit calculation reveals that the function y given by

$$y(x)=\sqrt[\mu-1]{\frac{\Gamma(j+1)}{\Gamma(j+1-n)}}\, x^j$$

with $j=n/(1-\mu)$ also solves the initial value problem. Thus, we indeed see that, in general, the uniqueness of the solution cannot be expected without the Lipschitz condition. ◆

7.6 Existence and Uniqueness Results for Caputo Fractional Differential Equations

Having established the fundamentals of a theory for fractional differential equations with Riemann-Liouville derivatives, we now come to the corresponding problem for Caputo operators. In view of the fact that the latter seem to be much more important than the former as far as applications outside of mathematics are concerned, we shall discuss this problem in a more detailed fashion. The main emphasis will be on initial value problems. In particular, this section will be devoted existence and uniqueness questions.

We begin once again with equations of the form

$$D_{*0}^n y(x)=f(x,y(x)), \tag{7.2a}$$

combined with appropriate initial conditions. As indicated in Section 7.5, these conditions have the form

$$D^k y(0)=y_0^{(k)}, \quad k=0, 1, \ldots, m-1, \tag{7.2b}$$

where as usual we have set $m=[n]$.

The first result is an existence result that corresponds to the classical Peano existence theorem for first order equations.

Theorem 7.22 Let $0<n$ and $m=\lceil n\rceil$. Moreover, let $y_0^{(0)}, y_0^{(1)}, \ldots, y_0^{(m-1)}\in\mathbf{R}$, $K>0$ and $h^*>0$. Define $G:=\{(x,y): x\in[0,h^*], |y-\sum_{k=0}^{m-1}x^k y_0^{(k)}/k!|\leqslant K\}$, and let the function $f: G\to\mathbf{R}$ be continuous. Furthermore, define $M:=\sup_{(x,z)\in G}|f(x,z)|$ and

$$h:=\begin{cases}h^*, & \text{if } M=0,\\ \min\{h^*, (K\Gamma(n+1)/M)^{1/n}\}, & \text{else.}\end{cases}$$

Then, there exists a function $y\in C[0,h]$ solving the initial value problem (7.2).

For the proofs of most of the theorems in this section, we will use the following lemma that adapts the statement of Lemma 7.7 to the present situation.

Lemma 7.10 Assume the hypotheses of Theorem 7.21. The function $y\in C[0,h]$ is a solution of the initial value problem (7.2) if and only if it is a solution of the nonlinear Volterra integral equation of the second kind

$$y(x)=\sum_{k=0}^{m-1}\frac{x^k}{k!}y_0^{(k)}+\frac{1}{\Gamma(n)}\int_0^x(x-t)^{n-1}f(t,y(t))\mathrm{d}t \tag{7.3}$$

with $m=\lceil n\rceil$.

Proof The proof that every continuous solution of the Volterra equation also solves the initial value problem is very close to the proof of the corresponding part of Lemma 7.7; we therefore leave the details to the reader.

For the other direction, we define $z(x):=f(x,y(x))$ and once again note that $z\in C[0,h]$ by our assumptions on y and f. Then, using the definition of the Caputo differential operator, the differential equation can be rewritten as

$$\begin{aligned}z(x)=f(x,y(x))=D_{*0}^n y(x)=D_0^n(y-T_{m-1}[y;0])(x)\\ =D^m J_0^{m-n}(y-T_{m-1}[y;0])(x).\end{aligned} \tag{7.4}$$

Since we are dealing with continuous functions, we may apply the operator J_0^m to both sides of the equation and find

$$J_0^m z(x)=J_0^{m-n}(y-T_{m-1}[y;0])(x)+q(x)$$

with some polynomial q of degree not exceeding $m-1$. Since z is continuous, the function $J_0^m z$ on the left-hand side of this equation has a zero of order (at least) m at the origin. Moreover, the difference $y-T_{m-1}[y;0]$ has the same property by construction, and therefore the function $J_0^{m-n}(y-T_{m-1}[y;0])$ on the right-hand side of our equation must have such an mth order zero too. Thus, the polynomial q has the same property, and we immediately deduce (since its degree is not more than $m-1$) that $q=0$. Consequently,

$$J_0^m z(x) = J_0^{m-n}(y - T_{m-1}[y;0])(x),$$

and by applying the Riemann-Liouville differential operator D_0^{m-n} to this equation, we find

$$y(x) - T_{m-1}[y;0](x) = D_0^{m-n} J_0^m z(x) = D^1 J_0^{1+n-m} J_0^m z(x) = D J_0^{1+n} z(x)$$
$$= J_0^n z(x).$$

Recalling the definitions of z and the Taylor polynomial $T_{m-1}[y;0]$, this is just the required Volterra equation. ◆

Proof (of Theorem 7.22) If $M=0$, then $f(x,y)=0$ for all $(x,y)\in G$. In this case, it is evident that the function $y:[0,h]\to\mathbf{R}$ with $y(x)=\sum_{k=0}^{m-1} y_0^{(k)} x^k/k!$ is a solution of the initial value problem (7.2). Hence we conclude, as required, that a solution exists in this case.

Otherwise, we apply Lemma 7.10 and see that our initial value problem (7.2) is equivalent to the Volterra equation (7.3). We thus introduce the polynomial T that satisfies the initial conditions, viz.

$$T(x) := \sum_{k=0}^{m-1} \frac{x^k}{k!} y_0^{(k)}, \tag{7.5}$$

and the set $U := \{y \in C[0,h] : \| y - T \|_\infty \leqslant K\}$. It is evident that U is a closed and convex subset of the Banach space of all continuous functions on $[0,h]$, equipped with the Chebyshev norm. Hence, U is a Banach space too. Since the polynomial T is an element of U, we also see that U is not empty. On this set U, we define the operator A by

$$(Ay)(x) := T(x) + \frac{1}{\Gamma(n)} \int_0^x (x-t)^{n-1} f(t, y(t))\,\mathrm{d}t. \tag{7.6}$$

Using this operator, the equation whose solvability we need to prove, viz. the Volterra equation (7.3), can be rewritten as

$$y = Ay,$$

and thus, in order to prove our desired existence result, we have to show that A has a fixed point. We therefore proceed by investigating the properties of the operator A more closely.

Our first goal in this context is to show that $Ay \in U$ for $y \in U$. To this end, we begin by noting that, for $0 \leqslant x_1 \leqslant x_2 \leqslant h$,

$$|(Ay)(x_1) - (Ay)(x_2)|$$
$$= \frac{1}{\Gamma(n)} \left| \int_0^{x_1} (x_1 - t)^{n-1} f(t, y(t))\,\mathrm{d}t - \int_0^{x_2} (x_2 - t)^{n-1} f(t, y(t))\,\mathrm{d}t \right|$$

$$=\frac{1}{\Gamma(n)}\left|\int_0^{x_1}((x_1-t)^{n-1}-(x_2-t)^{n-1})f(t,y(t))\mathrm{d}t+\int_{x_1}^{x_2}(x_2-t)^{n-1}f(t,y(t))\mathrm{d}t\right|$$

$$\leqslant\frac{M}{\Gamma(n)}\left(\int_0^{x_1}|(x_1-t)^{n-1}-(x_2-t)^{n-1}|\mathrm{d}t+\int_{x_1}^{x_2}(x_2-t)^{n-1}\mathrm{d}t\right). \tag{7.7}$$

The second integral in the right-hand side of (7.7) has the value $(x_2-x_1)^n/n$. For the first integral, we look at the three cases $n=1$, $n<1$ and $n>1$ separately. In the first case $n=1$, the integrand vanishes identically, and hence the integral has the value zero. Secondly, for $n<1$, we have $n-1<0$, and hence $(x_1-t)^{n-1}\geqslant(x_2-t)^{n-1}$. Thus,

$$\int_0^{x_1}|(x_1-t)^{n-1}-(x_2-t)^{n-1}|\mathrm{d}t=\int_0^{x_1}((x_1-t)^{n-1}-(x_2-t)^{n-1})\mathrm{d}t$$

$$=\frac{1}{n}(x_1^n-x_2^n+(x_2-x_1)^n)\leqslant\frac{1}{n}(x_2-x_1)^n.$$

Finally, if $n>1$, then $(x_1-t)^{n-1}\leqslant(x_2-t)^{n-1}$, and hence

$$\int_0^{x_1}|(x_1-t)^{n-1}-(x_2-t)^{n-1}|\mathrm{d}t=\int_0^{x_1}((x_2-t)^{n-1}-(x_1-t)^{n-1})\mathrm{d}t$$

$$=\frac{1}{n}(-x_1^n+x_2^n-(x_2-x_1)^n)\leqslant\frac{1}{n}(x_2^n-x_1^n).$$

A combination of these results yields

$$|(Ay)(x_1)-(Ay)(x_2)|\leqslant\begin{cases}\dfrac{2M}{\Gamma(n+1)}(x_2-x_1)^n, & \text{if } n\leqslant 1,\\[2ex] \dfrac{M}{\Gamma(n+1)}[(x_2-x_1)^n+x_2^n-x_1^n], & \text{if } n>1.\end{cases} \tag{7.8}$$

In either case, the expression on the right-hand side of (7.8) converges to 0 as $x_2\to x_1$ which proves that Ay is a continuous function. Moreover, for $y\in U$ and $x\in[0,h]$, we find

$$|(Ay)(x)-T(x)|=\frac{1}{\Gamma(n)}\left|\int_0^x(x-t)^{n-1}f(t,y(t))\mathrm{d}t\right|\leqslant\frac{1}{\Gamma(n+1)}Mx^n$$

$$\leqslant\frac{1}{\Gamma(n+1)}Mh^n\leqslant\frac{1}{\Gamma(n+1)}M\frac{K\Gamma(n+1)}{M}=K.$$

Thus, we have shown that $Ay\in U$ if $y\in U$, i. e., A maps the set U to itself.

Since we want to apply Schauder's fixed point theorem, all that remains now is to show that $A(U):=\{Au: u\in U\}$ is a relatively compact set. This can be done by means of the Arzelà-Ascoli theorem. For $z\in A(U)$, we find that, for all $x\in[0,h]$,

$$|z(x)|=|(Ay)(x)|\leqslant\|T\|_\infty+\frac{1}{\Gamma(n)}\int_0^x(x-t)^{n-1}|f(t,y(t))|\mathrm{d}t$$

$$\leqslant\|T\|_\infty+\frac{1}{\Gamma(n+1)}Mh^n\leqslant\|T\|_\infty+K,$$

which is the required boundedness property. Moreover, the equicontinuity property can be derived from (7.8) above. Specifically, for $0\leqslant x_1\leqslant x_2\leqslant h$, we have found in the case $n\leqslant 1$ that

$$|(Ay)(x_1)-(Ay)(x_2)|\leqslant\frac{2M}{\Gamma(n+1)}(x_2-x_1)^n.$$

Thus, if $|x_2-x_1|<\delta$, then

$$|(Ay)(x_1)-(Ay)(x_2)|\leqslant\frac{2M}{\Gamma(n+1)}\delta^n.$$

Noting that the expression on the right-hand side is independent of y, x_1 and x_2, we see that the set $A(U)$ is equicontinuous. Similarly, in the case $n>1$, we may use the mean value theorem to conclude that

$$\begin{aligned}|(Ay)(x_1)-(Ay)(x_2)|&\leqslant\frac{M}{\Gamma(n+1)}((x_2-x_1)^n+x_2^n-x_1^n)\\&=\frac{M}{\Gamma(n+1)}((x_2-x_1)^n+n(x_2-x_1)\xi^{n-1})\\&\leqslant\frac{M}{\Gamma(n+1)}((x_2-x_1)^n+n(x_2-x_1)h^{n-1})\end{aligned}$$

with some $\xi\in[x_1,x_2]\subseteq[0,h]$. Hence, if once again $|x_2-x_1|<\delta$, then

$$|(Ay)(x_1)-(Ay)(x_2)|\leqslant\frac{M}{\Gamma(n+1)}(\delta^n+n\delta h^{n-1})$$

and the right-hand side is once more independent of y, x_1 and x_2, proving the equicontinuity. In either case, the Arzelà-Ascoli theorem yields that $A(U)$ is relatively compact, and hence Schauder's fixed point theorem asserts that A has a fixed point. By construction, a fixed point of A is a solution of our initial value problem. ◆

We note two important special cases of Theorem 7.22. The first of these states that, under certain assumptions, the solution exists on the entire interval $[0,h^*]$ (and so for all x for which $f(x,y)$ is defined) and not only for a subinterval $[0,h]$ with some $h\leqslant h^*$.

Corollary 7.5 Assume the hypotheses of Theorem 7.22, except that the set G, i.e. the domain of definition of the function f on the right-hand side of the differential equation (7.2a), is now taken to be $G:=[0,h^*]\times\mathbf{R}$. Moreover, we assume that f is continuous and that there exist constants $c_1\geqslant 0$, $c_2\geqslant 0$ and $0\leqslant\mu<1$ such that

$$|f(x,y)|\leqslant c_1+c_2|y|^\mu \quad \text{for all } (x,y)\in G. \tag{7.9}$$

Then, there exists a function $y \in C[0,h^*]$ solving the initial value problem (7.2).

Proof We use the polynomial T defined in (7.5) in the previous proof. Since $\mu<1$, we may find some $K>0$ such that

$$c_1+c_2(K+\max_{x\in[0,h^*]}|T(x)|)^\mu \leqslant \frac{K\Gamma(n+1)}{h^{*n}}.$$

Using this value of K, we then restrict our function f to the set $G_K := \{(x,y): x\in[0,h^*], |y-T(x)|\leqslant K\}$ (this is the set that was denoted by G in Theorem 7.22). Then we see that

$$M := \sup_{(x,y)\in G_K}|f(x,y)| \leqslant c_1+c_2\sup_{(x,y)\in G_K}|y|^\mu$$

$$\leqslant c_1+c_2(K+\max_{x\in[0,h^*]}|T(x)|)^\mu \leqslant \frac{K\Gamma(n+1)}{h^{*n}}.$$

Thus, we may apply Theorem 7.22 with this value of K and the given h^* and see that $(K\Gamma(n+1)/M)^{1/n}\geqslant h^*$ which implies that

$$h^*=\min\left\{h^*, \left(\frac{K\Gamma(n+1)}{M}\right)^{1/n}\right\}=h. \blacklozenge$$

The second corollary to Theorem 7.22 asserts the existence of a solution on the entire half-axis $[0,\infty)$ under appropriate conditions.

Corollary 7.6 Assume the hypotheses of Corollary 7.7, except that the set G, i.e., the domain of definition of the function f on the right-hand side of the differential equation (7.2a), is now taken to be $G:=\mathbf{R}^2$. Then, there exists a function $y\in C[0,\infty)$ solving the initial value problem (7.2).

Proof Let $h^*>0$. Under our assumptions, we may apply Corollary 7.5 for this h^* and conclude that a continuous solution exists on $[0,h^*]$. Since h^* can be chosen arbitrarily large, we find that a continuous solution exists on $[0,\infty)$.

Now we come to a uniqueness theorem that corresponds to the well-known Picard-Lindelöf result. It can be seen as an analogue to the statement shown for Riemann-Liouville operators in the previous section (Theorem 7.21).

Theorem 7.23 Let $n>0$ and $m=[n]$. Moreover, let $y_0^{(0)}, y_0^{(1)}, \ldots, y_0^{(m-1)}\in\mathbf{R}$, $K>0$ and $h^*>0$. Define the set G as in Theorem 7.22 and let the function $f: G\to\mathbf{R}$ be continuous and fulfil a Lipschitz condition with respect to the second variable, i.e.,

$$|f(x,y_1)-f(x,y_2)|\leqslant L|y_1-y_2| \tag{7.10}$$

with some constant $L>0$ independent of x, y_1, and y_2. Then, denoting h as in Theorem 7.22, there exists a uniquely defined function $y\in C[0,h]$ solving the initial value problem (7.2).

Corollary 7.7 Assume the hypotheses of Theorem 7.23, except that the set G, i.e., the domain of definition of the function f on the right-hand side of the differential equation (7.2a), is now taken to be $G:=[0,h^*]\times\mathbf{R}$. Moreover, we assume that f is continuous and that there exist constants $c_1\geqslant 0$, $c_2\geqslant 0$ and $0\leqslant\mu<1$ such that

$$|f(x,y)|\leqslant c_1+c_2|y|^{\mu}\quad \text{for all } (x,y)\in G. \tag{7.11}$$

Then, there exists a function $y\in C[0,h^*]$ solving the initial value problem (7.2).

Corollary 7.8 Assume the hypotheses of Corollary 7.7, except that the set G, i.e., the domain of definition of the function f on the right-hand side of the differential equation (7.2a), is now taken to be $G:=\mathbf{R}^2$. Then, there exists a function $y\in C[0,\infty)$ solving the initial value problem (7.2).

Theorem 7.24 Let $n>0$ and $m=[n]$. Moreover, let $y_0^{(0)}, y_0^{(1)}, \ldots, y_0^{(m-1)}\in\mathbf{R}$ and $h^*>0$. Define the set $G:=[0,h^*]\times\mathbf{R}$ and let the function $f:G\to\mathbf{R}$ be continuous and fulfil a Lipschitz condition with respect to the second variable with a Lipschitz constant $L>0$ that is independent of x, y_1, and y_2. Then there exists a uniquely defined function $y\in C[0,h^*]$ solving the initial value problem (7.2).

In particular, this theorem is applicable to linear equations, i.e., equations of the form

$$D_{*0}^{n}y(x)=f(x)y(x)+g(x)$$

with certain functions $f, g\in C[0,h^*]$, because here we may choose $L=\|f\|_{\infty}<\infty$.

We obtain an immediate consequence.

Corollary 7.9 Let $n>0$ and $m=[n]$. Moreover, let $y_0^{(0)}, y_0^{(1)}, \ldots, y_0^{(m-1)}\in\mathbf{R}$ and $h^*>0$. Define the set $G:=[0,\infty)\times\mathbf{R}$ and let the function $f:G\to\mathbf{R}$ be continuous and fulfil a Lipschitz condition with respect to the second variable with a Lipschitz constant $L>0$ that is independent of x, y_1, and y_2. Then there exists a uniquely defined function $y\in C[0,\infty)$ solving the initial value problem (7.2).

We end this section by taking a look at this problem from a different point of view.

Theorem 7.25 Let $0<n<1$ and assume $f:[0,b]\times[c,d]\to\mathbf{R}$ to be continuous and satisfy a Lipschitz condition with respect to the second variable. Then, for each $x^*\in[0,b]$ and each $y^*\in[c,d]$, the differential equation

$$D_{*0}^{n} y(x)=f(x, y(x)) \tag{7.12}$$

subject to the condition

$$y(x^{*})=y^{*} \tag{7.13}$$

has at most one solution.

Thus, we have a uniqueness theorem for the solutions of a fractional differential equation of the usual form combined with a prescribed value of the unknown solution at a point that may differ from the starting point of the fractional differential operator. In the case $x^{*}=0$, this is just the standard initial condition that we had discussed thoroughly at the beginning of this section, but if $x^{*}>0$, then we have a significantly different problem that is sometimes called a **terminal value problem** because one usually is interested in the solution on the interval $[0, x^{*}]$, i. e. , one provides a condition on the unknown solution at the terminal point of the interval of interest.

Chapter 8

Dynamic Equations on Time Scales

The theory of time scales was introduced by Stefan Hilger in his Ph. D. thesis in 1988 supervised by Bernd Aulbach in order to unify continuous and discrete analysis. In this chapter, we introduce the basic theory and development of dynamic equations on time scales.

8.1 Basic Definitions

A **time scale** (which is a special case of a **measure chain**), is an arbitrary nonempty closed subset of the real numbers. Thus,

$$\mathbf{R},\quad \mathbf{Z},\quad \mathbf{N},\quad \mathbf{N}_0,$$

i. e. , the real numbers, the integers, the natural numbers, and the nonnegative integers are examples of time scales, as are

$$[0,1]\cup[2,3],\quad [0,1]\cup\mathbf{N},\quad \text{and the Cantor set},$$

while

$$\mathbf{Q},\quad \mathbf{R}\backslash\mathbf{Q},\quad \mathbf{C},\quad (0,1),$$

i. e. , the rational numbers, the irrational numbers, the complex numbers, and the open interval between 0 and 1, are not time scales. Throughout this chapter we'll denote a time scale by the symbol $\mathbf{T}$. We assume throughout that a time scale $\mathbf{T}$ has the topology that it inherits from the real numbers with the standard topology.

In this section, we introduce the basic notions connected to time scales and differentiability of functions on them, and we offer the above two cases as examples.

Definition 8.1 Let $\mathbf{T}$ be a time scale. For $t\in\mathbf{T}$, we define the **forward jump operator** $\sigma:\mathbf{T}\to\mathbf{T}$ by

$$\sigma(t):=\inf\{s\in\mathbf{T}:\ s>t\},$$

while the **backward jump operator** $\rho:\mathbf{T}\to\mathbf{T}$ is defined by

$$\rho(t):=\sup\{s\in\mathbf{T}:\ s<t\}.$$

In this definition, we put $\inf \varnothing = \sup \mathbf{T}$ (i. e., $\sigma(t)=t$ if $\mathbf{T}$ has a maximum t) and $\sup \varnothing = \inf \mathbf{T}$ (i. e., $\rho(t)=t$ if $\mathbf{T}$ has a minimum t), where $\varnothing$ denotes the empty set. If $\sigma(t)>t$, we say that t is **right-scattered**; while if $\rho(t)<t$, we say that t is **left-scattered.** Points that are right-scattered and left-scattered at the same time are called **isolated.** Also, if $t<\sup \mathbf{T}$ and $\sigma(t)=t$, then t is called **right-dense**, and if $t>\inf \mathbf{T}$ and $\rho(t)=t$, then t is called **left-dense.** Points that are right-dense and left-dense at the same time are called **dense.** The **graininess function** μ: $\mathbf{T}\to[0,\infty)$ is defined by

$$\mu(t):=\sigma(t)-t.$$

We define the set $\mathbf{T}^\kappa$ which is derived from the time scale $\mathbf{T}$ as follows: If $\mathbf{T}$ has a left-scattered maximum m, then $\mathbf{T}^\kappa=\mathbf{T}-\{m\}$. Otherwise, $\mathbf{T}^\kappa=\mathbf{T}$. In summary,

$$\mathbf{T}^\kappa=\begin{cases}\mathbf{T}\setminus(\rho(\sup \mathbf{T}),\sup \mathbf{T}], & \text{if} \quad \sup \mathbf{T}<\infty,\\ \mathbf{T}, & \text{if} \quad \sup \mathbf{T}=\infty.\end{cases}$$

Finally, if $f:\mathbf{T}\to\mathbf{R}$ is a function, then we define the function f^σ: $\mathbf{T}\to\mathbf{R}$ by

$$f^\sigma(t)=f(\sigma(t)) \quad \text{for all} \quad t\in\mathbf{T},$$

i. e., $f^\sigma=f\circ\sigma$.

Example 8.1 Let us briefly consider the two examples $\mathbf{T}=\mathbf{R}$ and $\mathbf{T}=\mathbf{Z}$.

(i) If $\mathbf{T}=\mathbf{R}$, then we have for any $t\in\mathbf{R}$,

$$\sigma(t)=\inf\{s\in\mathbf{R}:\ s>t\}=\inf(t,\infty)=t$$

and similarly $\rho(t)=t$. Hence, every point $t\in\mathbf{R}$ is dense. The graininess function μ turns out to be

$$\mu(t)\equiv 0 \quad \text{for all} \quad t\in\mathbf{T}.$$

(ii) If $\mathbf{T}=\mathbf{Z}$, then we have for any $t\in\mathbf{Z}$,

$$\sigma(t)=\inf\{s\in\mathbf{Z}: s>t\}=\inf\{t+1,t+2,t+3,\ldots\}=t+1$$

and similarly $\rho(t)=t-1$. Hence, every point $t\in\mathbf{Z}$ is isolated. The graininess function μ in this case is

$$\mu(t)\equiv 1 \quad \text{for all} \quad t\in\mathbf{T}.$$

Throughout this chapter we make the blanket assumption that a and b are points in $\mathbf{T}$. Often we assume $a\leqslant b$. We then define the interval $[a,b]$ in $\mathbf{T}$ by

$$[a,b]:=\{t\in\mathbf{T}: a\leqslant t\leqslant b\}.$$

Open intervals and half-open intervals ect. are defined accordingly. Note that $[a,b]^\kappa=[a,b]$ if b is left-dense and $[a,b]^\kappa=[a,b)=[a,\rho(b)]$ if b is left-scattered. ◆

Sometimes the following induction principle is a useful tool.

Theorem 8.1 (Induction Principle) Let $t_0 \in \mathbf{T}$ and assume that

$$\{S(t): t \in [t_0, \infty\}$$

is a family of statements satisfying:

(i) The statement $S(t_0)$ is true.

(ii) If $t \in [t_0, \infty)$ is right-scattered and $S(t)$ is true, then $S(\sigma(t))$ is also true.

(iii) If $t \in [t_0, \infty)$ is right-dense and $S(t)$ is true, then there is a neighborhood U of t such that $S(s)$ is true for all $s \in U \cap (t, \infty)$.

(iv) If $t \in (t_0, \infty)$ is left-dense and $S(s)$ is true for all $s \in [t_0, t)$, then $S(t)$ is true. Then $S(t)$ is true for all $t \in [t_0, \infty)$.

Remark 8.1 A dual version of the induction principle also holds for a family of statements $S(t)$ for t in an interval of the form $(-\infty, t_0]$. That is, to show that $S(t)$ is true for all $t \in (-\infty, t_0]$, we have to show that $S(t_0)$ is true, that $S(t)$ is true at a left-scattered t implies $S(\rho(t))$ is true, that $S(t)$ is true at a left-dense t implies $S(r)$ is true for all r in a left neighborhood of t, and that $S(r)$ is true for all $r \in (t, t_0]$, where t is right-dense implies $S(t)$ is true.

8.2 Differentiation

Now we consider a function $f: \mathbf{T} \to \mathbf{R}$ and define the so-called **delta** (or **Hilger**) **derivative** of f at a point $t \in \mathbf{T}^\kappa$.

Definition 8.2 Assume $f: \mathbf{T} \to \mathbf{R}$ is a function and let $t \in \mathbf{T}^\kappa$. Then we define $f^\Delta(t)$ to be the number (provided it exists) with the property that given any $\varepsilon > 0$, there is a neighborhood U of t (i. e., $U = (t-\delta, t+\delta) \cap \mathbf{T}$ for some $\delta > 0$) such that

$$|[f(\sigma(t)) - f(s)] - f^\Delta(t)[\sigma(t) - s]| \leqslant \varepsilon |\sigma(t) - s| \quad \text{for all} \quad s \in \boldsymbol{U}.$$

We call $f^\Delta(t)$ the **delta** (or **Hilger**) **derivative** of f at t.

Moreover, we say that f is **delta** (or **Hilger**) **differentiable** (or in short: **differentiable**) on $\mathbf{T}^\kappa$ provided $f^\Delta(t)$ exists for all $t \in \mathbf{T}^\kappa$. The function $f^\Delta: \mathbf{T}^\kappa \to \mathbf{R}$ is then called the (delta) derivative of f on $\mathbf{T}^\kappa$.

Some easy and useful relationships concerning the delta derivative are given next.

Theorem 8.2 Assume $f: \mathbf{T} \to \mathbf{R}$ is a function and let $t \in \mathbf{T}^\kappa$. Then we have the following:

(i) If f is differentiable at t, then f is continuous at t.

(ii) If f is continuous at t and t is right-scattered, then f is differentiable at t with

$$f^{\Delta}(t)=\frac{f(\sigma(t))-f(t)}{\mu(t)}$$

(iii) If t is right-dense, then f is differentiable at t iff the limit

$$\lim_{s\to t}\frac{f(t)-f(s)}{t-s}$$

exists as a finite number. In this case,

$$f^{\Delta}(t)=\lim_{s\to t}\frac{f(t)-f(s)}{t-s}.$$

(iv) If f is differentiable at t, then

$$f(\sigma(t))=f(t)+\mu(t)f^{\Delta}(t).$$

Proof (i) Assume that f is differentiable at t. Let $\varepsilon\in(0,1)$. Define

$$\varepsilon^{*}=\varepsilon[1+|f^{\Delta}(t)|+2\mu(t)]^{-1}.$$

Then $\varepsilon^{*}\in(0,1)$. By Definition 8.2, there exists a neighborhood U of t such that

$$|f(\sigma(t))-f(s)-[\sigma(t)-s]f^{\Delta}(t)|\leqslant\varepsilon^{*}|\sigma(t)-s| \quad \text{for all } s\in U.$$

Therefore, we have for all $s\in U\cap(t-\varepsilon^{*},t+\varepsilon^{*})$,

$$\begin{aligned}|f(t)-f(s)|&=|\{f(\sigma(t))-f(s)-f^{\Delta}(t)[\sigma(t)-s]\}\\&\quad-\{f(\sigma(t))-f(t)-\mu(t)f^{\Delta}(t)\}+(t-s)f^{\Delta}(t)|\\&\leqslant\varepsilon^{*}|\sigma(t)-s|+\varepsilon^{*}\mu(t)+|t-s||f^{\Delta}(t)|\\&\leqslant\varepsilon^{*}[\mu(t)+|t-s|+\mu(t)+|f^{\Delta}(t)|]\\&<\varepsilon^{*}[1+|f^{\Delta}(t)|+2\mu(t)]\\&=\varepsilon.\end{aligned}$$

It follows that f is continuous at t.

(ii) Assume f is continuous at t and t is right-scattered. By continuity,

$$\lim_{s\to t}\frac{f(\sigma(t))-f(s)}{\sigma(t)-s}=\frac{f(\sigma(t))-f(t)}{\sigma(t)-t}=\frac{f(\sigma(t))-f(t)}{\mu(t)}.$$

Hence, given $\varepsilon>0$, there is a neighborhood U of t such that

$$\left|\frac{f(\sigma(t))-f(s)}{\sigma(t)-s}-\frac{f(\sigma(t))-f(t)}{\mu(t)}\right|\leqslant\varepsilon$$

for all $s\in U$. It follows that

$$\left|[f(\sigma(t))-f(s)]-\frac{f(\sigma(t))-f(t)}{\mu(t)}[\sigma(t)-s]\right|\leqslant\varepsilon|\sigma(t)-s|$$

for all $s\in U$. Hence, we get the desired result

$$f^{\Delta}(t)=\frac{f(\sigma(t))-f(t)}{\mu(t)}.$$

(iii) Assume f is differentiable at t and t is right-dense. Let $\varepsilon>0$ be given. Since

f is differentiable at t, there is a neighborhood U of t such that

$$|[f(\sigma(t))-f(s)]-f^{\Delta}(t)[\sigma(t)-s]|\leqslant\varepsilon|\sigma(t)-s|$$

for all $s\in U$. Since $\sigma(t)=t$, we have that

$$|[f(t)-f(s)]-f^{\Delta}(t)(t-s)|\leqslant\varepsilon|t-s|$$

for all $s\in U$. It follows that

$$\left|\frac{f(t)-f(s)}{t-s}-f^{\Delta}(t)\right|\leqslant\varepsilon$$

for all $s\in U$, $s\neq t$. Therefore, we get the desired result

$$f^{\Delta}(t)=\lim_{s\to t}\frac{f(t)-f(s)}{t-s}.$$

(iv) If $\sigma(t)=t$, then $\mu(t)=0$ and we have that

$$f(\sigma(t))=f(t)=f(t)+\mu(t)f^{\Delta}(t).$$

On the other hand, if $\sigma(t)>t$, then by part (ii)

$$\begin{aligned} f(\sigma(t))&=f(t)+\mu(t)\cdot\frac{f(\sigma(t))-f(t)}{\mu(t)}\\ &=f(t)+\mu(t)f^{\Delta}(t), \end{aligned}$$

and the proof of part (iv) is complete.

Next, we would like to be able to find the derivatives of sums, products, and quotients of differentiable functions. This is possible according to the following theorem.

Theorem 8.3 Assume f, $g:\mathbf{T}\to\mathbf{R}$ are differentiable at $t\in\mathbf{T}^{\kappa}$. Then:

(i) The sum $f+g:\mathbf{T}\to\mathbf{R}$ is differentiable at t with

$$(f+g)^{\Delta}(t)=f^{\Delta}(t)+g^{\Delta}(t).$$

(ii) For any constant α, $\alpha f:\mathbf{T}\to\mathbf{R}$ is differentiable at t with

$$(\alpha f)^{\Delta}(t)=\alpha f^{\Delta}(t).$$

(iii) The product $fg:\mathbf{T}\to\mathbf{R}$ is differentiable at t with

$$(fg)^{\Delta}(t)=f^{\Delta}(t)g(t)+f(\sigma(t))g^{\Delta}(t)=f(t)g^{\Delta}(t)+f^{\Delta}(t)g(\sigma(t)).$$

(iv) If $f(t)f(\sigma(t))\neq 0$, then $\frac{1}{f}$ is differentiable at t with

$$\left(\frac{1}{f}\right)^{\Delta}(t)=-\frac{f^{\Delta}(t)}{f(t)f(\sigma(t))}.$$

(v) If $g(t)g(\sigma(t))\neq 0$, then $\frac{f}{g}$ is differentiable at t and

$$\left(\frac{f}{g}\right)^{\Delta}(t)=\frac{f^{\Delta}(t)g(t)-f(t)g^{\Delta}(t)}{g(t)g(\sigma(t))}.$$

Proof Assume that f and g are delta differentiable at $t\in\mathbf{T}^{\kappa}$.

(i) Let $\varepsilon>0$. Then there exist neighborhoods U_1 and U_2 of t with

$$|f(\sigma(t))-f(s)-f^{\Delta}(t)(\sigma(t)-s)|\leqslant\frac{\varepsilon}{2}|\sigma(t)-s| \quad \text{for all} \quad s\in U_1$$

and

$$|g(\sigma(t))-g(s)-g^{\Delta}(t)(\sigma(t)-s)|\leqslant\frac{\varepsilon}{2}|\sigma(t)-s| \quad \text{for all} \quad s\in U_2.$$

Let $U=U_1\cap U_2$. Then we have for all $s\in U$,

$$\begin{aligned}
&|(f+g)(\sigma(t))-(f+g)(s)-[f^{\Delta}(t)+g^{\Delta}(t)](\sigma(t)-s)|\\
&\quad=|f(\sigma(t))-f(s)-f^{\Delta}(t)(\sigma(t)-s)+g(\sigma(t))-g(s)-g^{\Delta}(t)(\sigma(t)-s)|\\
&\quad\leqslant|f(\sigma(t))-f(s)-f^{\Delta}(t)(\sigma(t)-s)|+|g(\sigma(t))-g(s)-g^{\Delta}(t)(\sigma(t)-s)|\\
&\quad\leqslant\frac{\varepsilon}{2}|\sigma(t)-s|+\frac{\varepsilon}{2}|\sigma(t)-s|\\
&\quad=\varepsilon|\sigma(t)-s|.
\end{aligned}$$

Therefore, $f+g$ is differentiable at t and $(f+g)^{\Delta}=f^{\Delta}+g^{\Delta}$ holds at t.

(iii) Let $\varepsilon\in(0,1)$. Define $\varepsilon^*=\varepsilon[1+|f(t)|+|g(\sigma(t))|+|g^{\Delta}(t)|]^{-1}$. Then $\varepsilon^*\in(0,1)$ and hence there exist neighborhoods U_1, U_2, and U_3 of t such that

$$|f(\sigma(t))-f(s)-f^{\Delta}(t)(\sigma(t))-s|\leqslant\varepsilon^*|\sigma(t)-s| \quad \text{for all} \quad s\in U_1,$$

$$|g(\sigma(t))-g(s)-g^{\Delta}(t)(\sigma(t)-s)|\leqslant\varepsilon^*|\sigma(t)-s| \quad \text{for all} \quad s\in U_2,$$

and

$$|f(t)-f(s)|\leqslant\varepsilon^* \quad \text{for all} \quad s\in U_3.$$

Put $U=U_1\cap U_2\cap U_3$ and let $s\in U$. Then

$$\begin{aligned}
&|(fg)(\sigma(t))-(fg)(s)-[f^{\Delta}(t)g(\sigma(t))+f(t)g^{\Delta}(t)](\sigma(t)-s)|\\
&\quad=|[f(\sigma(t))-f(s)-f^{\Delta}(t)(\sigma(t)-s)]g(\sigma(t))\\
&\qquad+[g(\sigma(t))-g(s)-g^{\Delta}(t)(\sigma(t)-s)]f(t)\\
&\qquad+[g(\sigma(t))-g(s)-g^{\Delta}(t)(\sigma(t)-s)][f(s)-f(t)]\\
&\qquad+(\sigma(t)-s)g^{\Delta}(t)[f(s)-f(t)]|\\
&\quad\leqslant\varepsilon^*|\sigma(t)-s||g(\sigma(t))|+\varepsilon^*|\sigma(t)-s||f(t)|\\
&\qquad+\varepsilon^*\varepsilon^*|\sigma(t)-s|+\varepsilon^*|\sigma(t)-s||g^{\Delta}(t)|\\
&\quad=\varepsilon^*|\sigma(t)-s|[|g(\sigma(t))|+|f(t)|+\varepsilon^*+|g^{\Delta}(t)|]\\
&\quad\leqslant\varepsilon^*|\sigma(t)-s|[1+|f(t)|+|g(\sigma(t))|+|g^{\Delta}(t)|]\\
&\quad=\varepsilon|\sigma(t)-s|.
\end{aligned}$$

Thus, $(fg)^{\Delta}=f^{\Delta}g^{\sigma}+fg^{\Delta}$ holds at t. The other product rule in part (iii) of this theorem follows from this last equation by interchanging the functions f and g.

For the quotient formula of part (v), we use parts (ii) and (iv) to calculate

$$\begin{aligned}\left(\frac{f}{g}\right)^{\Delta}(t)&=\left(f\cdot\frac{1}{g}\right)^{\Delta}(t)\\&=f(t)\left(\frac{1}{g}\right)^{\Delta}(t)+f^{\Delta}(t)\frac{1}{g(\sigma(t))}\\&=-f(t)\frac{g^{\Delta}(t)}{g(t)g(\sigma(t))}+f^{\Delta}(t)\frac{1}{g(\sigma(t))}\\&=\frac{f^{\Delta}(t)g(t)-f(t)g^{\Delta}(t)}{g(t)g(\sigma(t))}\end{aligned}$$

Corollary 8.1 Let α be constant and $m\in\mathbf{N}$.

(i) For f defined by $f(t)=(t-\alpha)^m$, we have

$$f^{\Delta}(t)=\sum_{\upsilon=0}^{m-1}(\sigma(t)-\alpha)^{\upsilon}(t-\alpha)^{m-1-\upsilon}.$$

(ii) For g defined by $g(t)=\frac{1}{(t-\alpha)^m}$, we have

$$g^{\Delta}(t)=-\sum_{\upsilon=0}^{m-1}\frac{1}{(\sigma(t)-\alpha)^{m-\upsilon}(t-\alpha)^{\upsilon+1}},$$

provided $(t-\alpha)(\sigma(t)-\alpha)\neq 0$.

Definition 8.3 For a function $f:\mathbf{T}\to\mathbf{R}$, we shall talk about the second derivative $f^{\Delta\Delta}$ provided f^{Δ} is differentiable on $\mathbf{T}^{\kappa^2}=(\mathbf{T}^{\kappa})^{\kappa}$ with derivative $f^{\Delta\Delta}=(f^{\Delta})^{\Delta}:\mathbf{T}^{\kappa^2}\to\mathbf{R}$. Similarly, we define higher order derivatives $f^{\Delta^n}:\mathbf{T}^{\kappa^n}\to\mathbf{R}$. Finally, for $t\in\mathbf{T}$, we denote $\sigma^2(t)=\sigma(\sigma(t))$ and $\rho^2(t)=\rho(\rho(t))$, and $\sigma^n(t)$ and $\rho^n(t)$ for $n\in\mathbf{N}$ are defined accordingly. For convenience, we also put

$$\sigma^0(t)=t,\ \rho^0(t)=t,\ f^{\Delta^0}=f,\quad\text{and}\quad \mathbf{T}^{\kappa^0}=\mathbf{T}.$$

Theorem 8.4 (Leibniz Formula) Let $S_{\kappa}^{(n)}$ be the set consisting of all possible strings of length n, containing exactly κ times σ and $n-\kappa$ times Δ. If

$$f^{\Lambda}\quad\text{exists for all}\quad \Lambda\in S_{\kappa}^{(n)},$$

then

$$(fg)^{\Delta^n}=\sum_{\kappa=0}^{n}\Big[\sum_{\Lambda\in S_{\kappa}^{(n)}}f^{\Lambda}\Big]g^{\Delta^{\kappa}}$$

holds for all $n\in\mathbf{N}$.

Definition 8.4 Let $t\in\mathbf{C}$ (i.e., t is a complex number) and $\kappa\in\mathbf{Z}$. The factorial function $t^{(\kappa)}$ is defined as follows:

(i) If $\kappa\in\mathbf{N}$, then

$$t^{(\kappa)}=t(t-1)\ldots(t-\kappa+1).$$

(ii) If $\kappa=0$, then

$$t^{(0)}=1.$$

(iii) If $-\kappa\in\mathbf{N}$, then

$$t^{(\kappa)}=\frac{1}{(t+1)(t+2)\ldots(t-\kappa)}$$

for $t\neq-1,-2,\ldots,\kappa$.

In general,

$$t^{(\kappa)}:=\frac{\Gamma(t+1)}{\Gamma(t-\kappa+1)} \tag{8.1}$$

for all $t,\kappa\in\mathbf{C}$ such that the right-hand side of equation (8.1) makes sense.

Definition 8.5 We define the binomial coefficient $\begin{pmatrix}\alpha\\ \beta\end{pmatrix}$ by

$$\begin{pmatrix}\alpha\\ \beta\end{pmatrix}=\frac{\alpha^{(\beta)}}{\Gamma(\beta+1)}.$$

8.3 Integration

In order to describe classes of functions that are "integrable", we introduce the following two concepts.

Definition 8.6 A function $f:\mathbf{T}\to\mathbf{R}$ is called **regulated** provided its right-sided limits exist (finite) at all right-dense points in $\mathbf{T}$ and its left-sided limits exist (finite) at all left-dense points in $\mathbf{T}$.

Definition 8.7 A function $f:\mathbf{T}\to\mathbf{R}$ is called **rd-continuous** provided it is continuous at right-dense points in $\mathbf{T}$ and its left-sided limits exist (finite) at left-dense points in $\mathbf{T}$. The set of rd-continuous functions $f:\mathbf{T}\to\mathbf{R}$ will be denoted in this book by

$$C_{rd}=C_{rd}(\mathbf{T})=C_{rd}(\mathbf{T},\mathbf{R}).$$

The set of functions $f:\mathbf{T}\to\mathbf{R}$ that are differentiable and whose derivative is rd-continuous is denoted by

$$C_{rd}^{1}=C_{rd}^{1}(\mathbf{T})=C_{rd}^{1}(\mathbf{T},\mathbf{R}).$$

Theorem 8.5 Assume $f:\mathbf{T}\to\mathbf{R}$.

(i) If f is continuous, then f is rd-continuous.

(ii) If f is rd-continuous, then f is regulated.

(iii) The jump operator σ is rd-continuous.

(iv) If f is regulated or rd-continuous, then so is f^{σ}.

(v) Assume f is continuous. If $g:\mathbf{T}\to\mathbf{R}$ is regulated or rd-continuous, then $f\circ g$ has that property too.

Definition 8.8 A continuous function $f:\mathbf{T}\to\mathbf{R}$ is called **pre-differentiable** with (region of differentiation) D, provided $D\subset\mathbf{T}^{\kappa}$, $\mathbf{T}^{\kappa}\setminus D$ is countable and contains no right-scattered elements of $\mathbf{T}$, and f is differentiable at each $t\in\mathbf{D}$.

Theorem 8.6 Every regulated function on a compact interval is bounded.

Proof Assume $f:[a,b]\to\mathbf{R}$ is unbounded, i. e., for each $n\in\mathbf{N}$, there exists $t_n\in[a,b]$ with $|f(t_n)|>n$. Since

$$\{t_n:\ n\in\mathbf{N}\}\subset[a,b],$$

there exists a convergent subsequence $\{t_{n_\kappa}\}_{\kappa\in\mathbf{N}}$, i. e.,

$$\lim_{\kappa\to\infty} t_{n_\kappa}=t_0 \quad \text{for some} \quad t_0\in[a,b]. \tag{8.2}$$

Note that $t_0\in\mathbf{T}$ since $\{t_{n_\kappa}:\kappa\in\mathbf{N}\}\subset\mathbf{T}$ and $\mathbf{T}$ is closed. By (8.2), t_0 cannot be isolated, and there exists either a subsequence that tends to t_0 from above or a subsequence that tends to t_0 from below, and in any case the limit of $f(t)$ as $t\to t_0$ has to be finite according to regularity, a contradiction.

The following mean value theorem holds for pre-differentiable functions and will be used to prove the main existence theorems for pre-antiderivatives and antiderivatives later on in this section. Its proof is an application of the induction principle.

Theorem 8.7 (Mean Value Theorem) Let f and g be real-valued functions defined on $\mathbf{T}$, both pre-differentiable with D. Then

$$|f^{\Delta}(t)|\leqslant g^{\Delta}(t) \quad \text{for all} \quad t\in D$$

implies

$$|f(s)-f(r)|\leqslant g(s)-g(r) \quad \text{for all} \quad r,s\in\mathbf{T},\ r\leqslant s.$$

Proof Let $r,s\in\mathbf{T}$ with $r\leqslant s$ and denote $[r,s)\setminus D=\{t_n:n\in\mathbf{N}\}$. Let $\varepsilon>0$. We now show by induction that

$$S(t):|f(t)-f(r)|\leqslant g(t)-g(r)+\varepsilon\Big[t-r+\sum_{t_n<t}2^{-n}\Big]$$

holds for all $t\in[r,s]$. Note that once we have shown this, the claim of the mean value theorem follows. We now check the four conditions given in Theorem 8.1.

(i) The statement $S(r)$ is trivially satisfied.

(ii) Let t be right-scattered and assume that $S(t)$ holds. Then

$$\begin{aligned}|f(\sigma(t))-f(r)|&=|f(t)+\mu(t)f^{\Delta}(t)-f(r)|\\&\leqslant\mu(t)|f^{\Delta}(t)|+|f(t)-f(r)|\\&\leqslant\mu(t)g^{\Delta}(t)+g(t)-g(r)+\varepsilon\left[t-r+\sum_{t_n<t}2^{-n}\right]\\&=g(\sigma(t))-g(r)+\varepsilon\left[t-r+\sum_{t_n<\sigma(t)}2^{-n}\right]\\&<g(\sigma(t))-g(r)+\varepsilon\left[\sigma(t)-r+\sum_{t_n<\sigma(t)}2^{-n}\right].\end{aligned}$$

Therefore, $S(\sigma(t))$ holds.

(iii) Suppose $S(t)$ holds and $t\neq s$ is right-dense, i. e. , $\sigma(t)=t$. We consider two cases, namely $t\in D$ and $t\notin D$. First of all, suppose $t\in D$. Then f and g are differentiable at t and hence there exists a neighborhood U of t with

$$|f(t)-f(\tau)-f^{\Delta}(t)(t-\tau)|\leqslant\frac{\varepsilon}{2}|t-\tau|\quad\text{for all}\quad\tau\in U$$

and

$$|g(t)-g(\tau)-g^{\Delta}(t)(t-\tau)|\leqslant\frac{\varepsilon}{2}|t-\tau|\quad\text{for all}\quad\tau\in U.$$

Thus,

$$|f(t)-f(\tau)|\leqslant\left[|f^{\Delta}(t)|+\frac{\varepsilon}{2}\right]|t-\tau|\quad\text{for all}\quad\tau\in U$$

and

$$g(\tau)-g(t)-g^{\Delta}(t)(\tau-t)\geqslant-\frac{\varepsilon}{2}|t-\tau|\quad\text{for all}\quad\tau\in U.$$

Hence, we have for all $\tau\in U\cap(t,\infty)$,

$$\begin{aligned}|f(\tau)-f(r)|&\leqslant|f(\tau)-f(t)|+|f(t)-f(r)|\\&\leqslant\left[|f^{\Delta}(t)|+\frac{\varepsilon}{2}\right]|t-\tau|+|f(t)-f(r)|\\&\leqslant\left[g^{\Delta}(t)+\frac{\varepsilon}{2}\right]|t-\tau|+g(t)-g(r)+\varepsilon\Big(t-r+\sum_{t_n<t}2^{-n}\Big)\\&=g^{\Delta}(t)(\tau-t)+\frac{\varepsilon}{2}(\tau-t)+g(t)-g(r)+\varepsilon(t-r)+\varepsilon\sum_{t_n<t}2^{-n}\\&\leqslant g(\tau)-g(t)+\frac{\varepsilon}{2}|t-\tau|+\frac{\varepsilon}{2}(\tau-t)+g(t)-g(r)\\&\quad+\varepsilon(t-r)+\varepsilon\sum_{t_n<t}2^{-n}\\&=g(\tau)-g(r)+\varepsilon\Big(\tau-r+\sum_{t_n<\tau}2^{-n}\Big),\end{aligned}$$

so that $S(\tau)$ follows for all $\tau\in U\cap(t,\infty)$.

For the second case, suppose $t\notin D$. Then $t=t_m$ for some $m\in\mathbf{N}$. Since f and g are pre-differentiable, they both are continuous and hence there exists a neighborhood U of t with

$$|f(\tau)-f(t)|\leqslant\frac{\varepsilon}{2}2^{-m}\quad\text{for all}\quad\tau\in U$$

and

$$|g(\tau)-g(t)|\leqslant\frac{\varepsilon}{2}2^{-m}\quad\text{for all}\quad\tau\in U.$$

Therefore

$$g(\tau)-g(t)\geqslant-\frac{\varepsilon}{2}2^{-m}\quad\text{for all}\quad\tau\in U$$

and hence

$$\begin{aligned}|f(\tau)-f(r)|&\leqslant|f(\tau)-f(t)|+|f(t)-f(r)|\\&\leqslant\frac{\varepsilon}{2}2^{-m}+g(t)-g(r)+\varepsilon\Big(t-r+\sum_{t_n<t}2^{-n}\Big)\\&\leqslant\frac{\varepsilon}{2}2^{-m}+g(\tau)+\frac{\varepsilon}{2}2^{-m}-g(r)+\varepsilon\Big(\tau-r+\sum_{t_n<t}2^{-n}\Big)\\&=\varepsilon2^{-m}+g(\tau)-g(r)+\varepsilon\Big(\tau-r+\sum_{t_n<t}2^{-n}\Big)\\&\leqslant g(\tau)-g(r)+\varepsilon\Big(\tau-r+\sum_{t_n<\tau}2^{-n}\Big)\end{aligned}$$

so that again $S(\tau)$ follows for all $\tau\in U\cap(t,\infty)$.

(iv) Now let t be left-dense and suppose $S(\tau)$ is true for all $\tau<t$. Then

$$\begin{aligned}\lim_{\tau\to t^-}|f(\tau)-f(r)|&\leqslant\lim_{\tau\to t^-}\Big[g(\tau)-g(r)+\varepsilon\Big(\tau-r+\sum_{t_n<\tau}2^{-n}\Big)\Big]\\&\leqslant\lim_{\tau\to t^-}\Big[g(\tau)-g(r)+\varepsilon\Big(\tau-r+\sum_{t_n<t}2^{-n}\Big)\Big]\end{aligned}$$

implies $S(t)$ as both f and g are continuous at t.

An application of Theorem 8.1 finishes the proof.

Corollary 8.2 Suppose f and g are pre-differentiable with D.

(i) If U is a compact interval with endpoints $r,s\in\mathbf{T}$, then

$$|f(s)-f(r)|\leqslant\{\sup_{t\in U^\kappa\cap D}|f^\Delta(t)|\}|s-r|.$$

(ii) If $f^\Delta(t)=0$ for all $t\in D$, then f is a constant function.

(iii) If $f^\Delta(t)=g^\Delta(t)$ for all $t\in D$, then

$$g(t)=f(t)+C \quad \text{for all } t\in \mathbf{T},$$

where C is a constant.

Proof Suppose f is pre-differentiable with D and let $r,\ s\in \mathbf{T}$ with $r\leqslant s$. Define

$$g(t):=\{\sup_{\tau\in[r,\ s]^{\kappa}\cap D}|f^{\Delta}(\tau)|\}(t-r) \quad \text{for} \quad t\in \mathbf{T}.$$

Then

$$g^{\Delta}(t)=\sup_{\tau\in[r,\ s]^{\kappa}\cap D}|f^{\Delta}(\tau)|\geqslant|f^{\Delta}(t)| \quad \text{for all} \quad t\in D\cap[r,s]^{\kappa}.$$

By Theorem 8.7

$$g(t)-g(r)\geqslant|f(t)-f(r)| \quad \text{for all} \quad t\in[r,s],$$

so that

$$|f(s)-f(r)|\leqslant g(s)-g(r)=g(s)=\{\sup_{\tau\in[r,s]^{\kappa}\cap D}|f^{\Delta}(\tau)|\}(s-r).$$

This completes the proof of part (i). Part (ii) follows immediately from part (i), and part (iii) follows from part (ii).

Theorem 8.8 (Existence of Pre-Antiderivatives) Let f be regulated. Then there exists a function F which is pre-differentiable with region of differentation D such that

$$F^{\Delta}(t)=f(t) \quad \text{holds for all} \quad t\in D.$$

Definition 8.9 Assume $f:\mathbf{T}\to\mathbf{R}$ is a regulated function. Any function F as in Theorem 8.8 is called a **pre-antiderivative** of f. We define the **indefinite integral** of a regulated function f by

$$\int f(t)\Delta t=F(t)+C,$$

where C is an arbitrary constant and F is a pre-antiderivative of f. We define the **Cauchy integral** by

$$\int_r^s f(t)\Delta t=F(s)-F(r) \quad \text{for all} \quad r,s\in\mathbf{T}.$$

A function $F:\mathbf{T}\to\mathbf{R}$ is called an **antiderivative** of $f:\mathbf{T}\to\mathbf{R}$ provided

$$F^{\Delta}(t)=f(t) \quad \text{holds for all} \quad t\in\mathbf{T}^{\kappa}.$$

Theorem 8.9 (Existence of Antiderivatives) Every rd-continuous function has an antiderivative. In particular, if $t_0\in\mathbf{T}$, then F defined by

$$F(t):=\int_{t_0}^{t} f(\tau)\Delta\tau \quad \text{for} \quad t\in\mathbf{T}$$

is an antiderivative of f.

Proof Suppose f is an rd-continuous function. By Theorem 8.5(ii), f is regulated. Let F be a function guaranteed to exist by Theorem 8.8 together with D, satisfying

$F(t_0)=x_0$ and

$$F^{\Delta}(t)=f(t) \quad \text{for all} \quad t\in D.$$

This F is pre-differentiable with D. We have to show that $F^{\Delta}(t)=f(t)$ holds for all $t\in \mathbf{T}^{\kappa}$ (this, of course, includes all points in $\mathbf{T}^{\kappa}\setminus D$). So let $t\in \mathbf{T}^{\kappa}\setminus D$. Then t is right-dense because $\mathbf{T}^{\kappa}\setminus D$ cannot contain any right-scattered points according to Definition 8. 8. Since f is rd-continuous, it is continuous at t. Let $\varepsilon>0$. Then there exists a neighborhood U of t with

$$|f(s)-f(t)|\leqslant\varepsilon \quad \text{for all} \quad s\in U.$$

Define

$$h(\tau):=F(\tau)-f(t)(\tau-t_0) \quad \text{for} \quad \tau\in \mathbf{T}.$$

Then h is pre-differentiable with D and we have

$$h^{\Delta}(\tau)=F^{\Delta}(\tau)-f(t)=f(\tau)-f(t) \quad \text{for all} \quad \tau\in D.$$

Hence

$$|h^{\Delta}(s)|=|f(s)-f(t)|\leqslant\varepsilon \quad \text{for all} \quad s\in D\cap U.$$

Therefore

$$\sup_{s\in D\cap U}|h^{\Delta}(s)|\leqslant\varepsilon.$$

Thus, by Corollary 8. 2, we have for $r\in U$,

$$\begin{aligned}|F(t)-F(r)-f(t)(t-r)|&=|h(t)+f(t)(t-t_0)-[h(r)+f(t)(r-t_0)]-f(t)(t-r)|\\&=|h(t)-h(r)|\\&\leqslant\{\sup_{s\in D\cap U}|h^{\Delta}(s)|\}\ |t-r|\\&\leqslant\varepsilon|t-r|.\end{aligned}$$

But this shows that F is differentiable at t with $F^{\Delta}(t)=f(t)$.

Theorem 8. 10 If $f\in C_{\mathrm{rd}}$ and $t\in \mathbf{T}^{\kappa}$, then

$$\int_t^{\sigma(t)} f(\tau)\Delta\tau=\mu(t)f(t).$$

Proof By Theorem 8. 9, there exists an antiderivative F of f, and

$$\begin{aligned}\int_t^{\sigma(t)} f(\tau)\Delta\tau&=F(\sigma(t))-F(t)\\&=\mu(t)F^{\Delta}(t)\\&=\mu(t)f(t),\end{aligned}$$

where the second equation holds because of Theorem 8. 1(iv).

Theorem 8. 11 If $f^{\Delta}\geqslant 0$, then f is increasing.

Proof Let $f^{\Delta}\geqslant 0$ on $[a,b]$ and let s, $t\in \mathbf{T}$ with $a\leqslant s\leqslant t\leqslant b$. Then

$$f(t)=f(s)+\int_s^t f^{\Delta}(\tau)\Delta\tau\geqslant f(s),$$

so that the conclusion follows.

Theorem 8.12 If a, b, $c\in\mathbf{T}$, $\alpha\in\mathbf{R}$, and f, $g\in C_{\mathrm{rd}}$, then

(i) $\int_a^b[f(t)+g(t)]\Delta t=\int_a^b f(t)\Delta t+\int_a^b g(t)\Delta t$;

(ii) $\int_a^b \alpha f(t)\Delta t=\alpha\int_a^b f(t)\Delta t$;

(iii) $\int_a^b f(t)\Delta t=-\int_b^a f(t)\Delta t$;

(iv) $\int_a^b f(t)\Delta t=\int_a^c f(t)\Delta t+\int_c^b f(t)\Delta t$;

(v) $\int_a^b f(\sigma(t))g^{\Delta}(t)\Delta t=(fg)(b)-(fg)(a)-\int_a^b f^{\Delta}(t)g(t)\Delta t$;

(vi) $\int_a^b f(t)g^{\Delta}(t)\Delta t=(fg)(b)-(fg)(a)-\int_a^b f^{\Delta}(t)g(\sigma(t))\Delta t$;

(vii) $\int_a^a f(t)\Delta t=0$;

(viii) If $|f(t)|\leqslant g(t)$ on $[a,b)$, then

$$\left|\int_a^b f(t)\Delta t\right|\leqslant\int_a^b g(t)\Delta t;$$

(ix) If $f(t)\geqslant 0$ for all $a\leqslant t<b$, then $\int_a^b f(t)\Delta t\geqslant 0$.

Proof These results follow easily from Definition 8.9 and Theorems 8.3 and 8.7.

Theorem 8.13 Let a, $b\in\mathbf{T}$ and $f\in C_{\mathrm{rd}}$.

(i) If $\mathbf{T}=\mathbf{R}$, then

$$\int_a^b f(t)\Delta t=\int_a^b f(t)\mathrm{d}t,$$

where the integral on the right is the usual Riemann integral from calculus.

(ii) If $[a,b]$ consists of only isolated points, then

$$\int_a^b f(t)\Delta t=\begin{cases}\sum\limits_{t\in[a,b)}\mu(t)f(t), & \text{if } a<b,\\ 0, & \text{if } a=b,\\ -\sum\limits_{t\in[b,a)}\mu(t)f(t), & \text{if } a>b.\end{cases}$$

(iii) If $\mathbf{T}=h\mathbf{Z}=\{h\kappa:\kappa\in\mathbf{Z}\}$, where $h>0$, then

$$\int_a^b f(t)\Delta t=\begin{cases}\sum_{\kappa=\frac{a}{h}}^{\frac{b}{h}-1} f(\kappa h)h, & \text{if } a<b,\\ 0, & \text{if } a=b,\\ -\sum_{\kappa=\frac{b}{h}}^{\frac{a}{h}-1} f(\kappa h)h, & \text{if } a>b.\end{cases}$$

(iv) If $\mathbf{T}=\mathbf{Z}$, then

$$\int_a^b f(t)\Delta t=\begin{cases}\sum_{t=a}^{b-1} f(t), & \text{if } a<b,\\ 0, & \text{if } a=b,\\ -\sum_{t=b}^{a-1} f(t), & \text{if } a>b.\end{cases}$$

Proof If $\mathbf{T}=\mathbf{R}$, then Theorem 8.2 (iii) yields that $f:\mathbf{R}\to\mathbf{R}$ is delta differentiable at $t\in\mathbf{R}$ iff

$$f'(t)=\lim_{s\to t}\frac{f(t)-f(s)}{t-s}$$

exists, i. e., iff f is differentiable (in the ordinary sense) at t. Thus, part (i) follows from the standard fundamental theorem of calculus.

We now prove (ii). First note that $[a,b]$ contains only finitely many points since each point in $[a,b]$ is isolated. Assume that $a<b$ and let $[a,b]=\{t_0, t_1, \ldots, t_n\}$, where

$$a=t_0<t_1<t_2<\ldots<t_n=b.$$

Then

$$\begin{aligned}\int_a^b f(t)\Delta t &= \sum_{i=0}^{n-1}\int_{t_i}^{t_{i+1}} f(t)\Delta t\\ &= \sum_{i=0}^{n-1}\int_{t_i}^{\sigma(t_i)} f(t)\Delta t\\ &= \sum_{i=0}^{n-1}\mu(t_i)f(t_i)\\ &= \sum_{t\in[a,b)}\mu(t)f(t),\end{aligned}$$

where the third equation above follows from Theorem 8.10. If $b<a$, then the result follows from what we just proved and Theorem 8.12 (iii). If $a=b$, then $\int_a^b f(t)\Delta t=0$ by Theorem 8.12(vii). Parts (iii) and (iv) are special cases of part (ii).

Definition 8.10 If $a\in\mathbf{T}$, $\sup\mathbf{T}=\infty$, and f is rd-continuous on $[a,\infty)$, then we define the improper integral by

$$\int_a^{\infty} f(t)\Delta t := \lim_{b\to\infty}\int_a^b f(t)\Delta t$$

provided this limit exists, and we say that the improper integral converges in this case. If this limit does not exist, then we say that the improper integral diverges.

Theorem 8.14 (Chain Rule) Assume $g:\mathbf{R}\to\mathbf{R}$ is continuous, $g:\mathbf{T}\to\mathbf{R}$ is delta differentiable on $\mathbf{T}^\kappa$, and $f:\mathbf{R}\to\mathbf{R}$ is continuously differentiable. Then there exists c in the real interval $[t,\sigma(t)]$ with

$$(f\circ g)^{\Delta}(t)=f'(g(c))g^{\Delta}(t). \tag{8.3}$$

Proof Fix $t\in\mathbf{T}^\kappa$. First we consider the case where t is right-scattered. In this case,

$$(f\circ g)^{\Delta}(t)=\frac{f(g(\sigma(t)))-f(g(t))}{\mu(t)}.$$

If $g(\sigma(t))=g(t)$, then we get $(f\circ g)^{\Delta}(t)=0$ and $g^{\Delta}(t)=0$ and so (8.3) holds for any c in the real interval $[t,\sigma(t)]$. Hence we can assume $g(\sigma(t))\neq g(t)$. Then

$$\begin{aligned}(f\circ g)^{\Delta}(t)&=\frac{f(g(\sigma(t)))-f(g(t))}{g(\sigma(t))-g(t)}\cdot\frac{g(\sigma(t))-g(t)}{\mu(t)}\\&=f'(\xi)g^{\Delta}(t)\end{aligned}$$

by the mean value theorem, where ξ is between $g(t)$ and $g(\sigma(t))$. Since $g:\mathbf{R}\to\mathbf{R}$ is continuous, there is $c\in[t,\sigma(t)]$ such that $g(c)=\xi$, which gives us the desired result.

It remains to consider the case when t is right-dense. In this case,

$$\begin{aligned}(f\circ g)^{\Delta}(t)&=\lim_{s\to t}\frac{f(g(t))-f(g(s))}{t-s}\\&=\lim_{s\to t}\left\{f'(\xi_s)\cdot\frac{g(t)-g(s)}{t-s}\right\}\end{aligned}$$

by the mean value theorem in calculus, where ξ_s is between $g(s)$ and $g(t)$. By the continuity of g, we get that $\lim_{s\to t}\xi_s=g(t)$, which gives us the desired result.

Theorem 8.15 (Chain Rule) Let $f:\mathbf{R}\to\mathbf{R}$ be continuously differentiable and suppose $g:\mathbf{T}\to\mathbf{R}$ is delta differentiable. Then $f\circ g:\mathbf{T}\to\mathbf{R}$ is delta differentiable and the formula

$$(f\circ g)^{\Delta}(t)=\left\{\int_0^1 f'(g(t)+h\mu(t)g^{\Delta}(t))\mathrm{d}h\right\}g^{\Delta}(t)$$

holds.

Proof First of all, we apply the ordinary substitution rule from calculus to find

$$f(g(\sigma(t)))-f(g(s))=\int_{g(s)}^{g(\sigma(t))} f'(\tau)\mathrm{d}\tau$$
$$=[g(\sigma(t))-g(s)]\int_0^1 f'(hg(\sigma(t))+(1-h)g(s))dh.$$

Let $t\in\mathbf{T}^{\kappa}$ and $\varepsilon>0$ be given. Since g is differentiable at t, there exists a neighborhood U_1 of t such that

$$|g(\sigma(t))-g(s)-g^{\Delta}(t)(\sigma(t)-s)|\leqslant\varepsilon^*|\sigma(t)-s|,$$

where

$$\varepsilon^*=\frac{\varepsilon}{1+2\int_0^1 |f'(hg(\sigma(t))+(1-h)g(t))|\,dh}$$

for all $s\in U_1$. Moreover, f' is continuous on $\mathbf{R}$, and therefore it is uniformly continuous on closed subsets of $\mathbf{R}$, and (observe also that g is continuous as it is differentiable) hence there exists a neighborhood U_2 of t such that

$$|f'(hg(\sigma(t))+(1-h)g(s))-f'(hg(\sigma(t))+(1-h)g(t))|\leqslant\frac{\varepsilon}{2(\varepsilon^*+|g^{\Delta}(t)|)}$$

for all $s\in U_2$. To see this, note also that

$$|hg(\sigma(t))+(1-h)g(s)-(hg(\sigma(t))+(1-h)g(t))=(1-h)|g(s)-g(t)|$$
$$\leqslant|g(s)-g(t)|$$

holds for all $0\leqslant h\leqslant1$. We then define $U=U_1\cap U_2$ and let $s\in U$. For convenience, we put

$$\alpha=hg(\sigma(t))+(1-h)g(s)\quad\text{and}\quad\beta=hg(\sigma(t))+(1-h)g(t).$$

Then we have

$$\left|(f\circ g)(\sigma(t))-(f\circ g)(s)-(\sigma(t)-s)g^{\Delta}(t)\int_0^1 f'(\beta)\,dh\right|$$
$$=\left|[g(\sigma(t))-g(s)]\int_0^1 f'(\alpha)\,dh-(\sigma(t)-s)g^{\Delta}(t)\int_0^1 f'(\beta)\,dh\right|$$
$$=\Big|[g(\sigma(t))-g(s)-(\sigma(t)-s)g^{\Delta}(t)]\int_0^1 f'(\alpha)\,dh$$
$$+(\sigma(t)-s)g^{\Delta}(t)\int_0^1(f'(\alpha)-f'(\beta))\,dh\Big|$$
$$\leqslant|g(\sigma(t))-g(s)-(\sigma(t)-s)g^{\Delta}(t)|\int_0^1|f'(\alpha)|\,dh$$
$$+|\sigma(t)-s||g^{\Delta}(t)|\int_0^1|f'(\alpha)-f'(\beta)|\,dh$$
$$\leqslant\varepsilon^*|\sigma(t)-s|\int_0^1|f'(\alpha)|\,dh+|\sigma(t)-s||g^{\Delta}(t)|\int_0^1|f'(\alpha)-f'(\beta)|\,dh$$

$$\leqslant \varepsilon^* |\sigma(t)-s| \int_0^1 |f'(\beta)| dh + [\varepsilon^* + |g^\Delta(t)|] |\sigma(t)-s| \int_0^1 |f'(\alpha)-f'(\beta)| dh$$

$$=\frac{\varepsilon}{2}|\sigma(t)-s|+\frac{\varepsilon}{2}|\sigma(t)-s|$$

$$=\varepsilon|\sigma(t)-s|.$$

Therefore, $f \circ g$ is differentiable at t and the derivative is as claimed above.

Theorem 8.16 (Chain Rule) Assume $\upsilon: \mathbf{T} \to \mathbf{R}$ is strictly increasing and $\tilde{\mathbf{T}} := \upsilon(\mathbf{T})$ is a time scale. Let $w: \tilde{\mathbf{T}} \to \mathbf{R}$. If $\upsilon^\Delta(t)$ and $w^{\tilde{\Delta}}(\upsilon(t))$ exist for $t \in \mathbf{T}^\kappa$, then

$$(w \circ \upsilon)^\Delta = (w^{\tilde{\Delta}} \circ \upsilon)\upsilon^\Delta.$$

Proof Let $0<\varepsilon<1$ be given and define $\varepsilon^* = \varepsilon[1+|\upsilon^\Delta(t)|+|w^{\tilde{\Delta}}(\upsilon(t))|]^{-1}$. Note that $0<\varepsilon^*<1$. According to the assumptions, there exist neighborhoods N_1 of t and N_2 of $\upsilon(t)$ such that

$$|\upsilon(\sigma(t))-\upsilon(s)-(\sigma(t)-s)\upsilon^\Delta(t)| \leqslant \varepsilon^* |\sigma(t)-s| \quad \text{for all} \quad s \in N_1$$

and

$$|w(\tilde{\sigma}(\upsilon(t)))-w(r)-(\tilde{\sigma}(\upsilon(t))-r)w^{\tilde{\Delta}}(\upsilon(t))| \leqslant \varepsilon^* |\tilde{\sigma}(\upsilon(t))-r|, \ r \in N_2.$$

Put $N = N_1 \cap \upsilon^{-1}(N_2)$ and let $s \in N$. Then $s \in N_1$ and $\upsilon(s) \in N_2$ and

$$\begin{aligned}
&|w(\upsilon(\sigma(t)))-w(\upsilon(s))-(\sigma(t)-s)[w^{\tilde{\Delta}}(\upsilon(t))\upsilon^\Delta(t)]| \\
&\quad = |w(\upsilon(\sigma(t)))-w(\upsilon(s))-(\tilde{\sigma}(\upsilon(t))-\upsilon(s))w^{\tilde{\Delta}}(\upsilon(t)) \\
&\qquad +[\tilde{\sigma}(\upsilon(t))-\upsilon(s)-(\sigma(t)-s)\upsilon^\Delta(t)]w^{\tilde{\Delta}}(\upsilon(t))| \\
&\quad \leqslant \varepsilon^* |\tilde{\sigma}(\upsilon(t))-\upsilon(s)| + \varepsilon^* |\sigma(t)-s| |w^{\tilde{\Delta}}(\upsilon(t))| \\
&\quad \leqslant \varepsilon^* \{ |\tilde{\sigma}(\upsilon(t))-\upsilon(s)-(\sigma(t)-s)\upsilon^\Delta(t)| + |\sigma(t)-s| |\upsilon^\Delta(t)| \\
&\qquad + |\sigma(t)-s| |w^{\tilde{\Delta}}(\upsilon(t))| \} \\
&\quad \leqslant \varepsilon^* \{ \varepsilon^* |\sigma(t)-s| + |\sigma(t)-s| |\upsilon^\Delta(t)| + |\sigma(t)-s| |w^{\tilde{\Delta}}(\upsilon(t))| \} \\
&\quad = \varepsilon^* |\sigma(t)-s| \{ \varepsilon^* + |\upsilon^\Delta(t)| + |w^{\tilde{\Delta}}(\upsilon(t))| \} \\
&\quad \leqslant \varepsilon^* \{ 1 + |\upsilon^\Delta(t)| + |w^{\tilde{\Delta}}(\upsilon(t))| \} |\sigma(t)-s| \\
&\quad = \varepsilon |\sigma(t)-s|.
\end{aligned}$$

This proves the claim.

Corollary 8.3 (Derivative of the Inverse) Assume $\upsilon: \mathbf{T} \to \mathbf{R}$ is strictly increasing and $\tilde{\mathbf{T}} := \upsilon(\mathbf{T})$ is a time scale. Then

$$\frac{1}{\upsilon^\Delta} = (\upsilon^{-1})^{\tilde{\Delta}} \circ \upsilon$$

at points where υ^Δ is different from zero.

Proof Let $w = \upsilon^{-1}: \tilde{\mathbf{T}} \to \mathbf{T}$ in the previous theorem.

Theorem 8. 17 (Substitution) Assume $\upsilon:\mathbf{T}\to\mathbf{R}$ is strictly increasing and $\tilde{\mathbf{T}}:=\upsilon(\mathbf{T})$ is a time scale. If $f:\mathbf{T}\to\mathbf{R}$ is an rd-continuous function and υ is differentiable with rd-continuous derivative, then if $a, b\in\mathbf{T}$,

$$\int_a^b f(t)\upsilon^{\Delta}(t)\Delta t=\int_{\upsilon(a)}^{\upsilon(b)}(f\circ\upsilon^{-1})(s)\tilde{\Delta}s.$$

Proof Since $f\upsilon^{\Delta}$ is an rd-continuous function, it possesses an antiderivative F by Theorem 3. 5, i. e. , $F^{\Delta}=f\upsilon^{\Delta}$, and

$$\begin{aligned}\int_a^b f(t)\upsilon^{\Delta}(t)\Delta t &= \int_a^b F^{\Delta}(t)\Delta t\\ &=F(b)-F(a)\\ &=(F\circ\upsilon^{-1})(\upsilon(b))-(F\circ\upsilon^{-1})(\upsilon(a))\\ &=\int_{\upsilon(a)}^{\upsilon(b)}(F\circ\upsilon^{-1})^{\tilde{\Delta}}(s)\,\tilde{\Delta}s\\ &=\int_{\upsilon(a)}^{\upsilon(b)}(F^{\Delta}\circ\upsilon^{-1})(s)(\upsilon^{-1})^{\tilde{\Delta}}(s)\,\tilde{\Delta}s\\ &=\int_{\upsilon(a)}^{\upsilon(b)}((f\upsilon^{\Delta})\circ\upsilon^{-1})(s)(\upsilon^{-1})^{\tilde{\Delta}}(s)\,\tilde{\Delta}s\\ &=\int_{\upsilon(a)}^{\upsilon(b)}(f\circ\upsilon^{-1})(s)[(\upsilon^{\Delta}\circ\upsilon^{-1})(\upsilon^{-1})^{\tilde{\Delta}}](s)\,\tilde{\Delta}s\\ &=\int_{\upsilon(a)}^{\upsilon(b)}(f\circ\upsilon^{-1})(s)\,\tilde{\Delta}s,\end{aligned}$$

where for the fifth equal sign we have used Theorem 8. 16 and in the last step we have used Corollary 8. 3.

Theorem 8. 18 (Taylor's Formula) Let $n\in\mathbf{N}$. Suppose f is n-times differentiable on $\mathbf{T}^{\kappa^n}$. Let $\alpha\in\mathbf{T}^{\kappa^{n-1}}$, $t\in\mathbf{T}$, and define the functions g_κ by

$$g_0(r,s)\equiv 1 \quad\text{and}\quad g_{\kappa+1}(r,s)=\int_s^r g_\kappa(\sigma(\tau),s)\Delta\tau \quad \text{for } \kappa\in\mathbf{N}_0.$$

Then we have

$$f(t)=\sum_{\kappa=0}^{n-1}(-1)^\kappa g_\kappa(\alpha,t)\, f^{\Delta^\kappa}(\alpha)+\int_\alpha^{\rho^{n-1}(t)}(-1)^{n-1}g_{n-1}(\sigma(\tau),t)f^{\Delta^n}(\tau)\Delta\tau.$$

Corollary 8. 4 (Taylor's Formula) Let $n\in\mathbf{N}$. Suppose f is n-times differentiable on $\mathbf{T}^{\kappa^n}$. Let $\alpha\in\mathbf{T}^{\kappa^{n-1}}$, $t\in\mathbf{T}$, and define the functions h_κ by

$$h_0(r,s)\equiv 1 \quad\text{and}\quad h_{\kappa+1}(r,s)=\int_s^r h_\kappa(\tau,s)\Delta\tau \quad \text{for } \kappa\in\mathbf{N}_0.$$

Then we have

$$f(t)=\sum_{\kappa=0}^{n-1}h_\kappa(t,\alpha)f^{\Delta^\kappa}(\alpha)+\int_\alpha^{\rho^{n-1}(t)}h_{n-1}(t,\sigma(\tau))f^{\Delta^n}(\tau)\Delta\tau.$$

Theorem 8.19 (Intermediate Value Theorem) Assume $x:\mathbf{T}\to\mathbf{R}$ is continuous, $a<b$ are points in $\mathbf{T}$, and

$$x(a)x(b)<0.$$

Then there exists $c\in[a,b)$ such that either $x(c)=0$ or

$$x(c)x^{\sigma}(c)<0.$$

Theorem 8.20 Let $a\in\mathbf{T}^{\kappa}$, $b\in\mathbf{T}$ and assume $f:\mathbf{T}\times\mathbf{T}^{\kappa}\to\mathbf{R}$ is continuous at (t,t), where $t\in\mathbf{T}^{\kappa}$ with $t>a$. Suppose that for all $\varepsilon>0$ there exists a neighborhood U of t (independent of τ) such that

$$|f(\sigma(t),\tau)-f(s,\tau)-f^{\Delta}(t,\tau)(\sigma(t)-s)|\leqslant\varepsilon|\sigma(t)-s| \quad \text{for all} \quad s\in U,$$

where f^{Δ} denotes the derivative of f with respect to the first variable. Then

(i) $g(t):=\int_a^t f(t,\tau)\Delta\tau$ implies $g^{\Delta}(t)=\int_a^t f^{\Delta}(t,\tau)\Delta\tau+f(\sigma(t),t)$;

(ii) $h(t):=\int_t^b f(t,\tau)\Delta\tau$ implies $h^{\Delta}(t)=\int_t^b f^{\Delta}(t,\tau)\Delta\tau-f(\sigma(t),t)$.

Proof We only prove (i) while the proof of (ii) is similar.

Let $\varepsilon>0$. By assumption, there exists a neighborhood U_1 of t such that

$$|f(\sigma(t),\tau)-f(s,\tau)-f^{\Delta}(t,\tau)(\sigma(t)-s)|\leqslant\frac{\varepsilon}{2(\sigma(t)-a)}|\sigma(t)-s| \quad \text{for all} \quad s\in U_1.$$

Since f is continuous at (t,t), there exists a neighborhood U_2 of t such that

$$|f(s,\tau)-f(t,t)|\leqslant\frac{\varepsilon}{2} \quad \text{whenever} \quad s,\tau\in U_2.$$

Now define $U=U_1\cap U_2$ and let $s\in U$. Then

$$\begin{aligned}
&\left|g(\sigma(t))-g(s)-\left[f(\sigma(t),t)+\int_a^t f^{\Delta}(t,\tau)\Delta\tau\right](\sigma(t)-s)\right|\\
&=\left|\int_a^{\sigma(t)} f(\sigma(t),\tau)\Delta\tau-\int_a^s f(s,\tau)\Delta\tau-(\sigma(t)-s)f(\sigma(t),t)-(\sigma(t)-s)\int_a^t f^{\Delta}(t,\tau)\Delta\tau\right|\\
&=\left|\int_a^{\sigma(t)}[f(\sigma(t),\tau)-f(s,\tau)-f^{\Delta}(t,\tau)(\sigma(t)-s)]\Delta\tau\right.\\
&\quad\left.-\int_{\sigma(t)}^s f(s,\tau)\Delta\tau-(\sigma(t)-s)f(\sigma(t),t)-(\sigma(t)-s)\int_{\sigma(t)}^t f^{\Delta}(t,\tau)\Delta\tau\right|\\
&=\left|\int_a^{\sigma(t)}[f(\sigma(t),\tau)-f(s,\tau)-f^{\Delta}(t,\tau)(\sigma(t)-s)]\Delta\tau\right.\\
&\quad\left.-\int_{\sigma(t)}^s f(s,\tau)\Delta\tau-(\sigma(t)-s)f(\sigma(t),t)-(\sigma(t)-s)\mu(t)f^{\Delta}(t,t)\right|\\
&=\left|\int_a^{\sigma(t)}[f(\sigma(t),\tau)-f(s,\tau)-f^{\Delta}(t,\tau)(\sigma(t)-s)]\Delta\tau\right.
\end{aligned}$$

$$+\int_s^{\sigma(t)} f(s,\tau)\Delta\tau-(\sigma(t)-s)f(t,t)\Bigg|$$

$$=\left|\int_a^{\sigma(t)}[f(\sigma(t),\tau)-f(s,\tau)-f^{\Delta}(t,\tau)(\sigma(t)-s)]\Delta\tau+\int_s^{\sigma(t)}[f(s,\tau)-f(t,t)]\Delta\tau\right|$$

$$\leqslant\int_a^{\sigma(t)}|f(\sigma(t),\tau)-f(s,\tau)-f^{\Delta}(t,\tau)(\sigma(t)-s)|\Delta\tau+\int_s^{\sigma(t)}|f(s,\tau)-f(t,t)|\Delta\tau$$

$$\leqslant\int_a^{\sigma(t)}\frac{\varepsilon}{2(\sigma(t)-a)}|\sigma(t)-s|\Delta\tau+\left|\int_s^{\sigma(t)}\frac{\varepsilon}{2}\Delta\tau\right|$$

$$=\frac{\varepsilon}{2}|\sigma(t)-s|+\frac{\varepsilon}{2}|\sigma(t)-s|=\varepsilon|\sigma(t)-s|.$$

$$\bar{\mathbf{T}}=\mathbf{T}\cup\{\sup\mathbf{T}\}\cup\{\inf\mathbf{T}\}.$$

If $\infty\in\bar{\mathbf{T}}$, we call ∞ **left-dense**, and $-\infty$ is called **right-dense** provided $-\infty\in\bar{\mathbf{T}}$.

For any left-dense $t_0\in\mathbf{T}$ and any $\varepsilon>0$, the set

$$L_\varepsilon(t_0)=\{t\in\mathbf{T}:0<t_0-t<\varepsilon\}$$

is nonempty, and so is $L_\varepsilon(\infty)=\{t\in\mathbf{T}:t>\frac{1}{\varepsilon}\}$ if $\infty\in\bar{\mathbf{T}}$. The sets $R_\varepsilon(t_0)$ for right-dense $t_0\in\bar{\mathbf{T}}$ and $\varepsilon>0$ are defined accordingly. For a function $h:\mathbf{T}\to\mathbf{R}$, we define

$$\liminf_{t\to t_0^-}h(t)=\lim_{\varepsilon\to0^+}\inf_{t\in L_\varepsilon(t_0)}h(t)\quad\text{for left-dense}\quad t_0\in\bar{\mathbf{T}},$$

and $\liminf_{t\to t_0^+}h(t)$, $\limsup_{t\to t_0^-}h(t)$, $\limsup_{t\to t_0^+}h(t)$ are defined analogously.

Theorem 8.21 (L'Hôspital's Rule) Assume f and g are differentiable on $\mathbf{T}$ with

$$\lim_{t\to t_0^-}f(t)=\lim_{t\to t_0^-}g(t)=0 \tag{8.4}$$

for some left-dense $t_0\in\bar{\mathbf{T}}$.

Suppose there exists $\varepsilon>0$ with

$$g(t)>0,g^{\Delta}(t)<0\quad\text{for all}\quad t\in L_\varepsilon(t_0). \tag{8.5}$$

Then we have

$$\liminf_{t\to t_0^-}\frac{f^{\Delta}(t)}{g^{\Delta}(t)}\leqslant\liminf_{t\to t_0^-}\frac{f(t)}{g(t)}\leqslant\limsup_{t\to t_0^-}\frac{f(t)}{g(t)}\leqslant\limsup_{t\to t_0^-}\frac{f^{\Delta}(t)}{g^{\Delta}(t)}.$$

Proof Let $\delta\in(0,\varepsilon]$ and put $\alpha=\inf_{\tau\in L_\delta(t_0)}\frac{f^{\Delta}(\tau)}{g^{\Delta}(\tau)}$, $\beta=\sup_{\tau\in L_\delta(t_0)}\frac{f^{\Delta}(\tau)}{g^{\Delta}(\tau)}$. Then

$$\alpha g^{\Delta}(\tau)\geqslant f^{\Delta}(\tau)\geqslant\beta g^{\Delta}(\tau)\quad\text{for all}\quad\tau\in L_\delta(t_0).$$

By (8.5) and hence

$$\int_s^t\alpha g^{\Delta}(\tau)\Delta\tau\geqslant\int_s^t f^{\Delta}(\tau)\Delta\tau\geqslant\int_s^t\beta g^{\Delta}(\tau)\Delta\tau\text{ for all }s,t\in L_\delta(t_0),\ s<t,$$

so that

$$\alpha g(t)-\alpha g(s)\geqslant f(t)-f(s)\geqslant\beta g(t)-\beta g(s) \text{ for all } s,t\in L_\delta(t_0),\ s<t.$$

Now, letting $t\to t_0^-$, we find from (8.4)

$$-\alpha g(s)\geqslant -f(s)\geqslant -\beta g(s) \quad \text{for all} \quad s\in L_\delta(t_0)$$

and hence by (8.5)

$$\inf_{\tau\in L_\delta(t_0)}\frac{f^\Delta(\tau)}{g^\Delta(\tau)}=\alpha\leqslant\inf_{s\in L_\delta(t_0)}\frac{f(s)}{g(s)}\leqslant\sup_{s\in L_\delta(t_0)}\frac{f(s)}{g(s)}\leqslant\beta=\sup_{\tau\in L_\delta(t_0)}\frac{f^\Delta(\tau)}{g^\Delta(\tau)}.$$

Letting $\delta\to 0^+$ yields our desired result.

Theorem 8.22 (L'Hôspital's Rule) Assume f and g are differentiable on $\mathbf{T}$ with

$$\lim_{t\to t_0^-} g(t)=\infty \quad \text{for some left-dense} \quad t_0\in\overline{\mathbf{T}}. \tag{8.6}$$

Suppose there exists $\varepsilon>0$ with

$$g(t)>0,\ g^\Delta(t)>0 \quad \text{for all} \quad t\in L_\varepsilon(t_0). \tag{8.7}$$

Then $\lim\limits_{t\to t_0^-}\dfrac{f^\Delta(t)}{g^\Delta(t)}=r\in\overline{\mathbf{R}}$ implies $\lim\limits_{t\to t_0^-}\dfrac{f(t)}{g(t)}=r$.

Proof First suppose $r\in\mathbf{R}$. Let $c>0$. Then there exists $\delta\in(0,\varepsilon]$ such that

$$\left|\frac{f^\Delta(\tau)}{g^\Delta(\tau)}-r\right|\leqslant c \quad \text{for all} \quad \tau\in L_\delta(t_0)$$

and hence by (8.7)

$$-cg^\Delta(\tau)\leqslant f^\Delta(\tau)-rg^\Delta(\tau)\leqslant cg^\Delta(\tau) \quad \text{for all} \quad \tau\in L_\delta(t_0).$$

We integrate as in the proof of Theorem 8.21 and use (8.7) to obtain

$$(r-c)\left(1-\frac{g(s)}{g(t)}\right)\leqslant\frac{f(t)}{g(t)}-\frac{f(s)}{g(t)}\leqslant(r+c)\left(1-\frac{g(s)}{g(t)}\right) \text{ for all } s,\ t\in L_\delta(t_0);\ s<t.$$

Letting $t\to t_0^-$ and applying (8.6) yields

$$r-c\leqslant\liminf_{t\to t_0^-}\frac{f(t)}{g(t)}\leqslant\limsup_{t\to t_0^-}\frac{f(t)}{g(t)}\leqslant r+c.$$

Now we let $c\to 0^+$ to see that $\lim\limits_{t\to t_0^-}\dfrac{f(t)}{g(t)}$ exists and equals r.

Next, if $r=\infty$ (and similarly if $r=-\infty$), let $c>0$. Then there exists $\delta\in(0,\varepsilon]$ with

$$\frac{f^\Delta(\tau)}{g^\Delta(\tau)}\geqslant\frac{1}{c} \quad \text{for all} \quad \tau\in L_\delta(t_0)$$

and hence by (8.7)

$$f^{\Delta}(\tau) \geqslant \frac{1}{c} g^{\Delta}(\tau) \quad \text{for all} \quad \tau \in L_{\delta}(t_0).$$

We integrate again to get

$$\frac{f(t)}{g(t)} - \frac{f(s)}{g(t)} \geqslant \frac{1}{c}\left[1 - \frac{g(s)}{g(t)}\right] \quad \text{for all} \quad s,\ t \in L_{\delta}(t_0);\ s < t.$$

Thus, letting $t \to t_0^-$ and applying (8.6), we find $\liminf_{t \to t_0^-} \frac{f(t)}{g(t)} \geqslant \frac{1}{c}$, and then, letting $c \to 0^+$, we obtain $\lim_{t \to t_0^-} \frac{f(t)}{g(t)} = \infty = r$.

References

[1] P. D. Ritger and N. J. Rose, *Differential Equations with Applications*, New York: Dover Publications, 2000.

[2] M. Tenenbaum and H. Pollard, *Ordinary Differential Equations: An Elementary Textbook for Students of Mathematics, Engineering, and the Sciences*, New York: Dover Publications, 1985.

[3] Arnol'd, *Ordinary Differential Equations*, New-York: Springer-Verlag, 1992.

[4] Liu Hetao, *Qualitative Theory of Differential Equations*, Hefei: University of Science and Technology of China Press, 2009.

[5] M. Braun, *Differential Equations and Their Applications*, New York: Springer-Verlag, 1998.

[6] C. H. Edwards and D. E. Penney, *Elementary Differential Equations*, Beijing: China Machine Press, 2005.

[7] D. G. Zill and M. R. Cullen, *Differential Equations with Boundary-Value Problems*, Brooks Cole, Cengage Learning, 2008.

[8] R. K. Nagle, E. B. Saff and A. D. Snider, *Fundamentals of Differential Equations and Boundary Value Problems*, New Jersey: Pearson Education, 2011.

[9] I. Podlubny, *Fractional Differential Equations*, New York: Academic Press, 1999.

[10] K. S. Miller and B. Ross, *An Introduction to the Fractional Calculus and Fractional Differential Equation*, New York: John-Wiley, 1993.

[11] A. A. Kilbas, H. H. Srivastava and J. J. Trujillo, *Theory and Applications of Fractional Differential Equations*, Amsterdam: Elsevier Science B. V., 2006.

[12] K. Diethelm, *The Analysis of Fractional Differential Equations*, New York: Springer-Verlag, 2010.

[13] M. Bohner and A. Peterson, *Dynamic Equations on Time Scales, An Introduction with Applications*, Boston: BirkhÄauser, 2001.

[14] M. Bohner and A. Peterson, *Advances in Dynamic Equations on Time Scales*, Boston: BirkhÄauser, 2003.

[15]王高雄等. 常微分方程. 3 版. 北京:高等教育出版社,2006.

[16]丁同仁,李承志. 常微分方程教程. 2 版. 北京:高等教育出版社,2004.